Decolonizing Science and Modernity in South Asia

"*Decolonizing Science and Modernity in South Asia* elucidates issues vital to the history, present, and future of South Asia. The volume celebrates the astounding diversity of the continent's historical and contemporary engagement with the sciences, technology, and indigenous and imported healing traditions and practices. Especially welcome are the perspectives on medical pluralism inclusive of Perso-Arabic medicine, the growth, and nature of the Indian scientific community, vernacular science, and the accommodations and adaptations of knowledge practices and technologies required under the challenges of colonialism and independence".

—Michael A. Osborne, *Oregon State University and University of California, Santa Barbara*

"An innovative collection of essays powerfully enriched with the experience and practice of decolonial life and thought. Readers will find that the vibrant field of South Asian Studies of science, technology, and medicine continues to provide vital answers in the attempt to comprehend the legacy of empire in the rise of disciplines, intellectual careers, and ways of knowing. The vernacular as also the alternation of modernity into plural forms is crucial as ways of taking stock of the sciences in South Asia as we face our present".

—Sujit Sivasundaram, *Author of 'Waves Across the South: A New History of Revolution and Empire', Faculty of History, University of Cambridge*

"This is an important collection of essays focussing on traditional roots and modern development of science, technology, and medicine in colonial South Asia. The essays written by both established and young scholars bring forth fresh perspectives for understanding and interrogating modernity and its complex relationship with colonialism. It will be of interest to scholars from a variety of disciplines".

—Sekhar Bandyopadhyay, *School of History, Philosophy, Political Science and International Relations, Victoria University of Wellington*

"This book skillfully charts how the difficult path between indigeneity and modernity has been navigated in the development of the sciences in India, offering thoughtful insights into the ways in which Indian scientists have innovated and challenged Western hegemony during the past two centuries".

—Crispin Bates, *School of History, Classics and Archaeology, University of Edinburgh*

"Sahara Ahmed and Suvobrata Sarkar have put together an innovative and exciting collection of essays, that challenge many received views on the place of science in the development of modern India. The authors provide new perspectives and insights on familiar topics and explore new topics, mining novel sources with new interdisciplinary approaches. The collection will be essential reading for understanding the place of science in modern India".

—Michael Worboys, *Centre for the History of Science, Technology and Medicine, University of Manchester*

Sahara Ahmed · Suvobrata Sarkar
Editors

Decolonizing Science and Modernity in South Asia

Questioning Concepts, Constructing Histories

Springer

Editors
Sahara Ahmed
Department of History
Rabindra Bharati University
Kolkata, West Bengal, India

Suvobrata Sarkar
Department of History
Rabindra Bharati University
Kolkata, West Bengal, India

ISBN 978-981-97-1828-3 ISBN 978-981-97-1829-0 (eBook)
https://doi.org/10.1007/978-981-97-1829-0

This Springer imprint is published by the registered company Springer Nature Singapore Pte Ltd.
The registered company address is: 152 Beach Road, #21-01/04 Gateway East, Singapore 189721, Singapore

Presented to our teachers
Professor Deepak Kumar
and
Professor Arun Bandopadhyay

Acknowledgements

The volume explores the production of new knowledge and changing understandings of science in South Asian terminologies and focusses on the individuals and institutions that translated 'Western' modernity and science in South Asia. Chronologically, our first debt is to our contributors who agreed to be a part of this volume. We would like to thank them all. Could it be possible to decentre the West and recentre South Asian modes of thinking? They induce us to explore new methodologies, research questions, and sources to conceptualize the theme. We also put on record our sincere thanks to the editor of the *Economic and Political Weekly* for the necessary permission to reproduce the article by Prof. Shiv Visvanathan.

The debt we owe to Prof. Deepak Kumar and Prof. Arun Bandopadhyay, our revered teachers, for helping us understand the context of the question of science and modernity in South Asia from a socio-historical perspective is unfathomable. Thanks are also due to Prof. Suranjan Das and Prof. Tirthankar Roy for their help and kindness.

We sincerely thank our colleagues in the department for their constant encouragement and support. Without the excellent institutional support, the volume would not see light of the day, for which we are earnestly grateful to our university administration. Our students and scholars have opened up new avenues of thought with their questions in the classroom and beyond. We thank them wholeheartedly.

Our debt to Mahek and Ayushmaan is too great to be enumerated. They endured the long hours that we devoted to the volume. Our families have been with us all the way.

We express our deepest gratitude to Ms. Satvinder Kaur and Ms. Sathya Subramaniam at Springer for their valuable editorial assistance and to our anonymous reviewers for their comments and encouragements.

Kolkata, India

Sahara Ahmed
Suvobrata Sarkar

Introduction

The hydraulic machine for supplying London with water is a stupendous work. By its means, an ample supply of water is raised from the river Thames, so as constantly to keep full of lofty reservoir, whence, by means of conduits and leaden pipes, it is conveyed all over the town, and even to the upper rooms of houses four stories high, to the great comfort and ease of the inhabitants...the English carry their passion for machines to such an extent, that machinery is introduced into their kitchens...

Mirza Abu Taleb, the erudite Lucknow-born scholar, thus wrote in his travelogue (1810)

This foreign language has been a real hindrance to rapid spread of literacy in the country. In an educational institution, it encourages cram and effectively damps all creative efforts. The first fifteen years of independence have not brought about any spectacular increase in literacy. If there be no rapid spread of modern ideas among our people, our attempts at industrial and technical progress would have to lean heavily on foreign support, and we would be always counted among the underdeveloped nations of the world.

Satyendranath Bose, the renowned scientist, while delivering his convocation address at the University of Calcutta (1962)

The present volume endeavours to analyse the appropriation of modern knowledge in South Asia. The contributors focus on how scientific, technical, and medical knowledge and practices were reshaped within a specific socio-historical context. Mirza Abu Taleb Khan (1752–1806) was one of the most prolific writers in the late 18th and early 19th centuries. His travelogue, *Masir-i Talibi fi Bilad-i Afrang*, based on his travels from February 1799 to August 1803, comprehensively covers the state of England before and after the Industrial Revolution. However, he was not alone—Munshi Itesamuddin (1730–1800) talked of the advanced navigational science of England and admired Oxford's octagonal observatory and the exhibits in the medical institute. He produced his *Shigurf Nama-i Velayet* (1765), much before Abu Taleb, based on his year-and-a-half stay in Britain.[1] The Muslim intelligentsia which was, at that stage, struggling for its identity against all odds, recognized the achievements of modern science and technology. During the first hundred years of British rule, scientific activities and continued display of new and improved technology became a regular feature under colonial relations. Indians were not averse to the adoption of new scientific theories and modern technologies. On several occasions, they had exhibited considerable appreciation for Western knowledge—Botanical Garden at Sibpur, for example, created considerable interest among the Calcutta elites, and many of whom, including the orthodox-minded Raja Radhakanta Deb (1784–1867), used to visit the garden to listen to the discussions by the then superintendent Nathaniel Wallich.[2]

Pre-colonial India was no tabula rasa. It had a vigorous tradition in different branches of science and medicine. However, gradual colonization brought forth a huge cultural collision. This encounter was initially disturbing, argues Deepak Kumar. It also had within the question of attitude towards each other, an uneasy acceptance, a quest for identity and, finally, the seeds of decolonization.[3] Under colonial milieu, the pattern and scope of relationship between Indian and the Western medical system took an altogether different shape. The *daktari* medicine, Bengali vernacular forms of Western medicine, have largely remained absent from the accounts of nineteenth-century South Asian medical history. Little detailed account, according to Projit Mukharji, has been available of the varied Bengalis who came to practice Western medicine at various strata of colonial establishment, ranging from the lowly hospital assistants to the grand Edinburgh-or London-trained MDs.[4] Radhagobindo Kar (1852–1918), Licentiate of the Royal College of Physicians (LRCP) from Edinburgh, was a teacher at the medical school in Calcutta and published extensively on various aspects of health and medicine in Bengali.[5] Unfortunately, he rarely appears, until recently, in the academic discussion on South Asian medical history. Considering the rapid progress of the history of medicine as it relates to India under British colonial rule, now at least he deserves a well-researched biography!

In the context of early nineteenth century, argued Tapan Raychaudhuri, a major intellectual exercise on the part of South Asians was to determine what they could

---

[1] Mushirul Hasan (ed.), 'Westward Bound: Travels of Mirza Abu Taleb', *Exploring the West: Three Travel Narratives*, New Delhi: Oxford University Press, 2009, pp. ix–x.

[2] Satpal Sangwan, 'Indian Response to European Science and Technology 1757–1857', *The British Journal for the History of Science*, Vol. 21, No. 2, 1988, p. 231 (211–232).

[3] Deepak Kumar, 'The Culture of Science and Colonial Culture, India 1820–1920', *The British Journal for the History of Science*, Vol. 29, No. 2, 1996, p. 195 (195–209).

[4] Projit Bihari Mukharji, *Nationalizing the Body: The Medical Market, Print and Daktari Medicine*, London and New Delhi: Anthem Press, 2012, p. 35.

[5] For example, once he elaborated upon the dreadful disease plague, 'This pamphlet is a short treatise on the disease known as the plague. In compiling this pamphlet, my aim has been to place before the vernacular medical profession as well as the public at large a short history of the disease, its causes, characteristics, and rational treatment, so far as I have been able to gather them from all available sources. It is well known that a good deal of misapprehension exists in the public mind regarding the disease itself, as also the measures adopted by the government for its prevention'. Radhagobinda Kar, *Plague: Ihar Sangkhipto Biboran O Etatsambandhe Adesh, Biggyapan O Niyomadi*, Kolkata: Gurudas Chatterjee, Bengal Medical Library, 1898, p. 1.

adopt from the West and, as important, what elements in Western life must be rejected at all cost.[6] A few decades later, by mid-nineteenth century, it converted into a cultural synthesis. Such idea gave them the best of both worlds—initially, it prepared them to absorb culture-shock and ultimately assured a possible opportunity to rise above the obstacles imposed by the foreign rulers.[7] The indigenous medical traditions and their interactions with the colonial health establishment are worth a mention here. The facilities afforded by state (allopathic) medicine were at no point of time sufficient to supplement the indigenous systems. An important implication of the limited reach of colonial medical facilities was that indigenous medicine had enough space to operate, particularly in rural and semi-urban areas. However, fears of possible marginalization due to the onslaught of Western medicine, the protagonists of indigenous medicine were forced to take a critical look at the state of their art. Their assessment of the situation, according to K. N. Pannikar, was a complex amalgamation of past pride, discontent with the present, and anxiety about the future—ultimately efforts to revitalize indigenous systems during the late nineteenth and early twentieth centuries.[8]

Another interesting dimension was, in the last decades of the nineteenth century, some of the individual 'native' scientists gathered sufficient strength to collaborate (and sometimes differ) with their metropolitan peers and fight colonial bureaucracy. The Royal Institution of Great Britain occupies a prominent place in the annals of science. Jagadis Chandra Bose (1858–1937), an Indian physicist, for the first time, spoke at this institution, of his discoveries of the optical properties of 'electric' waves, in January 1897. For the Royal Institution, Bose's lecture was, just one more entry in its record books. For science, the occasion was yet another addition to its vast corpus. For Bose and for India, Subrata Dasgupta emphasizes, it was a moment of profound history—'Western science, the science of Galileo and Newton, had finally taken root in India among Indians, and on that January evening in 1897, the West tacitly acknowledged that fact'.[9] However, Bose was not lone pioneer of modern Indian science. There were a small handful of others. In particular, there was his contemporary and colleague at Presidency College, Prafulla Chandra Ray (1861–1944), a chemist who discovered mercurous nitrite in 1896, and founded a chemical works—Bengal Chemical and Pharmaceutical Works Ltd. The next generation of Indian scientists—Meghnad Saha (1893–1956), Satyendranath Bose (1894–1974), and others—were taught at Presidency College by Bose and Ray. By the early 1940s, the Indian scientific community made its intellectual presence felt in the international scientific world. There were at least nine fellows of the Royal Society as well as a

---

[6] Tapan Raychaudhuri, *Europe Reconsidered: Perceptions of the West in Nineteenth-Century Bengal*, Delhi: Oxford University Press, 1988, p. xi.

[7] During 1860–80, Deepak Kumar has referred to a number of cultural essayists (Bankimchandra Chattopadhyay, Rajendralal Mitra, Maulvi Ubaidullah, among others) who tried to articulate modern scientific rationality in terms of indigenous traditions and requirements. Kumar, *Science and the Raj: A Study of British India*, New Delhi: Oxford University Press, 2nd edition, 2006, pp. 192–201.

[8] K. N. Panikkar, 'Indigenous Medicine and Cultural Hegemony', *Colonialism, Culture, and Resistance*, New Delhi: Oxford University Press, 2007, p. 173.

[9] Subrata Dasgupta, *Jagadis Chandra Bose and the Indian Response to Western Science*, Ranikhet: Permanent Black, paperback edition, 2009, pp. 2–3.

Nobel laureate in physics.[10] A prominent member of this group, Satyendranath Bose sent a letter accompanying his manuscript on Planck's blackbody to Albert Einstein in 1924:

> I have ventured to send you the accompanying article for your perusal and opinion. I am anxious to know what you think of it. You will see that I have tried to deduce the coefficient $8\pi v2/c3$ in Planck's law independent of the classical electrodynamics only assuming that the ultimate elementary regions in the phase-space has the content $h3$. I do not know sufficient German to translate the paper. If you think the paper is worth publishing, I shall be grateful if you arrange for its publication in *Zeitscrift fur Physik*.[11]

In July 1924, Einstein sent Bose a postcard that accepted his paper on radiation statistics and acknowledged its importance. The rest, as they say, is history!

More than past three decades, a number of studies have explored and critiqued science as colonial knowledge and as a tool of imperial dominance.[12] With the anti-diffusionist, constructionist, and global turns in the discipline more widely, recent researches have declined to see science as a set of ideas and Western construction, rather wanted to analysis sciences as practices and bodies of knowledge that embedded in particular spatial contexts, and practised by local actors in locally relevant political and cultural contexts.[13] The geographical scope of the essays collected in this book focussing on South Asia is designed to elucidate a shift in scholarly focus from the imperial sciences to the colonized subjects—not as the passive recipients of Western knowledge, but as practitioners and participants in global scientific, technological, and medical modernity.

---

[10] V. V. Krishna, 'The Emergence of the Indian Scientific Community', in Binay Kumar Pattnaik (ed.), *Sociology of Science and Technology in India*, New Delhi: Sage, 2014, p. 48.

[11] Quoted in Kameshwar C. Wali (ed.), *Satyendra Nath Bose: His Life and Times*, New Jersey and London: World Scientific, 2009, p. xx.

[12] George Basalla, 'The Spread of Western Science', *Science*, New Series, Vol. 156, No. 37, 1967, pp. 611–22; Roy Macleod, 'On Visiting the "Moving Metropolis": Reflections on the Architecture of Imperial Science', *Historical Records in Australian Science*, Vol. 5, No. 3, 1982, pp. 1–16; Daniel R. Headrick, *The Tools of Empire: Technology Transfer in the Age of Imperialism, 1850–1940*, New York: Oxford University Press, 1981; Michael Adas, *Machines as the Measure of Men: Science, Technology, and Ideologies of Western Dominance*, Ithaca, NY: Cornell University Press, 1989; Roy Macleod and Deepak Kumar (eds.), *Technology and the Raj: Western Technology and Technical Transfers to India 1700–1947*, New Delhi: Sage, 1995.

[13] David Wade Chambers and Richard Gillespie, 'Locality in the History of Science: Colonial Science, Technoscience, and Indigenous Knowledge', *Osiris*, Vol. 15, No. 1, 2000, pp. 221–240; David Arnold, *The New Cambridge History of India: Science, Technology and Medicine in Colonial India*, Vol. III:5,, Cambridge: Cambridge University Press, 2000; Gyan Prakash, *Another Reason: Science and the Imagination of Modern India*, New Delhi: Oxford University Press, 2000; Dhruv Raina, *Images and Contexts: The Historiography of Science and Modernity in India*, New Delhi: Oxford University Press, 2003; Pratik Chakrabarti, *Western Science in Modern India: Metropolitan Methods, Colonial Practices*, New Delhi: Permanent Black, 2004; Kapil Raj, *Relocating Modern Science: Circulation and the Construction of Scientific Knowledge in South Asia and Europe Seventeenth to Nineteenth Century*, Delhi: Permanent Black, 2006.

Decolonizing South Asian Past: Science and Modernity

Decolonization has emerged as a new meaning of colonialism. It originally referred to as the political process of withdrawal of European powers from their colonial possessions. However, one simply cannot think of political modernity without referring to certain European categories and concepts—Dipesh Chakrabarty rightly reminds us that '...all bear the burden of European thought and history'. The modern Indian educated middle-class warmly embraced science and humanism that the European Enlightenment propagated. The very critique of colonialism itself is unthinkable, argues Chakrabarty, 'except as a legacy, partially, of how Enlightenment Europe was appropriated in the subcontinent'.[14] Today, we all know that the ethos and function of colonialism can be extended to the realm of science as well. Like colonialism itself, colonial science[15] is more than a set of institutions or structures; it is an economic as well as a cultural intervention. Thus, decolonization now signifies the need to historicize the development and domestication of modern science in formal European colonial territories. It requires not simply an awareness of European science's hidden colonial past, but necessitates seeing science itself as colonial/imperial episteme. The task of doing history of science in the Global South need not be one simply tracing the mobility of Western knowledge, putting them in Afro-Asian contexts, and commenting on their behaviour in different environments, but taking seriously what the new knowledge means from the perspective of the local people. It requires not merely looking at how people respond to external agencies, as Clapperton Mavhunga emphasizes, but situating the latter's arrival, meanings, and materialization within the locals' techno-scientific *longue durée*. He further argues that it is important to explore what science and innovation might mean from Africa in lieu of outside introductions or influences.[16] One can ask similar questions in the context of South Asia. Can one see South Asians as intellectuals thinking about and doing science based on intellect? Also, can one identify use of vernaculars by South Asians as modes of theory, even if they engaged Western modes of thought and practice?

Recently, a handful of scholars have argued that South Asia should not be treated simply as a 'site' to which the conventional analytics and methodologies of the broader field of history of technology can be extended. Rather, it should be a location to unravel new analytics and methods—a direct engagement with the question of the modern and the place of technology in the construction of the modern.[17]

---

[14] Dipesh Chakrabarty, *Provincializing Europe: Postcolonial Thought and Historical Difference*, New Delhi: Oxford University Press, 2001, p. 4.

[15] Deepak Kumar's pioneering work reveals that the entire gamut of the relationship between science and colonization can be precisely expressed by the term *colonial science*. Kumar, *Science and the Raj*, op.cit., p. xii.

[16] Clapperton Chakanetsa Mavhunga, 'Introduction' (ed.), *What Do Science, Technology, and Innovation Mean from Africa*, Cambridge, Massachusetts and London: The MIT Press, 2017, p. 1–4.

[17] Prakash Kumar, 'Introduction: New Histories of Technology in South Asia', *Technology and Culture*, Vol. 60, No. 4, 2019, p. 933 (933–952). According to Kumar, the contributors aim to demonstrate how looking at the history of technology from South Asia changes what it means to do history of technology.

The meaning (s) and significance of modernity have been prominent in the South Asian historiography for quite some time. A number of interpretations arise from the literature on science and modernity in particular. One of these is whether a concept of modernity originating in the discourse of Western societies can be useful elsewhere. Similarly, is there a single modernity or many? To some scholars working on South Asian science, the notion of 'alternative modernities' has proved attractive as it allows the possibility that modernity can be understood in multiple ways. The use of this concept, as Warwick Anderson proposes, legitimates local adaptations and reworkings of Western knowledge.[18] For example, Projit Mukherji demonstrates how the incorporation of various medical technologies has affected the non-biomedical therapeutics, and, thus attained their own distinctive modernity. He argues that the use of modern stethoscope as a new therapeutic technology in Ayurvedic medicine has evidently been mentioned in ethnographic accounts in Bengal rather than in the conventional archives.[19] Thus, writing for science and modernity in a South Asian tradition is to contribute simultaneously in the task of decolonizing ways of knowing the past.

What then are or should be the common talking points of Asia-centred concerns in scholarly discussions on science, technology, and medicine? What are the best ways to conduct this objective in a geographically focussed study of science and modernity? Is conceptualization of science through a deliberate reversal of colonial constructions, therefore bringing to light identities, knowledge, and practices that have been often erased, can serve the purpose? The discipline of history of science in South Asia has been enriched with significant borrowings from many fields—area studies, cultural studies, ethnography, feminist studies, post-colonial studies, etc. Debating modernity in studies of science in South Asia is a unique example of this trend wherein area studies concerns have been reflected into studies of science, technology, and medicine in South Asia.[20] Science was more than just a practical or ideological project when it came to empire. The modern Western science was inextricably connected with colonialism, especially British colonialism. It was highly exploitative in character. In this context, Rohan Deb Roy has recently expressed his concern—'The empires may have virtually disappeared, but the cultural biases and disadvantages they imposed have not'. He talks about a 'cosmopolitan model' which

[18] Warwick Anderson, 'Introduction: Postcolonial Technoscience', *Social Studies of Science*, Vol. 32, Nos. 5–6, 2002, pp. 643–658.

[19] Projit Bihari Mukharji, 'Akarnan: The Stethoscope and Making of Modern Ayurveda, Bengal, c. 1894–1952', *Technology and Culture*, Vol. 60, No. 4, 2019, pp. 953–978. See also, Mukharji's *Doctoring traditions: Ayurveda, small technologies and braided sciences*, Chicago: University of Chicago Press, 2016.

[20] Prakash Kumar, Projit Bihari Mukharji and Amit Prasad, 'Decolonizing Science in Asia', *Verge: Studies in Global Asias*, Vol. 4, No. 1, 2018, p. 38 (24–43).

recognizes the true achievements and potential of scientists from outside the Western world.[21]

It would be helpful to link some of the logic to modern-day practices, so that current readers can identify the colonial nature of their scientific and technological activities.[22] Extraction of knowledge from interlocutors and unnamed locals is a practice that continues even today. A number of top-ranking Global North institutions run programmes in Afro-Asian countries, and then never acknowledge local participation in that project.[23] The laying down of long-distance telegraph cables that connected the empire[24] is similar to the laying down of Google's undersea cables around the world today. In many instances, technological innovation is driven from Silicon Valley and the rest of the world becomes an outpost for implementing, marketing, or attempting to regulate these technologies and their impact. Of course, individuals from former colonies now form a significant workforce in the Silicon Valley. The IITians who emerged as the biggest Indian names in Silicon Valley got their B.Tech. degrees no later than the early 1980s. It was also the early days of the computer revolution. Many IITians who came to the USA for advanced degrees in other branches of engineering, Ajantha Subramanian mentions, eventually switched to software.[25] The focus on decolonization is very timely and invites the readers to (re)frame the ways in which we think about techno-scientific advancement and success.

Emerging Disciplines and Science in Vernacular

Scholarship over the last few decades has pointed out how modernity has become a fundamental analytical category while discussing history of science in South Asia. The entangled nature of knowledge production attracts considerable attention[26]—the encounter between modern science and the South Asian knowledge systems dates back to the early modernity. If one carefully observes the interaction between the

[21] He rightly opines, 'Decolonization promises to make science more appealing by integrating its findings more firmly with questions of justice, ethics, and democracy'. Rohan Deb Roy, 'Science Still Bears the Fingerprints of Colonialism', *Smithsonian Magazine*, 9 April 2018. https://www.smithsonianmag.com/science-nature/science-bears-fingerprints-colonialism-180968709/ (Accessed on 12 February, 2023).

[22] Warwick Anderson, 'Racial Conceptions in the Global South', *Isis*, Vol. 105, No. 4, 2014, pp. 782–792.

[23] Virginia Gewin, 'Decolonization should extend to collaborations, authorship, and co-creation of knowledge', *Nature*, 22 November 2022. https://www.nature.com/articles/d41586-022-03822-1 (Accessed on 12 February, 2023).

[24] Deep Kanta Lahiri Choudhury, *Telegraphic Imperialism: Crisis and Panic in the Indian Empire, c. 1830–1920*, Basingstoke: Palgrave Macmillan, 2010.

[25] Ajantha Subramanian, *The Caste of Merit: Engineering Education in India*, Cambridge, Massachusetts and London: Harvard University Press, 2019, p. 286.

[26] Sheldon Pollock, 'Forms of Knowledge in Early Modern South Asia', *Comparative Studies of South Asia, Africa and the Middle East*, Vol. 24, No. 2, 2004, pp. 19–21.

intellectuals of two linguistic genres—Sanskrit and Persian—can find several examples of constellations in the fields of philosophy and science. Above all, one should consider the proliferation in knowledge production in the vernaculars during 1550 and 1750. Hence, Dhruv Raina argues, a shift of focus is required from the history of science to the history of knowledge.[27] The encounter between modern science and the multiple knowledge systems of South Asia, according to him, is epistemological in nature while at the same time part of the larger politics of knowledge. The process of making new cultural practices (of science) local can be done through (a) translating modern science into local languages of South Asia and (b) revisiting these traditional knowledge systems through the frame of modern science.[28]

Some of the most interesting encounters between the so-called traditional sciences of South Asia and modern science date back to the end of the eighteenth century which produced interesting results in different branches of science and mathematics. What was the nature of the interaction between traditional mathematics and a modern discipline like calculus? Master Ramchandra, mathematician and science teacher at the Delhi College had developed an alternative way of solving problems of elementary calculus in the nineteenth century. Although initially a few British orientalists sought to promote Ramchandra's book in English schools, the project was later aborted in India as subsequent government policy did not encourage cultural assimilation at the level of pedagogy.[29] The first chapter in the volume by Santanu Chacraverti situates Śubhaṅkarī, a tradition of computing and mathematical instruction, in its social, cultural, and pedagogic context of early modern Bengal. The term usually denotes a corpus of computational instruction enshrined in verses, most commonly used for mathematical instruction in the *pāthśālās*—the elementary schools in rural Bengal. On closer examination, one finds that the contents of the Śubhaṅkarī corpus can be divided into three categories—algorithms for practical computation; arithmetical and simple algebraic problems; and playing with and exploring the properties of numbers. This tradition was largely oral, with the quantity of written material being relatively small. More importantly, as claimed by the author, the terminology and the system of numerical notation used became almost undecipherable by the late twentieth century.

The appropriation of modern science by the indigenous elite in South Asia is discussed at length in the literature. It seeks to demonstrate how some of the ideas associated with modern science—reason, universalism, progress for example—were deployed by the elite in their imagination of nationhood and modernity. With their

---

[27] Dhruv Raina, 'Revisiting Social Theory and History of Science in Early Modern South Asia and Colonial India', *Extreme-Orient Extreme-Occident*, Vol. 36, No. 2, 2013, p. 202 (191–210).

[28] Raina talks about three terms apparently used interchangeably in the literature to refer the adaptation of the practices of modern science into the frame of the existing South Asian knowledge systems—'domestication', 'naturalization', and 'localization'. Dhruv Raina, 'The Naturalization of Modern Science in South Asia: A Historical Overview of the Process of Domestication and Globalization', in Jörgen Renn (ed.), *The Globalization of Knowledge in History*, Berlin: Max Planck Institute for the History of Science, 2012, pp. 345–366.

[29] Dhruv Raina and S Irfan Habib, *Domesticating Modern Science: A Social History of Science and Culture in Colonial India*, New Delhi: Tulika, 2004.

crucial role in the interface between new opportunities and the existing social realities, the interlocutors set up an agenda for modern science to take deeper roots in South Asian soil. They started interpreting modern science for their fellow countrymen and devised strategies for its acceptance under colonial conditions.[30] From Akshay Kumar Dutta (1820–1886), Rajendralal Mitra (1823–1891), Ramendra Sundar Trivedi (1864–1919) in Bengal to Balaji Prabhakar Modak (1847–1906) in Maharashtra and Ruchi Ram Sahni (1863–1948) in Punjab—with their various ideas of science, their critiques of indigenous society, and their strategies to deal with the situation—left a distinctive mark on the nature of adoption and adaptation of modern science in South Asia.[31] Kamlesh Mohan in her chapter evaluates Ruchi Ram Sahni's contribution towards spreading of scientific awareness among the Punjabis in pre-independence era. For a deeper understanding of his mission, the chapter decodes his perception of the relationship between science and religion. How far it was influenced by his association with Brahmo intellectuals in Punjab and his teachers in Calcutta? Like other contemporary Indian interlocutors, the author argues, Sahni had faced with two crucial issues—(a) tension between the practice of nationalism and cultivation of Western science within the colonial framework, and (b) conflict between the agenda of imperialist science and aspirations of the Indian people. Finally, she contextualizes Sahni's mission with the long-term project of modernizing Indian society.

Although modern science had been introduced in Indian schools and colleges from the early nineteenth century, the non-English reader's introduction to the theories, concepts, and practices of modern science originally took place through vernacular publications. Here, Charu Singh argues, the authors brought novel inventions, new technologies, and unfamiliar instruments into the cultural imaginary of South Asian readers. The vernacular publications emerged as a key medium for creating science-literate publics who otherwise had very limited exposure to scientific experiments in laboratories.[32] Despite a considerable growth of Hindi writings on the issues of science in the early twentieth century, it soon faced with the limitation of the language—in terms of the inadequate terminology in Hindi that could be used to define scientific concepts and practices. The third chapter by Sandipan Baksi is an attempt to trace the process that went into creating a dictionary of scientific terms in Hindi. It is based primarily on a survey of the content of one of the significant popular science periodicals of India—*Vigyan*. It emphasizes on the debates around the foundational principles for creating a glossary of scientific terms. The central point of these debates was the source of vocabulary to be used for constructing scientific terminology in Hindi. According to Baksi, these debates were indicative of

[30] J. Lourdusamy, *Science and National Consciousness in Bengal 1870–1930*, New Delhi: Orient Longman, 2004, p. 4.

[31] Suvobrata Sarkar (ed.), *History of Science, Technology, Environment, and Medicine in India*, London and New York: Routledge, 2022.

[32] Charu Singh, 'The shastri and the air-pump: Experimental fictions and fictions of experiment for Hindi readers in colonial north India', *History of Science*, Vol. 60, No. 2, 2022, pp. 232–254.

the larger political conflict coloured by linguistic nationalism framed in conjunction with a revivalist perspective.

Plural Healing and Multiplicity

While writing the introduction of a volume titled *Plural Medicine, Tradition and Modernity* (2002), Waltraud Ernst remarked on the rapid growth of 'colonial' or 'imperial' medicine as it relates to British India, while the study of 'indigenous' or 'folk' medicinal practices of South Asia received much less attention.[33] Since that time, several scholars ventured into this uncharted terrain and explored the underlying dynamics that led to the perceived marginalization of 'indigenous' medicine, the problematic nature of categorizations—'traditional' and 'modern' medicine, the scope and limitations of medical pluralism in diverse geographical and cultural settings, etc. The rise of subaltern and post-colonial theories eventually inclined social and cultural historians to consider the development of non-Western medical paradigms and indigenous medicine worthy subjects of historical investigation. These studies refuted the claim of Western medicine's epistemological and therapeutic superiority and argued that a number of profoundly effective and highly sophisticated 'traditional' systems of healing not only preceded the advent of modern Western medicine in South Asia, but also adapted successfully to the changing circumstances of a colonial situation.

The increasing use of therapeutics other than biomedicine, and the import, especially of alternative forms of medicine from South and Southeast Asia to America and Europe have initiated curiosity in indigenous medical traditions, such as Ayurveda, Unani, and Siddha. The pharmaceutical industry of Ayurvedic medicine is part of a larger process of transformation, bringing Indian forms of knowledge and practices under the influence of Western science and modernity. In this context, Madhulika Banerjee has shown how the growth of modern pharmaceutical production of Ayurveda has been shaped by the colonial history, encounter with biomedicine and its epistemology, the demands of globalized market and international trade regimes.[34] Recently, Rachel Berger has shown how Ayurveda during early twentieth century, a period of critical modernization, was defined by structures of colonial governmentality. He has further discussed how these experiments in governance were operational at the local level rather than a centralized planning.[35]

In the first chapter of this part, Saurav Kumar Rai argues that the distinctive colonial context of the late nineteenth and early twentieth century in India led to the

---

[33] Waltraud Ernst (ed.), *Plural Medicine, Tradition and Modernity, 1800–2000*, London and New York: Routledge, 2002.

[34] Madhulika Banerjee, *Power, Knowledge, Medicine: Ayurvedic Pharmaceuticals at Home and in the World*, Hyderabad: Orient Blackswan, 2009.

[35] Rachel Berger, *Ayurveda Made Modern: Political Histories of Indigenous Medicine in North India, 1900–1955*, New York: Palgrave Macmillan, 2013.

creation of various 'modernized traditions' around this time. This unique combination of oxymoronic traits was peculiar to colonial set-up manifesting 'the rule of colonial difference'. In the sphere of medical modernity, this peculiar phenomenon was substantially manifested through the emergence of 'modern' Ayurveda. In their revivalist quest, the Ayurvedic practitioners and proponents of the time dramatically reshaped and redefined it which was remarkably different from its pre-colonial version. It was very well reflected in Ayurvedic incorporation of the traits on the basis of which Western medicine was claiming its superiority (such as scientificity, standardization, professionalization, institutionalization, etc.). Yet at the same time, it attempted to contest the epistemological absolutism of Western medicine as well in a very subtle way. Thus, the author argues that even while the Ayurvedic revivalism was a product of (colonial) modernity, it does not necessarily symbolize the victory of modernism as the refurbished Ayurveda had been trying to derive its sanctity both from being 'traditional' as well as 'modern' at the same time.

Traditional medicines have generally been reduced to an inferior position and forced to justify itself as adequately 'modern' and 'scientific' even as these exercises created contradiction and controversy during colonial times. In South and Southeast Asia, as recently claimed by a group of scholars, these indigenous medicines have been forced to adopt the systematic disciplining and rationalizing processes of biomedicine and so on (licensing for example). They further argue that there are hierarchies between different types of alternative medicines—'some of which have been accorded different validity based on affinity with scientized modes of assessment'.[36] Although scholarly writings on indigenous medicine in colonial India have done justice to Ayurveda, there has been a paucity of literature on Unani over the same period. Guy Attewell's study fills this gap by providing a comprehensive account of the domains in which the Unani profession was in dialogue with itself, with other therapeutic practices—Ayurvedic, allopathic, and folk practices and traditions, and most significantly with the public—those who looked for its services, as practitioners of *tibb* attempted to define authentic and legitimate Unani knowledge and practice.[37]

Probing the area of popular healing of South Asia, recently another group of scholars have looked at the way that it is marginalized by the state and medical establishments while at the same time being very crucial in the everyday lives of the ordinary people. They have examined the domain of 'subaltern therapeutics' that both interacts with and resists statist and elite forms of medical practice.[38] Neshat Quaiser in his chapter in the present volume deals with the *bazaari*-subaltern Unani medical practices, demonstrating how contrary to the predominant view within the studies of South Asian medical practices, Unani has been a dispersed and discursive terrain since the beginning with simultaneous presence of the two broad paradoxical

[36] Paul T. Cohen, Chris Lyttleton and Thapin Phatcharanuruk, 'Western and Traditional Medicine in India, Myanmar and Thailand', *Sojourn: Journal of Social Issues in Southeast Asia*, Vol. 37, No. 2, 2022, p. 263 (262–289).

[37] Guy Attewell, *Refiguring Unani Tibb: Plural Healing in Late Colonial India*, Hyderabad: Orient Longman, 2007.

[38] David Hardiman and Projit Bihari Mukherji (eds.), *Medical Marginality in South Asia: Situating subaltern therapeutics*, London and New York: Routledge, 2012.

spheres of Unani practices, such as textual Unani on the one hand, and bazaari-subaltern Unani on the other in different geographical, social, and cultural locales. This chapter, however, forcefully argues that the relationship between the textual Unani and bazaari-subaltern Unani medical practices cannot be comprehended properly through a blanket totalizing elite-subaltern binary framework. Moreover, the concept of subaltern itself is to be re-examined contextually, particularly to account for the ways in which a textual tradition such as Unani itself can be rendered to a state of subalternity, particularly in post-colonial situation. Quaiser further examines the ways in which bazaari-subaltern Unani medical public sphere remained outside the textual Unani and statist medical imaginations and control, therefore, was declared inauthentic for its practitioners were considered untrained, and yet it catered the medical requirements of vast masses during the colonial and post-colonial periods. Illustrative cases are drawn from provinces of Bihar, Bengal, Tamil Nadu, Telangana and Delhi, United Provinces.

The bone-setters of Bosipur village in Burdwan district of West Bengal are traditional orthopaedics specializing in treating broken bones using indigenous medical knowledge. However, due to lack of patronage from both the government and the local elites (including patients), Arabinda Samanta has recently emphasized, they are marginalized. They are neglected in official healthcare schemes for a variety of social, economic, and political reasons.[39] Most of the primitive communities with their close association with the nature gradually developed healing techniques and methods with the available flora, fauna, and other naturally occurring substances. Apart from the so-called indigenous forms of medicine (Ayurveda, Unani, Siddha, etc.), there also existed several little known and un-institutionalized therapeutic practices—folk medicine for example. In this context, Rup Kumar Barman explores the sub-Himalayan Bengal (Jalpaiguri, Alipurduar, and Cooch Behar districts of West Bengal) and identifies the folk medicinal traditions of the indigenous communities in general, and of the Rajbanshis, a dominant community, in particular. Like the bone-setters of Burdwan district, the Rajbanshi healers are ordinary people engaged in different occupations (farmers, labours, domestic helps, even teachers). But unlike the bone-setters, one can find female practitioners of medicine among the community. Another point of departure from the bone-setters was the tendency to institutionalize their medicinal knowledge and practices. As Barman mentions about Rajmohan Eshor, the royal physician of the Cooch Behar state, who developed a special school for the cure of fracture and arthritis at Nishiganj. Later, his family members have built up several such clinics in different parts of West Bengal and Northeast India. Thus, a number of recent publications by social scientists indicate the continued engagement by the scholars with the indigenous medical systems of South Asia.

[39] Arabinda Samanta, 'Revisiting a Marginalized Community: The Bone-setters of Bengal', in Bipasha Raha and Subhayu Chattopadhyay (eds.), *Mapping the Path to Maturity: A Connected History of Bengal and the North-East*, New Delhi: Manohar, 2018, pp. 255–267.

Institutionalization and Professionalization of New Knowledge

The emphasis on 'original' research and the push to create conditions for young Indians to engage in scientific enterprise came from Indians themselves. In 1869, long before the opportunity for scientific research at the university level, Mahendra Lal Sircar (1833–1904), a graduate of the Calcutta Medical College, set forth before the citizens of Calcutta a plan for founding an institute equipped with a library and laboratories. He wrote, 'We want an Institution which will combine the character, the scope and objects of the Royal Institution of London and of the British Association for the Advancement of Science' and then added, 'I want freedom for this Institution. I want it to be entirely under our own management and control. I want it to be solely native and purely national'.[40] The Indian Association for the Cultivation of Science (1876) was the first attempt at an institutional manifestation of the relationship between scientific research and nationalism in the backdrop of colonialism. At the turn of the twentieth century, Chandrasekhara Venkata Raman (1888–1970) came to carry out his research at this institute, first in acoustics and later in optics, X-rays, and magnetism. Those researches culminated in Raman's discovery of the new 'scattering' phenomenon of light, named the Raman Effect, for which he received the Nobel Prize in physics in 1930.[41]

The first decade of the twentieth century saw intense debate on what Indians had received at the end of a century and a half of British rule: amelioration was sought through the slogans of *Swadeshi* (self-reliance) and *Swaraj* (self-rule). The new vision of India debated continuously in the years to come and the quest for techno-scientific knowledge preceded and facilitated the emergence of this vision. As Vice-Chancellor, Asutosh Mukherjee (1864–1924) had been working relentlessly to institutionalize science teaching and research at the University of Calcutta. He was keen to bring Meghnad Saha, a pioneer astrophysicist, to Calcutta University with a professorship in physics (1921):

> I have read with indescribable interest and pleasure the account of your work. I had already seen the notice in the Nature…I am now endeavouring to provide for you an independent position and a suitable laboratory when you return here and I have little doubt that my efforts

[40] M. L. Sircar, 'On the Desirability of National Institution for the Cultivation of Sciences by the Natives of India' (1869); quoted in Deepak Kumar, 'India', Hugh Richard Slotten, Ronald L. Numbers, and David N. Livingstone (eds.), *The Cambridge History of Science*, Vol. 8: Modern Science in National, Transnational, and Global Context, Cambridge: Cambridge University Press, 2020, p. 466.

The inauguration of the institute in 1876, according to Kumar, was no less important than the establishment of the Indian National Congress (1885), a political organization that was to spearhead the national movement.

[41] G. Venkataraman, *Journey into Light: life and science of C. V. Raman*, Bangalore: Indian Academy of Sciences, 1988.

> will be successful. I do not think you will have to regret your decision to stay in your own University rather than accept an appointment elsewhere.[42]

Outside the scope of the university system, the institution which made the strongest imprint of a nationalist science with its characteristics assertion on the ethics of 'struggle, renunciation and sacrifice', was the Bose Institute established by Jagadish Chandra Bose in 1917.[43] However, these are not isolated incidents. In his chapter, Suvobrata Sarkar traces the institutional history of electrical engineering in colonial Bengal taking into account the evolution of Bengal Engineering College, Sibpur, and the College of Engineering and Technology, Jadavpur. The arrival of electricity necessitated the introduction of new institutional facilities. With the growth of the technological system, a supportive culture also grew—trained manpower to handle the machines, better educational facilities, new business culture, etc. The CESC, a London-based power company, was instrumental in the generation and supply of electricity in Calcutta. From the beginning, they realized the need to create a pool of personnel (basically Indians) with elementary knowledge on electrical engineering. Thus, the Bengal Engineering College started providing basic instruction on the subject. The erudite Indians identified, on several occasions, the imperfect training of Indian engineers, including electrical. A nationalist initiative, College of Engineering and Technology, from its inception, used to provide advanced training in electrical engineering. Investment in education, especially techno-scientific, was now seen to be of great importance. The electrical power was to be the key to a modern industrial economy excited Bengali *bhadralok*. The chapter is basically on the education of electrical engineers as it had developed in the late years of the Raj, and the interface it had with entrepreneurship and industry in a colonial metropole—Calcutta.

While examining the reception of Freudian psychoanalysis in late nineteenth-century India, Dhruv Raina has shown how the setting up of a discipline in different cultural context opens up discussion on localization of knowledge—whether it involves the production of new curricula, the creation of new organizational structures, or even the production of new knowledge?[44] Oral Health was comparatively an uncharted domain in India in the late eighteenth century but before long dental health was deemed to be an integral part of medical science. The mouth, tongue,

---

[42] Asutosh Mukherjee to Meghnad Saha, May 11, 1921; quoted in Uma Dasgupta (ed.), 'Introduction', *Science and Modern India: An Institutional History, c. 1784–1947*, New Delhi: Pearson, 2011, p. liv.

The numerous science institutes established during 1784 and 1945, according to Dasgupta, exemplified how the separate ideologies and functions of colonialism and nationalism, the differences in the traditional and modern systems of knowledge, and an imperial administration following its own logic, all came together in the venture of organizing these science institutes in British India. Ibid, p. lx.

[43] Madhumita Majumdar, *Science and the Nationalist Agenda in Bengal, 1904–1947*, unpublished Ph.D. thesis, University of Calcutta, 2004, p. 120.

[44] Dhruv Raina, 'Travelling Both Ways: The Adaptation of Disciplines, Scientific Textbooks and Institutions', in Feza Gönergun and Dhruv Raina (eds.), *Science between Europe and Asia: Historical Studies on the Transmission, Adoption and Adaptation of Knowledge*, Dordrecht: Springer, 2011, pp. 168–170.

teeth, jaws and its shape, the art of dentistry, diagnosis of disease, and its treatment are an essential part of dental science. Although in common perception dentistry is limited to the teeth, in actuality it incorporates the shape of the mouth, jaws, face, and forehead. Sahara Ahmed traces the origin of the term dontology in the context of colonial India. The shape of the teeth, gradual development, and persistence of abnormalities comprise the crux of the matter of dontology. Similarities in subject matter have prompted the use of the two terms 'oral health' and 'dental health' interchangeably. 'Scientific dentistry' was introduced in India in the nineteenth century by the British and modern dental care and the study of this discipline were initiated a hundred years ago. Dental care was restricted to the extraction of teeth at the hospitals and dispensaries by medical men and assistants. Despite this, for long, dentistry predominantly remained in the hands of the unqualified. The year 1920 is a milestone in the history of dental care and public health with the establishment of the first dental college in Calcutta. Significantly, this perspective found expression in the endeavours of Dr. Rafiuddin Ahmed and the institute he established in Calcutta (1920), eventually came to be known as the R. Ahmed Dental Medical College and Hospital. Dr. Ahmed advocated the American model of dentistry in lieu of his training rather than the British model which honed its thrust on the study of basic medical subjects. The duration of the medical course was two years but through the years, it witnessed many transitions which would comprise the crux of the chapter.

The years between 1947 and 1966, covering the period from independence to the end of the third Five-Year Plan, provided an arena for the most acute debates over the content of industrial development. Essentially these debates centred on the form of ownership and control over the industrial undertakings which were already in operation and those which were to be established. During the period, Nasir Tyabji has anticipated that industrialization in India involved not only the establishment of new enterprises by individual capitalists, or even by the state, but also measures of social engineering by the government. State activity was essential to nurture the development of entrepreneurs with a tangible industrial frame of mind.[45] Dinesh C. Sharma argues that in the late 1960s and 1970s, electronics was an emerging sector of technology development and industrial production. It was seen as a sunrise industry, with potential applications of electronics in varied sectors such as defence, atomic energy, entertainment, communication, and education. All these factors attracted the governments, policy makers, and Indian industry to this field. The government's entry into electronics and computing sectors was driven by the desire to achieve self-reliance, ending multinational monopoly and dependence on imports. In the run-up to the formation of the Department of Electronics and soon afterwards, scientists and technocrats took upon themselves the task of setting up an electronics industry. They started looking for opportunities in R&D, design, production, export, and import. Electronics was considered strategic—at par with space and atomic energy, so they strived for national control over it. This gave rise

[45] Nasir Tyabji, *Forging Capitalism in Nehru's India: Neocolonialism and the State, c. 1940–1970*, New Delhi: Oxford University Press, 2015, pp. xiv–xv.

to the phenomenon of 'government entrepreneurship', whereby scientific departments planned commercial ventures for production of components, computers, and software. Several ventures were conceived at Indian Statistical Institute, Tata Institute of Fundamental Research, Department of Atomic Energy, and Department of Energy under Prasanta Chandra Mahalanobis (1893–1972), Homi Jehangir Bhabha (1909–1966), Vikram Sarabahi (1919–1971), and M. G. K. Menon (1928–2016). Most of them failed for lack of business and market orientation. The ones that took shape helped India develop certain capabilities and manpower in advanced areas of R&D and production. According to the author, this helped create nuclei to human capital and expertise that eventually helped the private sector in seeding the software revolution in the decades that followed.

Metropolitan Methods, Colonial Practices

The encounter between metropolitan sciences of the West triggered numerous projects of cultural redefinition and engagement with European modernity and Indian culture. Western science had to be reinvented in the vocabulary of a modernizing India. This process could be viewed and S. Irfan Habib and Dhruv Raina have argued, either in terms of the proclamation of the new or in terms of a course of cultural appropriation. The expansion of Western science was accelerated by the combined efforts of imperial bureaucrats, their scientific peers, and indigenous traditions. The local elites envisaged this encounter with European sciences as a path to revitalization. This reflective association itself continuously reconstituted modern science.[46] A large number of Indian interlocutors, belonging to different disciplines and walks of life, contributed to the new quest for scientific knowledge. Among those who were the first to take scientific research and teaching as their career were Pramathanath Bose (1855–1934), Jagadish Chandra Bose, and Prafulla Chandra Ray. Bose specialized in geology at the University of London and later joined the Geological Survey of India. In 1886, he wrote a pamphlet on 'Technical and Scientific Education in Bengal' and a decade later published three volumes on *A History of Hindu Civilization during British Rule*. Aggressive nationalism had converted a geologist into a historian. As Deepak Kumar opines from physical mapping, Bose shifted to cultural contours.[47]

V. V. Krishna in his chapter on the emergence of Indian scientific community explores the contribution of leading Indian scientists in laying the foundations for professionalization of science and advancement of scientific knowledge during the

---

[46] Habib and Raina have further argued that the assimilation of modern science began at the level of pedagogy. This was a complex process where science teachers had to contend with indigenous knowledge systems, and produce translations of modern science into vernaculars. This encounter opened up the gateways for a dialogue that generated interesting experiments both in pedagogy and science and was to enrich both. S. Irfan Habib and Dhruv Raina (eds.), 'Introduction', *Social History of Science in Colonial India*, New Delhi: Oxford University Press, 2007, pp. xiii–xl.

[47] Deepak Kumar, *Science and Society in Modern India*, Cambridge: Cambridge University Press, 2023, p. 78.

1870s and 1940s. He demonstrates leading scientists in this era such as Mahendra Lal Sircar, Prafulla Chandra Ray, Ashutosh Mukherjee, Chandrasekhara Venkata Raman, Meghnad Saha, and Satendyanath Bose, to name some, waged an intellectual struggle against the British colonial gate keepers who resisted local and indigenous initiatives in the professionalization of science. In collaboration with Indian political elite, systematic efforts were directed at laying the foundations for cultivation of science institutions from translation of science into local languages, popularization of science, promotion of science teaching in colleges, and higher institutions of learning to basic scientific research. Between the 1890s and 1940s half a dozen advance research specialist groups emerged in plant physiology, chemistry, physics, mathematics, and astronomy. In a large measure, the genesis of Indian science community had taken roots outside the confines of colonial science enterprises as these institutions were not conducive for the professionalization of science and the advancement of knowledge. Indian scientific intellectual struggle in laying the foundations of basic research during this period had dented the Centre-Periphery relations or dichotomy in science practised by British colonial scientific enterprise. Thus, the chapter challenges the basic assertion of key classical colonial thinkers, such as George Basalla, that independent scientific tradition in the colonial era emerged outside the confines of colonial science enterprises.

The first generation of Indian scientists struggled hard to inaugurate a scientific research system. The professed objective was to draw India closer to the international community of science. Collectively and individually, they had set in motion the efforts to build new institutions to do research at international levels and train new generations of competent scientists who would stay and work in India. Through their international networks, well before 1947 and immediately after that, they articulated new ideas and programmes, and these networks remained so impactful in India that the planning and development of science, as emphasized by Robert Anderson, was a major responsibility that non-specialists had to acknowledge and allow scientists a free hand.[48] On the other hand, while exploring the academic roots of India's nuclear research, Jahnavi Phalkey has pointed out that the leading Indian physicists were convinced of the imperative to create within India the facilities and capabilities required for international participation in nuclear physics research.[49] We need to appreciate how these scientists overcame the many obstacles they faced, to a great extent in the initial phases, and to understand the lessons their lives teach us.

The relationship between 'centres' of science (or scientific metropolises) and the geographical regions lying at the periphery has been a subject of some interest to historians and sociologists of science. The relationship between Britain and India during the heyday of the British Empire was coloured by the complicating fact that India was a colony of Britain. Although Indian scientists were fully aware that some

---

[48] Anderson argues that there would not have been a sustained atomic energy programme without a continuously evolving relationship between science and politics in South Asia. Robert S. Anderson, *Nucleus and Nation: Scientists, International Networks, and Power in India*, Chicago and London: The University of Chicago Press, 2010, pp. 2–17.

[49] Jahnavi Phalkey, *Atomic State: Big Science in Twentieth-Century India*, Ranikhet: Permanent Black, 2013.

other countries like Germany might have been ahead of Britain in certain branches of science such as physics and chemistry in the early decades of the twentieth century, Arnab Rai Choudhuri highlights in his chapter that there was still a special craving for British honours in India, since such honours would bring additional benefits in the colonial set-up. Many of the leading Indian scientists of that era openly expressed the view that nationalistic feelings drove their science, but they still wanted British recognitions for their research. Since the Knighthood was an official honour bestowed by the British Crown, many Indian scientists had an ambivalent attitude towards it. However, the Fellowship of the Royal Society was viewed as recognition by the scientific peers of Britain, and it was perhaps the most coveted honour for Indian scientists of that era. Rai Choudhuri concludes that it was well known that the FRS election process was not a level playing field for Indian scientists, and a few Indian scientists with high international reputation (such as Prafulla Chandra Ray) were never elected FRS even after being nominated repeatedly by their British peers.

In his pioneering study on the life histories of two Indian scientists during colonial times, Jagadish Chandra Bose and Srinivasa Ramanujan (1887–1920), Ashis Nandy has brilliantly constructed ‘a cultural psychology of scientific creativity’ through their encounters between individual creativity and social reality. Bose’s creativity was influenced by the synthesis of the dominant traditions of Western science and the indigenous ideals of knowledge and enquiry. In doing so, as claimed by Nandy, Bose not only created a personalized Indian perspective of world science, but also tried hard to recover the self-esteem of his subjugated countrymen.[50] The theme, Indian creativity in modern science, attracts considerable attention.[51] However, Deepanwita Dasgupta has raised a few significant questions—Why to study the scientific practices of a small non-Western community on the colonial margins? Being imitations of the dominant practices of Europe and America, what could be the philosophical relevance of Indian science?[52]

In June 1924, Satyendra Nath Bose sent Einstein a short four-page paper containing a new proof of Planck’s law. That proof involved two unusual moves—first, a consistent interpretation of blackbody radiation in terms of Einstein’s newly proposed light quantum theory, and secondly, the use of a new statistics that counted such light particles as indistinguishable. With Einstein’s quick endorsement of those moves, the paper won Bose—until then completely unknown to the international scientific community—a lasting place in quantum theory. Yet, a second paper written shortly thereafter with very similar strategies on the interaction of matter and radiation remained largely ignored, and this is usually taken to be the result of Einstein’s adverse comments on the paper. Having thus shown up briefly on the scene of international science, Bose disappeared quickly from the stage of quantum theory, and

---

[50] Ashis Nandy, *Alternative Sciences: Creativity and Authenticity in Two Indian Scientists*, New Delhi: Oxford University Press, 2nd edition, 1995.

[51] Subrata Dasgupta, ‘Jagadis Bose’s panvitalism as intellectual history’, *Indian Journal of History of Science*, Vol. 57, No. 1, 2022, pp. 26–36.

[52] Deepanwita Dasgupta, *Creativity from the Periphery: Trading Zones of Scientific Exchange in Colonial India*, Pittsburgh: University of Pittsburgh Press, 2021.

for the rest of his life devoted himself primarily to the work of building a scientific community in India, adding new topics of research in crystallography and organic chemistry, in addition to his original domain of theoretical physics. He also carefully nurtured a generation of women students, who became important researchers later in life. Bose's brief track records in quantum theory, and his two papers, taken together, according to Deepanwita Dasgupta, thus offer us a window on to the world of 'peripheral science'—the relatively unexplored region of scientific practice where we often see great creativity among people who were trained by somewhat unconventional methods, bypassing the usual channel of recognized mentors and institutions. If Bose indeed trained himself in such an unconventional manner, how did he do it? And having been able to do it once, why was he not able to replicate his success for a second time? Since many scientific communities in Asia and Latin America had gone through a similar peripheral phase, at least in the beginning, it is important to understand these processes, grasp their difficulties, but also see how such peripheral practices can often become the birthplace for great scientific ideas, ushering in new conceptual changes. The reflections on such processes, Dasgupta further claims in her chapter, can help us design an optimal practice for science in the twenty-first century.

Science and Multiple Modernities

Indians differed widely in their responses to Western science and its proselytizing message of modernity. A handful of Indian elites endorsed the call for transformation and identified enthusiastically with the modernizing project. To accept the Western set of institutions and values in ditto was problematic. As it would turn them into imperfect copies of Western ideals. Thus, according to David Arnold, a need to construct a particularly Indian modernity has been seen as the driving force behind colonial subjects' participation in debates on science and technology.[53] As Partha Chatterjee asked long back how could Indians accept and assimilate Western modernity while at the same time seeking to contest colonial authority.[54] One response is to see, as Gyan Prakash does, that Western science and technology as being reinterpreted in and for the colony usually by Indian erudite and popularizers of science. Indian elites,

[53] Arnold put forward three observations. First, the discussions on modernity are largely discussions of Indian pronouncements on science and technology. Second, the need to construct an Indian modernity is seen to be arisen in 'response' to the perceived modernity of the new knowledge brought by the colonial power. This occurred primarily through a strategy of revivalism, a search for elements of modernity in India's ancient past and its indigenous systems of learning. Third, ideas of modernity are viewed as an integral component of the Indian nationalist movement that arose in the late nineteenth century and came into its own in the twentieth century. David Arnold, *The New Cambridge History of India: Science, Technology and Medicine in Colonial India*, Cambridge: Cambridge University Press, 2000, pp. 15–18.

[54] Partha Chatterjee, *Nationalist Thought and the Colonial World: A Derivative Discourse*, Minneapolis: University of Minnesota Press, 2nd impression, 1993.

in translating Western scientific ideas and technologies into vernacular languages, engaged in a 'renegotiation of knowledge and power', setting themselves up as the champions of an Indian modernity.[55] However, even at the close of the British rule, it remained uncertain how far scientific modernity could facilitate a consensus within India and how far a science conversant by Indian values could gain acceptance from an international audience.

Indian modernity is not a single entity, patented by Britain, but is capable of multiple forms and any number of cultural and political variants. Recently, Kris Manjapra has examined the changing scope and quality of interactions between Garman and South Asian intellectuals during the late years of the British Raj. A particular alignment of international politics enabled new kinds of scholarly and scientific encounters across national and colonial boundaries.[56] In the 1910s and 1920s, Indians considered Germany as the great physics and chemistry laboratory of the world and these two disciplines alone could provide the high road to scientific and industrial development. Collaborations in the sciences, arts, and humanities between German and Indian minds produced some extraordinary results—Meghnad Saha met Albert Einstein, Girindrasekhar Bose (1887–1953) began a correspondence with Sigmund Freud, and Rabindranath Tagore (1861–1941) travelled to Germany to recruit scholars for his newly founded university in Santiniketan. In 1930, Einstein had two highly publicized encounters with Tagore, in which as Manjapra has put it, both thinkers addressed fundamental philosophical questions about epistemology and ethics.[57]

There is no conflict, as recently claimed by a scholar, between Tagore's poetic imagination and scientific perception. On several occasions, his poetic vision of nature around him is not only harmonious with scientific insight but, it appears, directly inspired by it and integrates it. Some of his poems seem to mesh with the physical sciences, astronomy, and cosmology in particular; others with the life sciences, rejoicing the order and energy in the living world of nature.[58] The originality of Tagore's philosophy attracted the leading European scientists (Arnold Sommerfeld, Werner Heisenberg, including Einstein) of his time, combatting with the deep philosophical problems unrestrained by their scientific discoveries. Tagore's most sustained interaction was with the eminent Indian scientists of the day—Jagadish Chandra Bose and Prasanta Chandra Mahalanobis. As a matter of fact, the poet and Bose became great friends and the latter dedicated his book *The Nervous Mechanism of Plants* (1926) to Tagore.[59] Shiv Visvanathan in his chapter looks at Rabindranath Tagore's relationship and interaction with two scientific legends, Patrick Geddes, the

---

[55] Gyan Prakash, *Another Reason*, op.cit., pp. 83–84.

[56] Kris Manjapra, *Age of Entanglement: German and Indian Intellectuals across Empire*, Cambridge, Massachusetts and London: Harvard University Press, 2014, pp. 2–9.

[57] Ibid, p. 126.

[58] Partha Ghose, 'Rabindranath and Science', in Sukanta Chaudhuri (ed.), *The Cambridge Companion to Rabindranath Tagore*, Cambridge: Cambridge University Press, 2020, p. 349.

[59] The poet responded enthusiastically at this, 'I realized this is where our truth lies, this light, this life—this is India's essence'. Quoted in Partha Ghose, ibid, p. 339.

Scottish biologist, and Jagadish Chandra Bose, and also between him and Mohandas Karamchand Gandhi (1869–1948). Each is an event on its own, but each telescope into the other to give an intriguing picture of a multifaceted man. The letters that Tagore wrote and received from the two scientist-intellectuals are also analysed. Tagore was no ordinary nationalist and went beyond the tired categories of the modern nation state. He wanted India to smell the West, taste it, and understand the differences within it. He realized that imperialism is only one phase of the West, that there were other Wests that one could talk to and conspire with.

Whereas some Indian intellectuals and nationalists envisaged science as both the agency and emblem of their modernity, others (led by non-other than Gandhi) rejected Western modernity with its alien sciences and inappropriate technologies. Gandhi condemned the West precisely for the same reasons for which it had prided itself for so long, that is, science and technology with all its achievements, consequences, and ramifications. He seldom used the terms 'science' or 'technology'. His concern was civilization and mechanization. He believed that it was machinery (and Manchester) that impoverished India![60] Contemporary scholarship has portrayed Gandhi's relationship with industrial civilization and technology as a complex one; as opposed to the more reductionist interpretations of his work suggesting that he was an anti-modernist and Luddite. The chapter, by Dhruv Raina, in three parts returns to a historical moment dating back to the first two decades of the twentieth century to situate Gandhi's reflections on industrialism and the 'West' in the context of a larger social movement that he was instrumental in shaping and that shaped his own thinking. The second part briefly summarizes contemporary meta-reflections within STS on technology, in order to argue that the binary dichotomies that purportedly characterize Gandhi's writings on the subject disappear when Gandhi begins to be seen as a social determinist on the question of industrialization and technology. And finally, the third part takes up a recuperation of Gandhi within the appropriate technology movement in the last quarter of the twentieth century in India to detail the issues wherein Gandhi continues to be relevant for the contemporary practice of science and technology.

The standard narrative of Indo-German academic exchange usually begins with the German Indology and its role in shaping a national consciousness in India. The initial German enthusiasm for India was opposed by a criticism of the 'value' of Indian culture and a reaffirmation of Western philosophical and religious superiority. This divergence meant, according to Angelika Malinar, either an emphasis on the value of India's spiritual and metaphysical traditions, or an appreciation of its rationalistic and scientific achievements.[61] Swami Vivekananda (1863–1902), who

---

[60] The first clear-cut exposition of Gandhi's views is to be found in *Hind Swaraj* (1909). For a meaningful discussion see, Deepak Kumar, *'Culture' of Science and the Making of Modern India*, New Delhi: Primus, 2023, pp. 290–302.

[61] Angelika Malinar, 'Ideas of Indian Philosophy in Nineteenth-century Germany: Vivekananda, Deussen, and Garbe', in Hans Harder and Dhruv Raina (eds.), *Disciplines and Movements: Conversations between India and the German-speaking World*, Hyderabad: Orient Blackswan, 2022, p. 129.

met German philosopher and Sanskritist Paul Deussen in 1896, while acknowledging their contribution in preserving the integrity of the ancient scriptures, argued that the German philosophers were unable to master the Indian philosophical texts entirely. The interpretations of Indian philosophy by prominent German philosophers of the nineteenth century can be served as a model to understand the progress of the Sanskrit studies and philosophy in the academia.[62] The final chapter in the volume by Nirmalya Narayan Chakraborty claims that with the arrival of colonialism in India, the term 'Comparative Philosophy' gained a new momentum. Comparative philosophy in colonial India acquired a different connotation, different from the comparative philosophy that has been there in practice in classical period.

The complete rupture in the age-old traditional knowledge system has got some interesting consequences for the Indian academic scenario, especially with regard to philosophy. The present chapter is an attempt to focus on how the very concept of philosophy is being debated on by some contemporary Indian philosophers. Chakraborty highlights how in this debate the participants freely use ideas and debates found in European philosophical tradition, but give these a very distinctive traditional Indian colour. Comparative philosophy gives rise to a cosmopolitan philosophy in the sense that different philosophical traditions converge and this convergence could be said to be inclusive in nature. This inclusiveness moves beyond mimicking the other. Internalizing the insights of both the classical Indian and contemporary European philosophical traditions, a novel creative philosophical repertoire has been generated. This post-colonial Indian philosophy defies any geographical labelling.

Coloniality, Modernity, and Decoloniality

'Modernity, of course, is not a decolonial concept, but coloniality is. Coloniality is constitutive, not derivative, of modernity. That is to say, there is no modernity without coloniality…' Thus, convincingly argued by two scholars recently.[63] The spate of discussions and publications on decolonization or decoloniality indicates an enduring and increasing interest in the subject.[64] In Asia, Africa, and Latin America, the varying character, depth, and length of colonial experience set the multiple modalities and spatiotemporal contexts for the efforts of decolonizing knowledge. This situation

---

[62] Malinga further argues that despite the attempts by Paul Deussen and Richard Garbe to launch Indian philosophy as a subject of study, while it was established as an area of specialization in Indology, it remained a marginal course in the philosophy department of German universities. Ibid, pp. 164–165.

[63] Catherine E. Walsh and Walter D. Mignolo (eds.), 'Introduction', *On Decoloniality: Concepts, Analytics, Praxis*, Durham and London: Duke University Press, 2018, pp. 3–4.

[64] Sekhar Bandopadhyay, *Decolonization in South Asia: Meanings of Freedom in Post-independence West Bengal, 1947–52*, Hyderabad: Orient Blackswan, 2009; also see, Sabyasachi Basu Ray Chaudhury, 'Governance of migration in South Asia: the need for a decolonial approach', in Tamar Mayer and Trinch Tran (eds.), *Displacement, Belonging, and Migrant Agency in the Face of Power*, London and New York: Routledge, 2022, pp. 78–91.

is understandable given the seriousness of the problems that were accompanying with formal colonialism—poverty, inequality, exploitation, and epistemic imperialism, among others. The East-West divide has been given way to the North-South split. Despite the rise to economic prominence of some former colonies, the political and intellectual power globally remained, until the rise of China, largely concentrated in the hands of former colonizers. One consequence of the centrality bestowed to decolonization, is embedded in its long colonial history. With scholars focussing on the enduring impact of colonization, the tendency is to look back into the colonial past for the roots of contemporary problems, and to the even deeper past (ancient?) for the solutions to them.

In an era of (pseudo?) revivalism, Deepak Kumar anticipates, history of science acquires special significance and makes insistence on scientific temper and definitive sources even more important. The study of history of science may provide a better understanding of the historical processes and forces.[65] Can the ethos and function of decolonization or decoloniality be extended to the realm of science and modernity? Is there any pattern or design? David Fairchild, an American scientist, mentioned several Indonesian (Oedam, head gardener at the Bogor Botanical Gardens, for example) who helped him during his visit to the then Dutch East Indies in 1895. One Henry Ogg Forbes, another scientist from Scotland who studied the flora and fauna in the Dutch Indies during 1878 and 1883, had similar observations about the local expertise.[66] Recently, a group of academicians have identified people like Oedam as intermediaries (go-betweens) connecting Western and indigenous knowledge in the eighteenth and nineteenth centuries. They argue that by paying more attention to knowledge collectors, translators, procurers, and others, we can better understand that the modern world emerged not in Europe but in global exchanges.[67] Although the role of intermediaries is important, this group is often missing in the narratives of colonial history. They remain unsung heroes of our knowledge production.[68]

For Fajri Siregar, the movement to decolonize science invites the scientific community to forcefully enter to the field of knowledge production largely dominated by the Global North and to make provision for local knowledge produced by local talents all across the globe. Thus, it talks about moving away from scientific absolutism originated from the dynamics of colonization. The social scientists can

---

65 Deepak Kumar, 'Why HISTEM and how to do it?', Kuruvila Zachariah Memorial Lecture, Presidency University, Kolkata, 7 November 2019.

66 Irawan Santoso Suryo Basuki, 'Why the role of native Indonesians in developing science is often overlooked during colonialism?' *The Conversation*, 27 July, 2020. https://theconversation.com/why-the-role-of-native-indonesians-in-developing-science-is-often-overlooked-during-colonialism-143263 (Accessed on 15 June, 2023).

67 Simon Schaffer, Lissa Roberts, Kapil Raj, and James Delbourgo (eds.), *The brokered world: go-betweens and global intelligence, 1770–1820*, Sagamore Beach, MA: Watson Publishing International, 2009.

68 Several Indians without academic training in engineering and sciences excelled in the field of modern technology. However, they remain absent in the official narratives and can only be found in the contemporary vernacular publications. Suvobrata Sarkar, *Let there be Light: Engineering, Entrepreneurship and Electricity in Colonial Bengal, 1880–1945*, Cambridge: Cambridge University Press, 2020, pp. 203–212.

play a pivotal role in the process, but not many Indonesian social scientists, Siregar laments, have been involved in the decolonizing science, an important discourse.[69] A prominent Indian sociologist, Benoy Kumar Sarkar's (1887–1949) campaign for a materialistic orientation of Indian culture brings us to the last point we wish to emphasize—how did South Asians writing in the twentieth-century envisage terms like science and modernity? Sarkar argued (1918):

> The real and only cause of the parting of ways between the East and the West, nay, between the medieval and the modern, was the discovery of steam, or rather its application to production and transportation. The steam engine effected an industrial revolution during the first three decades of the nineteenth century. It is this revolution which has ushered in the 'modernism' of the modern world in social institutions, science, and philosophy, as well as brought about the supremacy of Eur-America over Asia…[70]

Thus, the volume interrelates a wide range of perspectives from the lived experiences of coloniality and decolonial thought and practices in different local histories from South Asia.

Kolkata, India

Sahara Ahmed
sahara.ahmed@rbu.ac.in

Suvobrata Sarkar
suvobrata.history@rbu.ac.in

[69] Fajri Siregar, 'Decolonizing science: it's high time for Indonesia to break free from overreliance on Western theories', *The Conversation*, 24 March 2022. https://theconversation.com/decolonising-science-its-high-time-for-indonesia-to-break-free-from-overreliance-on-western-theories-179474 (Accessed on 15 June 2023).

[70] Benoy Kumar Sarkar, *Hindu Achievements in Exact Sciences: A Study in the History of Scientific Development*, Bombay, Calcutta, and Madras: Longmans, Green and Co., 1918, p. 7.

Contents

Editors and Contributors

About the Editors

Sahara Ahmed is Professor at the Department of History, Rabindra Bharati University, Kolkata, India. She obtained her Ph.D. from University of Calcutta. Her research interests include histories of ecology, environment, and sustainable development, health and medicine in colonial and post-colonial contexts. She is the author of *Woods, Mines and Minds: Politics of Survival in Jalpaiguri and the Jungle Mahals, 1860–1970* (Primus, 2019). Her most recent publications—*Epidemics and the Indigenous Tribes: Sub-Himalayan Bengal and the Jungle Mahals*, in Poonam Bala and Russel Viljoen (eds.), *Epidemic Encounters, Communities in the Colonial World* (Lexington Books, 2023), and *Designing Scientific Mining: Evolution and Implementation, c. 1860s–1960s*, in Suvobrata Sarkar (ed.), *History of Science, Technology, Environment, and Medicine in India* (Routledge, 2022).

Suvobrata Sarkar teaches at the Department of History, Rabindra Bharati University, Kolkata, India. He obtained his Ph.D. from Jawaharlal Nehru University, New Delhi. His research explores history of technology in the context of the nineteenth- and twentieth-century South Asia. Sarkar is the author of *Let there be Light: Engineering, Entrepreneurship and Electricity in Colonial Bengal, 1880–1945* (Cambridge University Press, 2020), in addition to several articles and book chapters. He has also edited *History of Science, Technology, Environment and Medicine in India* (Routledge, 2022).

Contributors

Sandipan Baksi is a historian of science, with a focus on the development of agricultural science and technology in India during the colonial and post-colonial period.

He is particularly interested in studying the varied Indian engagements with development of science and technology in general and scientific agriculture in particular. He currently works as Director at the Foundation for Agrarian Studies, Bengaluru, India.

Rup Kumar Barman is Professor at the Department of History of Jadavpur University, Kolkata, India. He was also formerly Head of the Department of History of Jadavpur University (2016–20) and Vice-Chancellor of Murshidabad University (2023). Professor Barman has authored multiple books and research papers in various national and international academic journals. He is currently engaged in studies on state, border, citizenship, Dalit discourse, folk medicine, and environmental history. Some of his publications include *Nature, Communities and State: An Environmental History of Bengal and Assam* (Abhijeet Publications, 2019); *Raidak, a Transitional River: From Bhutan to Bangladesh to India* (Mittal, 2021), *River, Society, and Culture* (Primus, 2023) and *Enclaves of the India-Bangladesh Border* (Routledge, 2023).

Santanu Chacraverti is an activist associated with DISHA. An erstwhile academic, he now works on a broad spectrum of social and environmental issues, with an eye on the interface between traditional technology and modern scientific knowledge. He has researched and written mainly on the evolution of ideas, and on livelihood issues in an ecological context. His unpublished doctoral dissertation—*Ramendrasundar Trivedi and Bengal's Response to Modern Western Science*—deals with Ramendra Sundar's epistemology of science in analytical detail. His monograph on Bengal's *Subhankari* lore, published by the Asiatic Society, provides a detailed overview of the tradition. His *The Sundarbans Fishers: Coping in an Overly Stressed Mangrove Estuary* deals with the small-scale fishing community in the Sundarban delta and has been published by the ICSF Trust. The latter has also published his *Profits and Perils of Farming Fish*. He has authored several research reports, and his articles and book reviews have been published in various anthologies, journals, and newspapers.

Nirmalya Narayan Chakraborty is Professor of Philosophy, Rabindra Bharati University, Kolkata, India. Earlier he served as Vice-Chancelor of Rabindra Bharati University. A former Commonwealth fellow at the University of Glasgow, Scotland (2004–05), he also served as Visting Professor at University of Naples 'L'Orientale', Italy (2009), Assam University (2008), and University of Hyderabad (2007), India. He contributed chapters titled 'Tagore and the Idea of Emancipation' in Sukanta Chaudhuri (ed.), *The Cambridge Companion to Rabindranath Tagore* (2020); and 'Methodology in Indian Philosophy', in Purushottam Bilimoria (eds.), *Routledge Handbook of Indian Philosophy* (2018). His most recent publication is *A History of Navya-Nyaya Scholarship in Bengal* (Motilal Banarasidas, 2023).

Arnab Rai Choudhuri obtained Ph.D. from University of Chicago in 1985 and joined Indian Institute of Science in 1987, retiring from there recently as Professor of Physics. A theoretical plasma astrophysicist keenly interested in the history of science, he is known for developing the currently accepted flux transport dynamo model of the 11-year sunspot cycleAws. He is the author of *The Physics of Fluids and Plasmas* (Cambridge University Press) and *Astrophysics for Physicists* (Cambridge

University Press). He is an elected fellow of all three science academies of India and the World Academy of Sciences. He was the second Indian to win the Chandrasekhar Prize of Plasma Physics.

Deepanwita Dasgupta is currently Associate Professor of Philosophy at the University of Texas at El Paso. She did her Ph.D. from the University of Minnesota in 2010 on the problem of creativity in small and marginal scientific communities that generally begin their career as self-trained outsiders. She finally published a book on this theme, titled, *Creativity from the Periphery* (University of Pittsburgh Press, 2021). This book tells the story of a small physics community in India during the 1920s and their contributions in optics and quantum mechanics. Her recent research interests focus on the idea of wisdom in science, and whether it is possible to revive wisdom inquiry and natural philosophy once again in our present research environment. She has recently been a fellow both at the Pittsburgh Center for Philosophy of Science and at the University of York, UK, on this wisdom project.

Venni V. Krishna is currently Honourary Professor, School of Humanities and Languages at the University of New South Wales Sydney, Australia. He has been elected as Fellow of the Royal Society, NSW in 2019. He has a Ph.D. from the University of Wollongong, Australia and has more than 30 years of research, teaching and consultancy experience in science and technology policy studies, history and sociology of science and technology, innovation studies and science, technology and developing world in leading academic and research institutions in India, Australia, Singapore, China, and Canada. He was Professor in Science Policy and Chair, Centre for Studies in Science Policy, School of Social Sciences, Jawaharlal Nehru University, New Delhi, for more than 20 years. He held visiting faculty positions at the National University of Singapore; McGill University, Montreal, Canada; Maison des Science De l Homme, Paris; United Nations University, Japan; Western Sydney University, Sydney and Tsinghua University, Beijing. His latest publication includes *Universities in the National Innovation Systems: Experiences from Asia Pacific* (Routledge, 2018) and *Chinese Universities in the National Innovation System* (Routledge, 2020) He is Editor-in-Chief of the journal *Science, Technology and Society* (Sage).

Kamlesh Mohan is an Emeritus Professor of modern history at Panjab University, Chandigarh, India. Besides her several research papers in national and international journals, her major publications include *Science and Technology in Colonial India* (Aakar, 2016), *Towards a Gender History: Images, Identities, and Roles of North Indian Women* (Aakar, 2006), *Militant Nationalism in Punjab, 1919–1935* (Manohar, 1985; Routledge Edition, forthcoming). Her co-edited book titled *Pandemics and Literature: Regional and Global Perspectives* is forthcoming from Routledge. She was Visiting Professor at Maison de Sciences de I' Homme, Paris, twice (1998 and 2002).

Neshat Quaiser is Associate Researcher with the Centre de Sciences Humaines (CSH), Delhi, India. His research interest includes: Critical Social Theory; Historical Sociology of South Asian medical traditions with a focus on Unani medicine; Muslims, Islam, Sharia, and Communalism in Indian subcontinent; Culture and

Society; Historical Sociology of Urdu language and literature; Colonialism and colonial politics; Post-colonial Law and Legal Theory; Religion, Nationalism, and Peasant Politics in Indian subcontinent with reference to Muslim politics and society; and Inter-religious public debate during the colonial period.

Saurav Kumar Rai has done his M.Phil. and Ph.D. from the History Department, University of Delhi, on social history of health and medicine. His monograph titled *Ayurveda, Nation and Society: United Provinces, c. 1890–1950* is scheduled to be published by Orient BlackSwan in 2023 under the prestigious series of York University 'New perspectives in South Asian History'. He has several articles to his credit in reputed journals including *Indian Economic and Social History Review, History and Sociology of South Asia, Gandhi Marg* and *Social Scientist*. He is also an author of an undergraduate history textbook titled *Debating Modern Indian History* (Manak Publications, 2021). At present, he is working as Research Officer at the Gandhi Smriti and Darshan Samiti, New Delhi.

Dhruv Raina is currently Professor at the Jawaharlal Nehru University, New Delhi, where he teaches the history and philosophy of science. He studied physics at the Indian Institute of Technology, Mumbai, and received his Ph.D. in the philosophy of science from Göteborg University. His research has focussed upon the politics and cultures of scientific knowledge in South Asia both in historical and contemporary contexts, the history and historiography of mathematics, and the philosophy of sciences and social sciences.

Dinesh C. Sharma is a science journalist and researcher based in New Delhi. His areas of interest include science and society, the history of technology in post-colonial India and innovation. His book—*The Outsourcer: The Story of India's IT Revolution* (MIT Press, 2015)—was awarded the Computer History Museum Prize by the Society for History of Technology. He has been a Fellow of the New India Foundation and Jawaharlal Nehru Memorial Fund, and a teaching fellow at JNU. He was Science Editor at Mail Today and Founding Managing Editor of India Science Wire. He holds Ph.D. in health communication from BHU and a postgraduate degree from Osmania University.

Shiv Visvanathan is a sociologist who did his Ph.D. from Delhi School of Economics and worked at the Centre for Study of Developing Societies and the Ambani Institute of Computer Technology. He is the author of four books, including *Organizing for Science* (Oxford University Press, 1985) and *A Carnival for Science* (Oxford University Press, 1997). He was Henry Luce Professor at Smith College Massachusetts and held visiting professorships at Stanford, Goldsmith, Maastricht, Pretoria. He was also a visiting professor at NID Ahmedabad, IIT Chennai, and the Raman Institute. He is currently a professor at Jindal University and an adjunct professor at the National Institute of Advanced Study, Bangalore. He also lectures and researches on film, corruption, and violence and has published over a thousand newspaper articles. He is currently working on issues regarding anthropocene and peace.

Part I
Emerging Disciplines and Science in Vernacular

Chapter 1

Monies, Measures, and Arithmetic(s): a Brief Glimpse into the Śubhaṅkarī Tradition

Santanu Chacraverti

1.1 A Personal Prelude

Through the mists of early childhood comes the memory of toying with cowrie shells. Possibly, most people had a few at home. Today, they are more difficult to come by.

The small smooth glossy shells seem to have been designed to captivate. Indeed, the word "porcelain" originates from the Italian "*porcellana*", meaning *cowrie shell*. However, in those days of blissful ignorance, we merely knew cowrie shells as interesting looking objects that had some connection to rituals, and with which we played the indoor game *Golokdhām*. I vaguely recall the sense of the game but not the rules. Another well-known game was *Daś P̃aciś*.[1]

The term kaṛi, meaning the cowrie shell,[2] became the metaphor for money and measure, manifested in numerous Bengali usages, with meanings that included money, quantity, wealth, and cheapness. The word found its way into personal names like Tinkari and Panchkari. However, it was only much later that I came to learn that the cowrie shell was an amazingly successful and resilient currency, which had played a crucial role in trade between regions, countries, and continents, including the African slave trade!

But, what does all this have to do with Śubhaṅkarī—a tradition of computing and mathematical instruction? Everything! For one, in the Bengal area and in

[1] For a clue to diacritics, see the upcoming 'note on orthography'.

[2] The term 'cowrie' seems to have been derived from the corresponding term in Bengali, Hindi, or Marathi. See, for example, Wilfrid Jackson (1916); and *Hobson-Jobson: A Glossary of Anglo-Indian Words and Phrases and of Kindred Terms*, London: John Murray, 1903, p. 269.

S. Chacraverti (✉)
Society for Direct Initiative for Social and Health Action (DISHA), Kolkata, India
e-mail: santanuchacraverti@dishaearth.org

S. Ahmed and S. Sarkar (eds.), *Decolonizing Science and Modernity in South Asia*,
https://doi.org/10.1007/978-981-97-1829-0_1

some other parts India, the way cowries were counted became the *paradigm of counting*, enshrined in the Śubhaṅkarī tradition! Other interesting aspects followed from this, as we shall see.

1.2 This Essay—Content and Structure

This essay aspires, among other things, to be a brief, introductory, and simplified sketch of the Śubhaṅkarī tradition of elementary mathematical instruction, famously enshrined in verses. The Śubhaṅkarī lore included both practical and more purely mathematical components. The tradition emerged during the Mughal period. It was very active in Bengal up to the early decades of the nineteenth century. Indeed, its practical components continued to be useful even into the twentieth century, because, while cowries had died out as currency, the units of money and measure remained organized on traditional lines until the establishment of decimal metrics in the 1950s.

The present tale tells about the tradition by trying to trace its intellectual threads within the multilayered texture of history, culture, society, and economic activity. It is presented in seven parts. Part I presents a few elementary facts about traditional Indian units and about Śubhaṅkarī. Part II is about the socio-historical context of the tradition. It looks at the most important site of Śubhaṅkarī instruction, the pāṭhśālā, primarily, though not entirely, through the familiar yet still refreshing lens of the Adam Reports. Part III introduces the reader to the traditional units of weight and measure of Bengal and provides a first peep into practical Śubhaṅkarī computation. This soon calls for a look into the monetary units. That comes in the next part, Part IV—devoted to money and related things (for example, trying to clear up the cobweb of confusion surrounding an apparently *timeless exchange rate* of 1280 cowries to a silver taṅkā). Part V provides further exposure to practical computations. Part VI moves to a more textbookish realm—the problems of basic arithmetic and algebra. Part VII concludes the tale, through a short discussion on the significance and demise of Śubhaṅkarī.

1.2.1 A Note on Orthography

Special symbols have been avoided in the case of proper nouns, i.e. person names, book names, place names, and caste names. Thus, *Bankura*, *Bardhaman*, and *Bankimchandra*, *Kayastha*, *Bagdi*, *Vaishnava*, *Pala*, *Sena*, *Bharatchandra*, *Bidyasundar*, and even *Subhankar* appear in their usual Romanized forms. So do terms that continue to be in common use today, such as the 'rupee' and 'paisa'.

For other words and terms from Bengali (and Sanskrit) we have used the usual letters of the IAST convention. Due to uncertainty about what should be the IAST for the *voiced retroflex flap* (represented in Bengali by the symbol "ড়"), we have used the symbol ɽ (the IPA symbol for that sound) and, for similar reasons, for the nasal twang imparted by *candrabindu* in Bengali, the tilde has been used (thus, *p̃a* in this article denotes the nasal version of *pa*).

In transliterating Bengali technical terms, the orthography follows the standard Bengali spelling. In transliterating the Śubhaṅkarī verses, *the orthography follows the modern standardized Bengali pronunciation and not the original spelling,* as the spellings varied across manuscripts and the pronunciation varied (and varies) across districts.

1.3 Part I: Śubhaṅkarī Preliminaries

1.3.1 Counting Fundamentals: In the Beginning Was the Unit

The stamp of society on knowledge is best seen in our choice of units.

A measure of length or weight is intrinsically arbitrary. With some change in the social and practical context, another could have been chosen in its place. This is most dramatically revealed in the use of *notional* units—for example, the *mill* (1/10th of a cent) used to measure money in the US, the 'rupee current' used in measuring money by East India Company accountants, and GMs ("notional Gold Mohurs") used as a measure of lawyer's fees in certain law courts in India. We shall encounter other notional units later.

However, it is important to recognize that not only the so-called notional units, but, any unit, whatsoever, is essentially notional—a matter of culture, convenience, and choice.

1.3.2 Of Measure and Measuring

But, while the unit, and therefore measure, is relative, *measuring is not.* This is because, once the unit is chosen, the properties of *real numbers* take over. As long as we are in the domain of real numbers, these principles seem inescapable. Are the principles also eternal in the Pythagorean-Platonic sense? Or are we just in touch with the fundamental patterns underlying our cognition? Interesting questions, but not quite our concern in this essay!

Between, measure and measuring, between units and calculation, there exists the realm of *methods employed to calculate*. Here, again, culture and context dominate. We find methods employed varying in form and tone across cultures. But, inter-method interpretation is possible and takes place regularly. We shall be doing a little of that here also. This again points to the existence of something fundamental underlying numbers and numerical operations.

1.3.3 Of Decimal Number Notation and Non-decimal Units

Our familiarity with the decimal system has resulted in a prejudice in favour of *ten*. However, the number *ten* has *no* intrinsic advantages as compared to, say, *twelve*.

The actual achievement of the ancient Indians was not in selecting *ten* as the base number. The achievement lay in developing a *place value* system with a separate symbol for zero. Such a system can be used in favour of any base number, as those acquainted with, say, the binary notation know well.

Then, why does the decimal base seem so natural to us? One, we are used to it. More importantly, our modern units of length, area, mass, and weight are organized along decimal lines. This harmony between the base of number notation and system of units is a huge advantage in computation.

However, the weights and measures *were not organized on decimal lines in India*. This, of course, was *also the situation* in the rest of the world until modern times. (Even today, our measurement of time is not organized on decimal lines, as every school kid learns the hard way.)

In India, in the ratios obtaining between the traditional units of weight and measure, 4 and its multiples dominated. In fact, had the base of the Indian system been four, eight, or sixteen (and particularly the number twelve), the situation would have been ideal.[3] However, consistency between domains is not a dominant feature of human culture. *This is where Śubhaṅkarī tradition of computation becomes relevant.*

1.3.4 Practical Computation in the Śubhaṅkarī Tradition

The Śubhaṅkarī tradition evolved methods to negotiate with traditional units of weight, measure, and currency organized on different ratios without forsaking a base-10 number system. One knowing the rules presented in the verses could undertake practical computation without engaging in complicated calculations. The verses, occasionally no more than doggerels, were called *āryā*s. The instruction was imparted in the *pāṭhśālā*s—the traditional elementary schools in rural Bengal, but also at home. Often, many children did not understand their use while memorizing them but grew to appreciate their value when learning to apply them in daily life. Therefore, true mastery of such verses was often acquired much later in life.[4]

[3] The general advantage of base twelve over base ten is that ten has only by 2 and 5 as its non-trivial divisors, but twelve has 2, 3, and 4, and 6 as its non-trivial divisors. This leads to substantial advantages in expressing fractions and its divisibility by 4 leads to obvious advantages with traditional Indian units.

[4] Lalitmohan Bandyopadhyay, "Shubhankar", *Bharatavarsha*, Fifth Year, Vol. 1, no. 5, Kartik, 1324 BS, pp. 670–71.

1.3.5 Orality

One purpose of versifying the problems was being able to retain them in memory, sharing them orally, and solving, at least some of them, without the aid of pen and paper. This is why Śubhaṅkarī were often referred to as *mānasāṅka*. This brings out the oral aspect of the tradition.

1.3.6 The Not-So-Practical Challenges

In addition to practical algorithms, there were verses of other types—concerned with honing arithmetic skills and trying to impart to students a fascination for numbers and their properties. Śubhaṅkarī verses describing arithmetic problems were occasionally exchanged among older children and adolescents as standard riddles.[5]

1.3.7 Sources Explored in the Present Research

The living tradition had largely died out by the early twentieth century. Thus, the basic research on which this paper draws was compelled to make sense of the tradition from original manuscripts, published manuscripts, from reports of the tradition in nineteenth and twentieth century literature, and from arithmetic textbooks that included elements of the tradition until the decimal system was officially accepted in the 1950s. We shall cite the sources as they come up in the course of the narrative.

1.4 Part II: The Socio-historical Context of Śubhaṅkarī

1.4.1 The Śubhaṅkarī Tradition in Time and Space

Śubhaṅkarī supposedly derives its name from a certain Subhankar Das or Subhankar Roy. His identity is uncertain. The Bishnupur area, present in the Bankura district in West Bengal, has developed strong claims as the birthplace of *the* eponymous Subhankar.[6] Nevertheless, the so-called Śubhaṅkarī verses bear the name of several authors.

[5] Day (1878), p. 92.

[6] For a review of the controversies and confusions regarding an eponymous Subhankar, see, for example, Chacraverti (2007), pp. 24–37.

Many Śubhaṅkarī manuscripts have been found in the Bengal area and Bengali nationalism has automatically appropriated the Śubhaṅkarī as a Bengali tradition.[7]

However, according to Sukumar Sen, computational verses, generally stamped or associated with the name of Subhankar, are not only a feature of Bengal *but also of Assam*. There, these were usually called 'Kāyathali' verses, as account keeping was largely the prerogative of the Kayastha caste. The name of Subhankar Das is a familiar one in Assam also. Manuscripts have been found in Assam with verses bearing the names of Subhankar, "Subhankar"-Gaurdas, Raghupati, Bakul Kayastha, Hriday Kayastha, and so on. One such manuscript was dated as 1730–31. Often the language and content of some of the verses found in Assamese manuscripts are very similar to some found in Bengali manuscripts.[8] Thus, as in the case of Navya Nyaya or Vaiṣṇava poetry, so in the case of Śubhaṅkarī also, we are looking at an instance of common Eastern Indian heritage. It would be interesting to get similar information from Bihar or Odisha, but, except for some verbal reports, such information has not been forthcoming.[9]

The earliest known couplet of Śubhaṅkarī, argued Sukumar Sen, carries the unmistakable stamp of Eastern Abahaṭṭha, the language that was the direct ancestor of the tongues of eastern India. On these grounds, he was inclined to trace the origins of the tradition to the fifteenth century. Be that as it may, the language of the manuscripts available to us suggests that the main corpus developed during the seventeenth and eighteenth centuries and many manuscripts belong to the late 18th and early nineteenth centuries. *Moreover, the extant manuscripts are from the western districts of the Bengal area and, hence, reflect the units prevailing in these areas.*

1.4.2 The Bengali Society and Elementary Education—Through Adam's Lens, Though not Entirely

The verses contain occasional hints about the social context of the Śubhaṅkarī. However, for Bengal, we also have a valuable guide in the reports prepared by William Adam on the state of education in Bengal in the early nineteenth century. When Adam did his educational surveys (1835–38), many elements of precolonial society—including much of the social character of the villages, the paṭhśālā, its pedagogy, and the syllabi taught—had lingered. Adam had a discerning eye.

[7] Chacraverti (2010).

[8] Sen 1963 (1965), pp. 482–83.

[9] The present author remembers his father, who grew up in Bihar, mentioning the presence of Śubhaṅkarī type verses in the Bihari tradition and the Late Ramesh Chandra Das of Utkal University mentioning the presence of similar verses in the Odiya tradition. However, studies on ethnomathetical traditions of Odisha and Bihar seem hard to come by.

Moreover, he was alert to how his own presence as a *sahib* seeking information for the Company government might affect the quality of the information elicited. Therefore, his reports are particularly valuable.

Adam prepared three reports from 1835 and 1838.[10] The first report drew on other official and unofficial reports. The next two reports were based on Adam's own investigations on specific areas and provided in-depth information.

The second report was on Rajshahi district. It is a pity that Adam could not examine the situation in any other district of eastern Bengal at first hand!

The third report was on the district and the city of Murshidabad, the districts of Birbhum and Bardhaman in western Bengal, and the districts of Gaya and Tirhoot in Bihar. (There is also some information on Midnapore provided by the then Acting Joint Magistrate of the district.) By the way, Adam does not use the term paṭhśālā, but refers to them as "indigenous elementary schools" and, occasionally, as "elementary Bengali schools", to distinguish them from elementary schools where Arabic or Persian were taught.

Our rendition of the portrait sketched by Adam will be a thumbnail. Nevertheless, we shall try to indicate some features of the society and education in Bengal drawing mostly on the reports of Adam.

- All thanas in each district, and many, if not most, villages seem to have had at least one paṭhśālā—sites of elementary instruction in reading, writing, and mathematics.
- *There was no state-sponsored programme* of popular education. The paṭhśālās grew up in response to a social need. The students often paid for the education with such fees (in cash or kind) as they could manage and, often, the schools were supported by the charity of private patrons. In a few cases, teachers with independent incomes offered the education free of cost.
- At the basic level, elementary arithmetic was taught, including multiplication tables up to 20. In *all* paṭhśālās offering the *full* course, *agricultural and commercial accounts* were taught at the advanced level. Students usually had to choose one or the other and the decision was based on family background and possible choice of future profession.
- In most schools, Śubhaṅkarī verses were taught, though many verses were composed by authors other than Subhankar. However, Adam says that such verses did not feature in the curriculum of all schools. This would seem to suggest that, notwithstanding the predominance of metrical compositions in pedagogy, the principles and algorithms of agricultural and commercial accounts were often taught *without the aid of verses*.
- *Writing was not neglected*. From the earliest to the most mature stage, the student was made to write out what he had learned and this, Adam felt, was superior to the mode of elementary instruction in his home country.[11]

---

[10] Adam (1941).

[11] William Adam, *Reports*, Anathnath Basu, op. cit., pp. 145–46.

- Among the teachers, the Kayastha caste, traditionally, the caste of scribes and accountants, were far better represented than any other caste, with Brahmans ranking a poor second.
- By no means, *all* or even *most* children in a village having a paṭhśālā attended the latter. Most students were from the upper castes. Children from lower caste families, particularly poor lower caste families, tended to be far fewer in number.[12]
- Muslims were noticeably *underrepresented* in the paṭhśālās even in districts with Muslim majority or very large proportion of Muslims.[13] The poor representation was noticeable as most Muslims in Bengal were Bengali, and that paṭhśālā was the only formal educational establishment where instruction was in the regional vernacular.
- Poromesh Acharya thinks that Adam does not mention *maktab*s—meant for elementary education of Muslim children. Therefore, Acharya has drawn on other sources to convincingly demonstrate that maktabs were essentially elementary schools of Arabic learning, where children were taught to read and recite the Quran at the basic level. Perhaps Bengali was also taught in an occasional maktab, but unlike paṭhśālās, the maktabs were not meant to be schools of general elementary education in the vernacular.[14] This explains why Muslim families seeking proper vernacular education for their children dispatched them to the paṭhśālās. (It doesn't seem that Adam has entirely missed mentioning maktabs. He called them "elementary Arabic schools" and seems to have had poor opinion of them[15]—in contrast to *elementary Persian schools*, which, he felt, were generally superior to the paṭhśālās. His problem seems to be that he has failed to always distinguish between maktabs and *madrasa*s.)
- Well-off and well-educated families often preferred to teach their kids at home. However, in general, learning at the paṭhśālā was usually considered superior to learning at home.[16] Therefore, particularly among the Hindus, caste status and poverty were chief deterrents to attending the paṭhśālā.
- Notwithstanding the poorer presence of very low caste children *as compared to* students from the upper castes, Adam notes that unlike in the elementary schools in Bihar, *the paṭhśālās in Bengal had a considerable presence of children from the lower castes*. In fact, there were children even from the very lowest castes studying in the paṭhśālās in some of the districts that Adam carefully studied.[17] In fact, he also notes that the interest of lower caste families in

[12] Adam notes this repeatedly in his second report and third report.

[13] As is evident in the second report (on Rajshahi) and the report on Murshidabad.

[14] Acharya (1989), pp. 147–92.

[15] For example, William Adam, *Reports*, Anathnath Basu, op. cit., pp. 152–53.

[16] William Adam, *Reports*, Anathnath Basu, op. cit., pp. 158–59.

[17] For an analytical discussion of the caste statistics of children in the paṭhśālās of Burdwan, Birbhum, and Murshidabad as presented by Adam see Poromesh Acharya, op. cit., p. 111–12, and Santanu Chacraverti, *Śubhaṅkarī*, op. cit., pp. 128–53.

sending their kids to school seemed to be increasing. He thought that this was due to the salubrious influence of the British rule. However, it is possible that he *was witnessing a trend that had been going on for some time*. Indeed, there were traditional trends in Bengali society positively oriented towards taking at least vernacular instruction beyond caste barriers, as noted by Poromesh Acharya.[18]

- Quite a few paṭhśālās had *lower caste teachers*, with some teachers coming from the '*lowest*' castes.
- This is not to suggest that the upper castes were, in general, positively inclined towards the educational empowerment of the people, but to suggest that elementary education of the traditional type was not necessarily seen as the privilege of certain specific castes, nor was extending such education to the lower castes *necessarily* seen as a threat to upper caste status.
- Adam does not seem to have found any girl students in the paṭhśālās he surveyed. Further, the overall scenario of social apathy towards and even serious social resistance to providing even elementary education to girls that Adam portrays does not seem to conflict with what other sources say. Adam admits that some girls in privileged, particularly, Zamindar families received some education in reading writing and keeping of accounts *in their homes*, to ensure they be able to protect their own rights after marriage.[19] He also admits that girls in Bengali Vaishnava families often received some education to allow them to read their religious literature.[20] Incidentally, in the Śubhaṅkarī verses also, there do not seem to be noticeable references to girls getting systematic instruction—except for a solitary *āryā* addressed to a '*Kayasther bala*' (*Kayastha lass*) indicating computational instruction at least for Kayastha girls.

1.5 Part III: Śubhaṅkarī—A First Peep into Practical Computation

1.5.1 The Content of Śubhaṅkarī

The content of the Śubhaṅkarī manuscripts can be divided into three categories.

1. Algorithms for practical computation;
2. Arithmetical and simple algebraic problems;
3. Playing with numbers and sheer computational fun.

The verses, however, was often terse and enigmatic, often not meant to be learnt from manuscripts but from teachers, elders, peers, and by actual application.

[18] Poromesh Acharya, *Banglar Deshaja Shikkhadhara*, op. cit. pp. 88–111.

[19] He says this in connection with his report on Natore, Rajshahi. William Adam, *Reports*, Anathnath Basu, op. cit., p. 188.

[20] William Adam, *Reports*, Anathnath Basu, op. cit., p. 189.

1.5.2 The First Category—Practical Verses

The practical material consisted of verses that helped one to negotiate successfully with traditional units of weight and measure. This was exceedingly important, as practical computation could be a serious problem.

1.5.3 The Computational Challenge

Bengal, like other places in the subcontinent, had a decimal system of numerical notation as it evolved in India. However, as in other parts of India, the weights, measures, and monetary units were proportioned not on 10 (or its powers) but on other numbers, mostly 4 and 16.

Thus, computation could be a headache, similar to what one can face while operating with sexagesimal measures of time and angle, but more severe, as the proportions often varied as one went up or down the scale.

1.5.4 Some Units of Weight

Let us now take a look at some units of weight that were used in Bengal in the immediate pre-colonial period.

Measures of weight:

1 *man* = 40 *ser*
1 *ser* = 4 *pauā*
1 *pauā* = 4 *chaṭāk*
1 *chaṭāk* = 4 *kāccā*

Another crucial unit of weight was known as the *tolā*. It was a more ancient unit than the ser. It was related to the *ser* by being defined as its 1/80th part. We also have the following list:

1 *tolā* = 12 *māṣā*
1 *māṣā* = 8 *rati*
1 *rati* = 4 *dhān*

The fundamental thing about the weights (and other measures) is that these were not standardized in the sense of modern weights. Even in the Bengal area, if measured against another standard measure, say *troy pounds* or *grains* (used by the British), they would vary from one district of Bengal to the next.

However, while the values varied, the ratios between the weights tended to be the same at least over a large area such as a province. Thus, in most areas of the Bengal region, particularly its western parts, 40 sers made a *man*, and 16 chaṭāks equalled a ser. Therefore, as the exact value of the *man* varied, so did that of a ser, a chaṭāk, or a kāccā. Moreover, we know that the traditional 'sicca' rupee from the Murshidabad mint weighed 1 tolā. The British found it to weigh 179.666 grains, which gives us a grain measure for a tolā.[21] This translates to 11.64216096 gm for a tolā. As 80 tolā = 1 ser, therefore, we get a value of 931.373 gm to a ser and corresponding values for the other units of weight. (The above have been given to indicate how those values would translate to our modern units. In the 1830s, the Company Government slightly modified the values to line them up with their own weights, so that the tolā was defined as 180 grains and result in a ser of 933.005 gm.[22] In 1956, the Indian government declared a ser of 933.10 gm.[23] All these changes raise some questions of measurement, but, we shall avoid them in this essay.)

Let us now move on to the units of length.

1.5.5 Linear Measures or Units of Length

What is known as the traditional standard Bengal measure, widely used in land surveying before 1956 is as follows:

6 ānguli = 1 chaṭāk (4 ½ inches—not to be confused with the homonymous unit of weight)
2 chaṭāk = 1 bighat (9 inches)
2 bighat = 4 chaṭāk = 1 hāt (cubit = 18 inches)
4 hāt = 16 chaṭāk = 1 kāṭhā (6 ft)
20 kāṭhā = 320 chaṭāk = 1 bighā[24]

From the above it follows that 1 kāṭhā = 1/20 bighā and 1 chaṭāk = 1/320 bighā.

In measuring land, one usually did not bother with *ānguli*s and *bighat*s, but focused on chaṭāk, kāṭhā, and bighā.

The mention of the chaṭāk, kāṭhā, and bighā in a table of *linear* measures might sound surprising, as most of us know them as units of land area. However, these are primarily units of length and, only secondarily, though not less importantly, also units of land area. Thus, the same terms were used in two senses, and everybody understood from the context (and occasional indicators added to the terms) in what sense the terms were being used.

---

[21] Prinsep (1840), p. 6 and pp. 103–04.

[22] James Prinsep, *Useful Tables*, op. cit., p. 67.

[23] Verman and Kaul (1970), p. 487.

[24] See, for example, *Manual of Mensuration and Surveying for Mandals and Patwaris in Assam*, Shillong: Department of Land Records and Agriculture, Assam, 1900, p. 5.

1.5.6 Of Square and Oblong Units

One bighā area is simply 1 bighā length × 1 bighā length. Employing modern usage, we can call it 1 sq bighā. This is fine for us as we are accustomed to thinking of area in terms of square units—sq m, sq ft, sq km, and so on.

However, traditionally, Indians were also open to units that could be conceived as oblong rectangles! Thus, unlike the bighā, 1 kāṭhā area was not 1 kāṭhā length squared. It was rather 1 kāṭhā length × 1 bighā length. Now, we know that 1 kāṭhā length = 1/20 bighā length. Therefore, 1 kāṭhā area = 1/20 bighā length × 1 bighā length = 1/20 bighā area. (Hence, we have a beautiful symmetry: 1 kāṭhā length = 1/20 bighā length and 1 kāṭhā area = 1/20 bighā area.)

Similarly for the chaṭāk! 1 chaṭāk length = 1/16th of 1 kāṭhā length and 1/320th of 1 bighā length. Further, 1 chaṭāk area was and is defined as being 1 chaṭāk length × 1 bighā length. Therefore, 1 chaṭāk area = 1/320th bighā length × 1 bighā length = 1/320 bighā area. Since, 1 kāṭhā area = 1/20th of 1 bighā area, therefore, 1 chaṭāk area = 1/16 kāṭhā area. Therefore, we have another great symmetry: 1 chaṭāk length = 1/16th of 1 kāṭhā length and 1 chaṭāk area = 1/16th of 1 kāṭhā area. Thus, the ratios between chaṭāk, kāṭhā, and bighā *areas* are exactly the same as the ratios between chaṭāk, kāṭhā, and bighā *lengths*.

In the above, we have covered an essential piece of ground.

1.5.7 More on Squares and Oblongs—Reiteration and Amplification

In the Western system, whether it is the British system or the SI system, the basic unit of area is inevitably a square—ft^2, mi^2, m^2, mm^2, km^2, etc.

One consequence of this modern approach is that the relation between a foot and inch (1:12) differs from the relation between ft^2 and $inch^2$ (1:144) and the relationship between metre and centimetre (1:100) differs from the one between m^2 and cm^2 (1:10,000).

In the Indian system, many of the basic units are defined in terms of oblongs. We have seen the examples above.

Now, if we apply the Indian approach to Western units, we can multiply any two British units or SI units to produce legitimate units of area. For example, multiplying 1 foot with 1 inch will produce an area of 1 foot-inch, a legitimate unit in terms of the Indian approach. Similarly, we could multiply 1 Km with 100 m, producing 100 mKm—again another perfectly legitimate unit in terms of the Indian approach.

In fact, this system of mixed-units makes multiplication much easier by avoiding fractional manipulation—hugely convenient in the absence of computing machines or computational wizards. Ashok Jhunjhunwala in his book discusses how an Indian carpenter working with a tape showing feet and inches, used the foot-inch as an unit to measure his areas, in the process avoiding complicated multiplications.[25] Clearly, he is invoking the Indian system.

We shall see how this works in the case of land areas, from an example given below.

1.5.8 *Delving Deeper*

Now, let us delve a little more into the units of land areas in Bengal. The following table sums up the essential information. Please note that the information here is based on extant manuscripts (Table 1.1).

Table 1.1 From length to area

Sl. No.	Length × Length	Area	Dimensions The exact dimensions indicated here are known as Bengal measures. They are not necessarily applicable to everywhere outside Bengal and not applicable even to some parts of Bengal; further, the equating with inches or ft is only approximate and corresponds to the practice of the *amin*s during both colonial and post colonial periods
1	1 chaṭāk × 1 chaṭāk = 4.5 inches × 4.5 inches = 3/8 ft × 3/8 ft	1 kāk (i.e. 1 square chaṭāk)	1/16 sq cubits or 20.25 sq inches or 9/64 sq ft)
2	1 chaṭāk × 1 kāṭhā = 1 chaṭāk × 16 chaṭāk or 4.5 inches × 6 ft = 3/8 f t × 6 ft	1 gaṇḍā = 16 kāk = 324 sq inches = (18 inches) squared	2.25 sq ft or 1 sq cubit. (Note, this gaṇḍā area is very different from the one that was prevalent elsewhere. There, the gaṇḍā might mean an area somewhat larger than a kāṭhā.)

(continued)

[25] Jhunjhunwala (1993), pp. 1–3. Not all Indian carpenters use this system, as one knows from experience. Perhaps, the more traditional ones did.

Table 1.1 (continued)

Sl. No.	Length × Length	Area	Dimensions The exact dimensions indicated here are known as Bengal measures. They are not necessarily applicable to everywhere outside Bengal and not applicable even to some parts of Bengal; further, the equating with inches or ft is only approximate and corresponds to the practice of the *amins* during both colonial and post colonial periods
3	1 chaṭāk × 1 bighā = 1 chaṭāk × 320 chaṭāk = 3/8 ft × 120 ft	1 chaṭāk area (occasionally referred to as kālir chaṭāk) more commonly as simply chaṭāk) = 20 gaṇḍā	45 sq ft
4	1 kāṭhā × 1 kāṭhā = 6 ft × 6 ft	1 dhul = 16 gaṇḍā	36 sq ft
5	1 kāṭhā × 1 bighā = 1 kāṭhā × 20 kāṭhā = 20 sq kāṭhā = 6 ft × 120 ft	This is 1 kāṭhā area. This is = 16 chaṭāk area and also = **20 dhul area** 1 kāṭhā area is occasionally referred to as 1 kālir kāṭhā	720 sq ft
6	1 bighā × 1 bighā	1 bighā area = 20 kāṭhā area 1 bighā is occasionally referred to as 1 kālir Bighā	14,400 sq feet = 20 kālir kāṭhā = 320 kālir chaṭāk

1.5.9 Three Important Aspects of the Units

Immediately notice four things.

First, here we are dealing with area dimensions in Bengal. The exact dimensions easily varied, as mentioned above. However, the ratios mostly did not! In most areas, the bighā was equal to 20 kāṭhā. Indeed, the Ā-īn-i-Akbarī refers to the kāṭhā area as the 'biswah', denoting twentieth of a bighā![26] And, everywhere, the kāṭhā was equal to 16 chaṭāk. Therefore, as the size of bighā varies, so did the sizes of the kāṭhā and chaṭāk.

Secondly, beginning with 1 Chaṭāk length and moving upwards from there, we find that multiplying any two units of length generates a legitimate unit of area in the system. The table below will bear this out (Table 1.2).

[26] Allāmī (1949), p. 67.

Table 1.2 Multiplying the units of length with each other

X	Chaṭāk	Kāṭhā	Bighā
Chaṭāk	Kāk	Gaṇḍā	Chaṭāk (area)
Kāṭhā	Gaṇḍā	Dhul	Kāṭhā (area)
Bighā	Chaṭāk (area)	Kāṭhā (area)	Bighā (area)

Thirdly, notice that there is no unit of area smaller than the Kāk. This stands to reason in the realm of land areas, for 1 Kāk measures 20.25 sq inches, which is about 0.14 sq ft and 0.013 sq m, sizes hardly of any significance in land measurement.

Fourthly, notice the numerical relationship between the land areas:

16 kāk = 1 gaṇḍā 16 chaṭāk = 1 kāṭhā
16 gaṇḍā = 1 dhul 20 kāṭhā = 1 bighā
20 gaṇḍā = 1 chaṭāk

Thus, we are looking essentially at ratios of 1:16 and 1:20, in tune with units of cowrie counting. This, understandably, was hugely convenient.

We have initiated ourselves for taking a dip into our first Śubhaṅkarī āryā.

1.5.10 The Kuṛobā Āryā

This is perhaps the most famous Śubhaṅkarī āryā. We are providing a more updated and comprehensive version,[27] as the manuscript versions are often truncated and can also involve complicated measurement issues, out of place in an introductory essay.

At this point, the verse should pose no problem at all for the reader.

The āryā with its English translation reads as follows:

Āryā	Translation
Kuṛobā kuṛobā kuṛobā lijje Kāṭhāy kuṛobā kāṭhāy lijje Kāṭhāy kāṭhāy dhul parimāṇ Biś dhul hay kāṭhār pramāṇ Dhul bāki thāke jadi kāṭhā nile par Ṣolo diye pure tāre sārā gaṇḍā dhar Chaṭāk dharite habe chaṭāk bighāy Gaṇḍā dhari lote habe chaṭāk kāṭhāy Chaṭāke chaṭāke hole kāk dhari labe Ekun korile pare kāli thik pābe	1. A *Bighā* multiplied by a *Bighā* gives a *Bighā* 2. A *Kāṭhā* multiplied by a *Bighā* gives a *Kāṭhā* 3. A *Kāṭhā* multiplied by itself gives a *Dhul* 4. Twenty *dhuls* make a *Kāṭhā* 5. If having converted *dhuls* to *kāṭhās*, some *dhuls* still remain 6. Multiply those by sixteen, converting them into gaṇḍās 7. A *Chaṭāk* multiplied by a *Bighā* gives a *Chaṭāk* 8. A *Chaṭāk* multiplied by a *Kāṭhā* gives a *Gaṇḍā* 9. A *Chaṭāk* multiplied with itself gives a *Kāk* 10. If you proceed in this manner, your computation will be correct

[27] From Matubbar (1995), pp. 439–40.

We can now recognize how the verse is mostly telling us how multiplying the various units of length yield the various units of area.

Lines 3 and 4 define a dhul area and relate it to the kāṭhā area. Compare with rows 4 and 5 of Table 1.1. Interestingly, the Ā-īn-i-Akbarī recognizes the dhul as the *biswanasah*.

Lines 5 and 6 tell us something interesting. If there are a sufficiently large number of dhuls, convert them into kāṭhā (obviously, by dividing by 20). If some dhuls still remain, just multiply the remaining dhuls by 16 converting them into gaṇḍās (as 16 gaṇḍā make a dhul). That is, no Dhul should remain in the final result. We shall see why.

The other lines should hold no mystery for us.

1.5.11 Calculating

Let us assume that we have a rectangular piece of land that is 4 bighā 16 kāṭhā 12 chaṭāk long and 2 bighā 12 kāṭhā 8 chaṭāk wide. How do you find the area of this piece? In the modern system, you convert both the lengths into a single unit, say a bighā, and then multiply. In the above, the length works out to be $4^{67}/_{80}$ bighā and the breadth translates $2\ ^{5}/_{8}$ bighā. So we would have to multiply $4\ ^{67}/_{80}$ bighā with $2\ ^{5}/_{8}$ bighā. In the modern system, without resorting to the help of a mechanical, digital, or human computer, this can be done in two ways. First, by multiplying the two numbers in the abovementioned fractional form or second, by converting the numbers into decimal fraction and multiplying. Both the procedures would involve almost inevitable use of writing material and some proficiency with fractional calculation.

The traditional system, though somewhat long drawn, is far easier because *it completely avoids the use of fractional manipulation*. Let us see.

The multiplication below displays how it was done. (Please refer to Table 1.2 for refreshing the memory regarding how the various area units were generated.) Note that we are multiplying the entries in the column on the left with the row at the top, to get the initial results. In case of units with the same names, we have placed "(a)" after the area unit to distinguish it from the length unit (Table 1.3).

Hereafter, the system proceeds in a fairly predictable manner. We can separate out the units of each type from the above array and add the similar units and convert smaller to larger units.

Table 1.3 Calculation: The first steps

Multiply	2 bighā	12 kāṭhā	8 chaṭāk
4 bighā	8 bighā(a)	48 kāṭhā(a)	32 chaṭāk(a)
16 kāṭhā	32 kāṭhā(a)	192 dhul	128 gaṇḍā
12 chaṭāk	24 chaṭāk(a)	144 gaṇḍā	96 kāk

Cells with similar units shaded similarly

Thus, we will get at first

8 bighā(a); 48 kāṭhā(a) + 32 kāṭhā(a); 32 chaṭāk(a) + 24 chaṭāk(a); 192 dhul; 128 gaṇḍā + 144 gaṇḍā; 96 kāk.

If the reader works on the above, remembering to *convert all multiples of 20 dhuls into* kāṭhā(a), s/he should come up with the following figures:

12 bighā(a), 13 kāṭhā(a), 5 chaṭāk(a), 18 gaṇḍā, 12 dhul

The 6th line of the verse tells us what to do with the 12 dhul still remaining. 16 gaṇḍā make a dhul. Therefore, we multiply 12 dhul by 16 and get 192 gaṇḍā. Again, 20 gaṇḍā make a chaṭāk(a). Again, we need to convert.

Doing the necessary we would arrive at the final result: 12 Bighā(a), 13 Kāṭhā(a), 15 chaṭāk(a), and 10 gaṇḍā.

Notice that although the steps are many, each step is simple and requires nothing more than application of the simplest operations of arithmetic and no use at all of fractional operations. Some of it could be done verbally, as it often was. Only for lengthy operations did one resort to *kāli karā*—using pen and paper.

Further, since the units of area are related to one another in multiples of 4 (and occasionally 16 and 20), and this proportional pattern resonates with the ratio between money units in traditional Bengal, working out the price or rent of a particular unit of area of land from rates mentioned for any unit of area is extremely easy. However, before we go into that, or delve into the verses involving units of weight, we must understand the nature of money in pre-colonial Bengal.

1.6 Part IV: Of Shells and Silver

1.6.1 Prejudice About Money

Across Europe and some other parts of the Old World, the approach to money seems to have suffered from what be called the "numismatic prejudice"—the belief that only metallic currency was truly money. This resulted, for example, in the tragic overlooking of the use of seashells as a vital monetary innovation, and resulted in the alarming neglect of cowries in archaeological finds.[28] It is only in 2008 that a historian (Niall Ferguson) could unflinchingly write:

> There was in fact no reason other than historical happenstance that money was for so long equated in the Western mind with metal.[29]

[28] Deyell (2020), p. 186.

[29] Niall Ferguson, *The Ascent of Money: A Financial History of the World*, New York: The Penguin Press, p. 27.

1.6.2 Waking up to Cowries

The term cowrie (or 'cowry') primarily denotes a family of sea snails. The cowrie family is subdivided into several genera and these, in their turn, into species. Cowries had numerous uses, including ritual and ornamental. However, the cowrie species most popularly used as currency is known as the *Monetaria moneta* (earlier, more familiar as *Cypraea moneta*).

Although there were some earlier investigations, the major breakthrough in understanding the economic importance of cowries was the analytically decisive work by Jan Hogendorn and Marion Johnson (1986).[30]

In this book, the authors showed that

(i) The largest and economically decisive source of *moneta* or the money cowries was the Maldives
(ii) These shells were money in all senses—e.g., they were seen to obey the Gresham's Law and abide by the Quantity Theory of Money
(iii) As money, cowries suffer from the limitation of tremendous cheapness, which makes them an unsatisfactory storage medium for large values
(iv) However, each form of money has one limitation or the other
(v) The tremendous advantage of cowries is that, unlike metallic coins, they cannot be debased or counterfeited
(vi) Moreover, their great cheapness makes them ideal small change
(vii) No wonder, they were used in abandon and with great success in commerce over large parts of the world
(viii) Consequently, the lack of academic interest in their role is due to shortcomings in academic mindsets rather than any limitations of the cowries.

Nevertheless, the pace at which the importance of cowries gained recognition has been intolerably slow.

From 2010 onwards, a number of scholars, notably John Deyell and Bin Yang, have examined the role of cowries with serious attention.[31] And, the relatively recent publication, *From Mountain Fastness to Coastal Kingdoms*,[32] has seen a team of scholars arguing the importance of the non-metallic money, particularly cowries, in Indian history, with special focus on Eastern India, particularly the Bengal area, showing *how the prevalence of non-metallic forms of money was not synonymous to economic backwardness.*

---

[30] Hogendorn and Johnson (1986).

[31] See, for example, Deyell (2010); Yang (2011).

[32] John Deyell and Rila Mukherjee (eds.) *From Mountain Fastness to Coastal Kingdoms*, op. cit.

1.6.3 Counting Cowries

Before proceeding further, let us examine how cowries were counted.

In Eastern India, and some other areas in India, cowries were traditionally counted in the following manner:

4 kaṛā (cowrie) = 1 gaṇḍā
5 gaṇḍā = 1 buṛi
4 buṛi = 1 pan
4 paṇ = 1 cauk
4 cauk = 1 kāhan

Thus, 1280 items cowries = 1 kāhan.

This manner of counting occasionally extended to other items as well. Thus, 1 gaṇḍā or 1 buṛi of any item meant 4 and 20 of those items respectively. Hence, the units of cowrie often denoted just numbers in the abstract, to be employed as to count any countable item.

Terms	gaṇḍā	buṛi	paṇ	Cauk	Kāhan
Numbers	4	20	80	320	1280

Notice that all the numbers above are divisible by 4 and the higher ones divisible by 16 and 20. We have already heard the rhythm of 4 in the units of weight and particularly in land measurement. *This is why doing the math while buying and selling in cowries was easy and perhaps why the above mode of counting emerged and persisted in this country.*

1.6.4 The Strange Faith in a Timeless Exchange Rate

There is apparently a tendency in the Indian tradition to accept the standard silver coin as being equal in value to 1 kāhan or 1280 cowries. Even notable historians, oblivious to economic common sense, have often taken this equation to represent historical reality. A suggestion floated by some contributors in the anthology mentioned above,[33] seems to shed some light as to why such an error might have crept in. Sayantani Pal has clarified that across large parts of Bengal in the Pala-Sena period references to silver currency *very often* meant not actual silver currency but notional or nominal silver money—i.e. actual payments were made in cowries.[34] Deyell, in the same volume, clearly suggested, apparently in agreement with D.C.

[33] John Deyell and Rila Mukherjee, op. cit.

[34] Sayantani Pal, 'Media of Exchange under the Palas and the Senas as Reflected in their Inscriptions', in John Deyell and Rila Mukherjee, op. cit., pp. 53–76.

Sircar, that in the Pala-Sena context, what was referred to as the purāṇa (silver coin) was most often an accounting unit merely denoting 1280 cowries.[35] This is possibly one source of the error that 1280 cowries amounted to an actual silver coin.

This is not to say that actual metallic money—either in the form of officially imprinted coins as from Harikela or Samatata area, or silver or gold dust in varying quantities—was not used as money in different parts of Bengal in the post-Gupta period.[36] However, all those forms of money must have exchanged with cowries at the prevailing rate of exchange, about which we have no information.

However, let us move on to sixteenth to seventeenth century Bengal—when the Śubhaṅkarī took shape.

1.6.5 The Silver Taṅkā and the Cowrie in Bengal Economy

From the fifteenth century, the silver taṅkā was the established official currency and was available in considerable quantities, at least in the urban areas and, usually, for all major transactions.[37] This continued during the Mughal period—only the new coins now had different legends.

On the other hand, cowries dominated small change in Bengal (and Odisha). Of course, most substantial purchases involved some amount of the taṅkā coins. Moreover, all major prices and, of course, all revenues and duties, were reckoned in the taṅkā. However, occasionally, large payments (including that of land revenue) were also made in cowries, particularly in the rural areas. For example, even in the second half of the eighteenth century, the entire revenue of Sylhet was collected in cowries.[38] This was somewhat inconvenient, as the cheapness of cowries rendered making large payments cumbersome. However, cowries were ideal as very small change. Thus, the 'big problem of small change' was solved with the precision of a liner brush.

In general, Mughal India seems to have had a *relative* dearth of copper. And in Bengal, possibly due to cowrie predominance, both copper currency and fractional denominations of the taṅkā were almost non-existent during both the Sultanate and Mughal periods.[39] In fact, this scarcity turned out to be a blessing as it helped Bengal avoid the confusions ensuing from triple currency. (There was the gold coin. However, although its symbolic value was high, its role was insignificant in usual economic transactions.)

---

[35] John Deyell, "Hard Money and 'Cashless' Economies: Medieval Bengal and the Greater Asian World", in John Deyell and Rila Mukherjee, op. cit. p. 190.

[36] Referred to, for example, in Ghosh (2012–13).

[37] See, for example, Deyell (2010).

[38] Haraksingh (1973), p. 25, pp. 50–51.

[39] John Deyell, 'Cowries and coins: The dual monetary system of the Bengal Sultanate', 71–72. Chatterjee (2019), p. 1.

1.6.6 Lower Value Currency Elsewhere in India

The Mughals, in deference to Indian custom and the manner of cowrie counting, structured most of their coin denominations in the ratios of 4 and 16. However, they had a liking for decimal units. Thus, we find Abū L-Faẓl mentioning coins equal to one-tenth and one-twentieth part of the Rupaiyā.[40] However, the Indian custom possibly prevailed to make them less than popular. Thus, Tavernier, who travelled in India in the mid-seventeenth century, mentions Indians having for their silver money *only* the rupee, the half, the quarter, the eighth, and the sixteenth,[41] something familiar to those acquainted with Mughal numismatics.

1.6.7 The Ever-Varying Value of the Cowrie

Belying the myth of a fixed rate of the cowrie to the silver coin, the rate constantly fluctuated. This is attested by travellers to all parts of India and definitely to eastern India. Tomé Pires, who visited Bengal in the second decade of the sixteenth century reports that 7 kāhan cowries exchanged for a silver tanqat (taṅkā). Tavernier's data from the seventeenth century suggests very different figures. Alexander Hamilton's data regarding Bengal in the early decades of the eighteenth century shows that 2560 to 2880 cowries exchanged for a 'rupee current'.[42] And again, we have other sets of figures from Andrew Sterling regarding Odisha from 1760s to 1803 and from Francis Buchanan regarding Bengal from around 1807.[43]

1.6.8 The Fractions of the Taṅkā, and of the Kaṛā as Its 1/1280th Part

With the Mughal conquest in the late sixteenth century, the Mughal rupaiyā gradually took over and became the new taṅkā. The new coin was in the same weight range as the prevailing Sultani taṅkā and the Sher Shahi rupaiyā. Anyway, Bengal had none of the lower denomination Mughal coins available elsewhere and managed its smaller transactions only in cowries. However, it generated notional fractions corresponding to those values. The notional sixteenth part of the taṅkā was

---

[40] Allāmī (1989), p. 32.

[41] Tavernier (1925), p. 22.

[42] A 'rupee current' was a notional money of account that the East India Company used. It was valued at 2 shillings. In the early decades of the eighteenth century, it was a little less than the Murshidabad sicca rupee in value.

[43] Pires and Rodrigues (1944), pp. 93–94; Jean Baptiste Tavernier, *Travels*, op. cit., p. 24; Hamilton (1995), p. 7; Sterling (1846), pp. 59–60; Buchanan (1833), p. 321.

Table 1.4 The usual notional fractions of a taṅkā

1 kaṛā or bat (also called *1 pākā kaṛi*) = 1/1280 portion of a taṅkā	4 kaṛā = 1 gaṇḍā = 1/320th of a taṅkā (also called pākā gaṇḍā)	5 gaṇḍā = 1 pākā buṛi also called 1 pāi = 1/64th of a taṅkā	4 pāi = 1 ānā = 1/16th of a taṅkā	4 ānā = 1 siki (1/4 of a taṅkā

Table 1.5 The usual units of cowrie counting

1 kaṛi (cowrie) = 1/1280th of a kāhan	4 cowries = 1 gaṇḍā = 1/320th of a kāhan	5 gaṇḍā = 1 buṛi = 1/64th of a kāhan	4 buṛi = 1 paṇ = 1/16th of a kāhan	4 paṇ = 1 cauk = 1/4th of a kāhan

the ānā, the eighth was two ānā, the quarter was 4 ānā or a siki and the half was 8 ānā or 2 siki (during the Company currency regimen, later known as *ādhuli*). Moreover, one-fourth of an ānā was called the pāi, which, elsewhere, was called the paisa. (The precolonial notional Bengali pāi is not be confused with the coin called "pāi" struck by the colonial government and valued at $^1/_3$rd of a paisa.)[44]

Moreover, Bengal, hugely dependent on the cowries, did something still more radical. It *notionally* fractionated the taṅkā primarily into 1280 parts, mimicking the divisions of the kāhan. The $^1/_{1280}$th part of the taṅkā was called a *kaṛā*. Now, etymologically, the kaṛā means kaṛi (i.e. cowrie). Therefore, to distinguish the $^1/_{1280}$th part of the taṅkā from the usual cowrie, the former was known as the pākā kaṛi (stable or secure cowrie). The actual cowrie was often called the kãcā kaṛi (unstable cowrie),[45] as its value fluctuated.

The usual notional fractions of the taṅkā were as follows:

The reader would have noticed that Table 1.4 is homologous in structure to the units of cowrie counting (Table 1.5).

And this is why the arithmetic tables tell us that 1280 kaṛā make a taṅkā, which subsequent scholars understood *to mean that 1280 cowries actually exchanged for a taṅkā*![46]

We hope this suffices to dispel the myth surrounding 1280.

Incidentally, there developed a tradition of indicating these fractions—the $^1/_{1280}$th, the $^1/_{320}$th, $^1/_{64}$th, $^1/_{16}$th, $^1/_4$th—and their multiples. These were not the kind of fractional notation that we know, but elemental symbols such as these:

/, |, ৵, and ৸

These symbols could be used to indicate the fractions of money, land, weights, etc. and, with appropriate placement techniques, would indicate which item was being mentioned.

[44] This also began as a notional currency of the Company accountants before becoming an actual currency, though not a popular one.

[45] Ghosh (1893), p. 18.

[46] See, for example, Chakravarti (2016), p. 365.

We won't be using this notation here. In what follows, in order to avoid confusion, we shall add a subscript **m** at the end of the fractions of the taṅkā, to indicate that these are fractions of **m**etallic money. Thus, we shall refer to the pākā gaṇḍā as gaṇḍā_m, and the kaṛā as kaṛā_m, so as not to leave any chance of confusing with the corresponding units of cowrie counting.

Was there any utility of breaking up the taṅkā in that manner? There was. Let us see, how that worked?

1.7 Part V: The Kaṛā, Counting, and a Further Taste of Practical Śubhaṅkarī Arithmetic

1.7.1 The Utility of Fractionating the Taṅkā

Tomé Pires informed us that in sixteenth century Bengal *7 kāhan cowries* exchanged for a taṅkā.

Recalling that the ratio of siki, ānā, pāi, pākā gaṇḍā, and kaṛā to the taṅkā is exactly the same as the cauk, pan, buṛi, gaṇḍā, and kaṛi to the kāhan, we immediately get Table 1.6.

Let us assume someone wants to buy land at the rate of 3 taṅkā a kāṭhā (those were cheap days). He wants to buy a piece of land measuring 7 kāṭhā and 14 chaṭāk. What would be the price in cowries?

First, the price of 7 kāṭhā(a) land = 15 taṅkā = 15 × 7 kāhan cowries.

Next, we have 14 chaṭāk(a) land.

Just remember that 16 chaṭāk(a) = 1 kāṭhā(a), as 1 taṅkā = 16 ānā, making the necessary conversions amazingly easy—again proving the utility of fractionating the taṅkā as in the above.

The reader might try her hand in working out the value of the entire piece of land in cowries. Once one does similar computations a few times, this becomes amazingly easy.

Table 1.6 The taṅkā-cowrie exchange rate witnessed by Tomé Pires

Taṅkā and its fractions	1 taṅkā	$^1/_4$ taṅkā or 1 siki	$^1/_{16}$ taṅkā or 1 ānā	¼ ānā or 1 pāi	$^1/_5$ pāi or 1 pākā gaṇḍā	¼ pākā gaṇḍā or 1 kaṛā
Number of cowries available for each value in the above	7 kāhan (7 × 1280)	7 cauk (7 × 320)	7 paṇ (7 × 80)	7 buṛi (7 × 20)	7 gaṇḍā (7 × 4)	7

1.7.2 The Dhul Problem—A Digression

Recall getting rid of the dhul?

The dhul and chaṭāk(a) are fairly close in size, 1 dhul being 4/5 of a chaṭāk(a). Then why discriminate against the dhul in the final statement of area?

Actually, 20 dhul make a kāṭhā(a) and 400 dhul make a bighā(a). Thus in order to deduce the value of a dhul from the value of a bighā(a) we have to undertake division by 400, a division, which, as the reader may easily check, results in a less happier combination of fractions. Similarly in order to deduce the *price* of a *dhul* from the *price* of a *kāṭhā* we have to divide by 20. This is a happier procedure than the division by 400. But, it is less so than the division by 16. If we think about it a little, we will realize that chaṭāk(a) and gaṇḍā values are far more convenient than dhul values.

1.7.3 Further Divisions

As small a unit as it was, the kaṛā or bat (as a notional fraction of the taṅkā) had further subdivisions. These are captured by the following verse:

Verse	Translation
Kāk chaturthe batek jāni *Tin krānte bat bākhāni* *Naba danti kariā sār* *Sātāś jabe bat bichār* *Āśi tile batek kar* *Lekhār guru Subhankar*	The *kaṛā* is equal to 4 *kāk*. It is also equal to 3 *krānti*, and equal to 9 *danti*, also equal to 27 *jab*, and further, equal to 80 *til* Thus 1 *kāk* = ¼ *kaṛā*, 1 *krānti* = 1/3 *kaṛā*, 1 *danti* = 1/3 *krānti* = 1/9 *kaṛā*, 1 jab = 1/3 *danti* = 1/27 *kaṛā* 1 til = 1/20 *kāk* = 1/80 *kaṛā* It goes without saying that these are very small fractions of the silver taṅkā

Thus, in keeping accounts, one could use, apart from the previously mentioned fractions of the rupee, the following *notional* denominations (notice the subscript **m** placed at the end of the denominations indicating that the latter are divisions of the taṅkā.):

1 Kaṛā_m = 1/1280 taṅkā
1 kāk_m = 1/5120 taṅkā
1 krānti_m = 1/3840 taṅkā
1 danti_m = 1/11520 taṅkā
1 jab_m = 1/34560 taṅkā
1 til_m = 1/102400 taṅkā

In the case of kāk, of course, the subscript **m** helps also to distinguish it from the kāk of land area.

Even cowries were notionally fractionated. The actual cowrie was notionally fractionated into 4 notional kāks and 80 notional tils. We shall denote these as kāk_c and til_c, to indicate that they are fractions of the cowrie.

1.7.4 Why These Strange Fractions?

The answer to this question is simple enough.

There is a verse that provides information on what would the price per kāṭhā(a) would be for each category of price per bighā(a).[47] For brevity, we shall not cite the verse but present the information it contains right away.

The verse tells us, when the price per bighā(a) of a piece of land is given in so many taṅkā, ānā, pāi, gaṇḍā_m, and kaṛā_m, what would be the price per kāṭhā(a).

The values per kāṭhā(a) may be found from the following table:

Per *bighā(a)*	Per *kāṭhā(a)*
For each taṅkā of the value	16 gaṇḍā_m
For each ānā	1 gaṇḍā_m
For each pāi	1 kaṛā_m
For each gaṇḍā_m	16 til_m
For each kaṛā_m	4 til_m

The first two entries suffice to demonstrate how this works. Remember, 1 bighā(a) = 20 kāṭhā(a). Therefore, if 1 bighā(a) = 1 taṅkā, then the price of 1 kāṭhā(a) would be $^1/_{20}$th of 1 taṅkā. The table of money units shows that 1 taṅkā = 320 gaṇḍā_m. Therefore, 1/20 of a taṅkā = 16 G_m. Similarly, the relationship between bighā(a) and kāṭhā(a) is very conveniently echoed in the relationship between ānā and gaṇḍā_m. This accounts for the second entry. Now, if from the price (or revenue) of a certain plot of land, the price or revenue per Bighā(a) works out to be *a tākā, b ānā, c paisa,* and *d* gaṇḍā_m, then, with the aid of the above (versified) table, we find the amount per Kāṭhā(a) to be 16 *a* gaṇḍā_m + *b* gaṇḍā_m, *c Kaṛā,* and 16*d* til_m. Depending on the values of a, b, c, and d, these in turn may need to be translated further to higher units. With this, we have the price or revenue per Kāṭhā(a) expressed in appropriate value of the taṅkā and its fractions. Now, since the fractional units of the taṅkā most often had no suitable numismatic representation, one would need to *convert the fractional values into cowries.*

And, predictably, *you had verses telling you how to go about it.* For example, when one could buy so many kāhan, so many cauk, so many paṇ etc. cowries for a taṅkā, one might need to find out how many cowries one could get for an ānā (without needing to undertake the division by 16).

[47] Asiatic Society manuscript 3749A, verse no. 8.

One such verse tells us exactly how to do that. It reads:

Kāhane laibe paṇ choke nibe buṛi
Gaṇḍāy laibe kāk paṇe p̃āc koṛi
Kaṛāte laibe pañca tilér likhan
Bhriguram Das kahe ānār dharan[48]

Here, we are talking about the number of cowries. Therefore, the terms 'gaṇḍā', 'kāk', and 'til' in the verse denote units of actual cowrie counting (the gaṇḍā, the kāk_c, and til_c).

What the verse tells us might be explained in simple algebraic terms as follows:

If the price of cowries is such that for 1 taṅkā you get X kāhan, Y chauk, Z paṇ, A gaṇḍā and B cowries, then for 1 ānā you would get X paṇ, Y buṛi, 5Z cowries, A kāk_c and B til_c cowries. This, of course, comes from dividing each of the denominations, kāhan, paṇ, etc. by 16. Once this verse was memorised one would not have to make the division and would be able to tell the result straight away.

However, the result includes the units such as kāk_c and til_c, which, as fractions of the cowrie, are fictitious entities. However if the number of kāk_c (in this case A kāk_c) and the number of til_c (in this case B til_c) add up to whole numbers of cowries (even approximately) then these can be taken into account. These procedures are however, not mentioned in the above verse and are being inferred as obvious practical procedures.

There is another āryā that describes the number of cowries available for a gaṇḍā_m, given the number of cowries available for a taṅkā. It may be translated as follows:

> For each *kāhan* carefully take a *gaṇḍā*/Remember children to take a *kāk*$_c$ for a *pan*/Know for certain that you must take a *til*$_c$ for each *gaṇḍā*/This, according to *Śubhankar*, is the value of a *pākā gaṇḍā*. (See footnote 8)

The pākā gaṇḍā is gaṇḍā_m. The principles and procedures involved in understanding and applying this *Āryā* are similar to the one above, except, instead of 16, the division undertaken is by 320, as one is deriving the purchase of a gaṇḍā_m from that of a taṅkā.

1.7.5 Other Computations

We have been discussing land areas. There are verses that allowed one to find the price of smaller units of weight from that of higher units. These were very similar to those on land areas as in the case of units of weight also, the ratios were similar to that obtaining between the *tākā* and its various fractions.[49] There were

---

[48] Asiatic Society manuscript 3749A, verse no. 24.

[49] For example, we might recall that 1 man = 40 ser; 1 ser = 4 pauā; 1 pauā = 4 chaṭāk; and 1 chaṭāk = 4 kāchchā.

also verses concerned with deriving the daily wage from the monthly wage, the daily wage given the yearly wage and vice-versa, and the monthly wage given the yearly wage. Further, there were verses dealing with *māthat*—a zamindari tax levied as a proportion of the rent payable and other verses dealing with interest—e.g., working out the interest per day from the interest per month.[50] In all these cases, the basic arithmetic operations were similar to the ones we have already seen and, hence, we leave them out of the discussion.[51]

1.8 Part VI: Arithmetic, Algebra, and Fun with Numbers

The second category of Śubhaṅkarī verses, as we have mentioned at the beginning, consisted of rhymed versions of arithmetic and simple algebraic problems. For example, we have verses posing multiplication problems for students. We shall not burden the reader with the original language and present translations.

Example 1

A Pir has his estate in Mecca city
His estate measures a ***hundred crore*** *bighā*
Hundred *blades make a bundle*
Each blade has a ***hundred*** *grains.*
The Pir's blessing ensures uniform yield
Each bighā yields seven kāhan bundles
"How many grains of rice are there"?
Asks the Pir to his scribe.[52]

Ignoring the incongruity of rich rice yields in the deserts of Mecca, let us note that a crore is 10^7, and 1 *kāhan* measures 1280. Therefore, we have (100×10^7) bighā $\times (7 \times 1280)$ bundles per bighā $\times$ 100 grains per bundle $= 7 \times 1280 \times 10^{11}$ grains $= 896 \times 10^{12}$ grains.

Example 2

In Indra's heavenly abode stands the Parijāt tree
Each day ten million new flowers blossom on it
The price of each flower is one and a quarter man gold
How many flowers bloom and how much is their total gold value in four yugas

[50] Some such verses are cited in Madhusudan Deb, *Shubhankari: Indigenous Bengali Arithmetic Including A Complete System of Mental Arithmetic*, Calcutta: Sarbamangala Library, no date, and in the possibly the finest published anthology of Śubhaṅkarī verses, namely, Palit (2001).

[51] For an analytical discussion, see Santanu Chacraverti, *Śubhaṅkarī*, op. cit., pp. 93–98.

[52] Visvabharati Manuscript no. 17, last page.

This is simple multiplication, except that it involves very large numbers. For example, if the total length each yuga is taken to be common figure of 12 years, then four yugas amount to 48 years then the final number is somewhat less. However, if one the combined span of four yugas is understood to be the classical 432,000 years, we are operating within a different order of magnitude.

Then there were some simple arithmetic problems.

There was a temple very tall
The Wind God's son drowned it in rage
Half sank in mud, one-third in water
One-tenth lay rolling in the moss
Fifty-two yards[53] *continued to stand*
What is the total height, asks Śubhankar...[54]

Various versions of the same verse occur in different manuscripts, many of which indicate the method of solution. This was the standard arithmetic one—determining the fractional part of the total height that continued to stand. This would be $1 - ({}^{1}/_{2} + {}^{1}/_{3} + {}^{1}/_{10})$. This is ${}^{1}/_{15}$. And, that is 52 yards. Therefore, the total height would be $52 \times 15 = 780$ yards.

And another,

A thief tries to steal the Queen's jewel necklace.
He succeeds in taking half the jewels. One-fourth falls to the ground.
One-sixth remains on the Queen's neck.
The King happens to get thirty-two.
How many jewels were there in the necklace?[55]

Here, the answer is given but not the method of solution. However, it is obvious that one expected the usual arithmetic method would be employed.

1.8.1 Slightly More Difficult Problems

An interesting *Āryā* reads:

A goat for a tākā, a cow for a siki
A buffalo for tākā five
Hundred animals in hundred tākā

[53]The specific term used of course is not yard but 'gaj', which, however, had almost the same length as a yard.

[54]Visvabharati Manuscript no. 17, last page. This seems to be an extremely common verse. It occurs in both the Asiatic Society manuscripts (3749A p. 9 and 3749 pp. 9–10) and also cited by Dineshchandra Sen op. cit. p. 901.

[55]Visvabharati Manuscript no. 17, p. 9.

Doth announce Sadashiv[56]

The unstated question is respectively how many **goats**, **cows,** and **buffaloes** must be there so that:

The total number of animals is 100
The sum of animal prices add up to 100 tākā

Remember that siki is 0.25 tākā.

Writing **g** for number of goats, **c** for number of cows and **b** for number of buffalo, we have two distinct equations:

$$g + c + b = 100 \ldots \tag{1.1}$$

$$1g + 0.25c + 5b = 100 \ldots \tag{1.2}$$

Notice that we have 3 unknown variables but only two distinct equations. In such situations, we say that the equations are *undetermined.* If the equations are consistent and the variables denote real numbers, then the number of real solutions will be infinite. However, when the variables can only denote natural numbers, e.g., the number of animals, then, thankfully, we can have a finite set of solutions. Now, let us see if we can find the solutions to the above equations.

Subtracting the second row from the first, we have $0.75\ c - 4\ b = 0$. Multiplying this by 4 we have $3c - 16b = 0$. Therefore, we have $3c = 16b$. Therefore, $c = {}^{16}/_{3}\ b$. Therefore, Eq. (1.1) reads $g + {}^{16}/_{3}\ b + b = 100$. Therefore, $g + {}^{19}/_{3}\ b = 100$. Let us provisionally take $b = 3$. Then $g + ({}^{19}/_{3}) \times 3 = 100$. Therefore, $g + 19 = 100$. Then $g = 81$. Now, $c = {}^{16}/_{3}$ b. Therefore, $c = 16$. Therefore, we have one set of solutions $g = 81$, $c = 16$, and $b = 3$. This satisfies Eq. (1.1). We see that it also satisfies Eq. (1.2) $[81 + (0.25 \times 16) + (5 \times 3) = 100]$. Therefore, this is a satisfactory set of solutions. We can have another such set by just doubling b and c [$b = 6$, $c = 32$, and $g = 62$]. Further exploration along the same lines produces five satisfactory solution sets, as follows:

Goats	Cows	Buffaloes
81	16	3
62	32	6
43	48	9
24	64	12
5	80	15

[56] Sen (1993), p. 897.

A more difficult problem reads:

The royal maid comes to the market and asks
Help me to purchase two and a half ser for one kāhan
Listen to the rates for each
Ghee is worth one kāhan per ser and molasses worth one pan per ser
A ser of flour is worth one buṛi
Quickly deliver lest the king gets angry.[57]

That is, what amount of ghee, molasses, and flour must be bought so that the total amounts to 2 ½ ser while the total price was 1 kāhan?

Once again there are three unknowns and two distinct equations. We leave the problem for the reader. Mahendranath Mukhopadhyay, a scholar who had the opportunity of interviewing the traditional teachers of the paṭhśālās, says that the method involved a kind of intelligent trial and error. He explores some of these methods in his work.[58]

The technique of solving these is never discussed in the manuscripts. This, as Mahendranath indicated, *was taught to the students directly.*

1.8.2 Fun with Numbers

The third category of Śubhaṅkarī consisted of pure computational fun. One often used a riddle-like language, meaningful to all reared in the culture. Often, one played around with extraordinarily large numbers, indicating a typical Indian fascination with immense magnitudes. On other occasions, the numbers dealt with had interesting properties explored in the verses. For example, there is an understandable fascination with primes and with using them to generate numbers with interesting features and symmetries. Let us see some examples.

1.8.3 An Example of Riddle-Like Language

Even those born in the 1960s, often remember memorizing a verse on words and terms that stood for *various numbers from one to ten.*

However, the lexicon was not rigid. There were minor variations involving synonymous expressions or other Puranic associations.

---

[57] Hemendranath Palit, op cit., p. 53.

[58] Mahendranath Mukhopadhyay, *Asthita Samadahan*, Bankura: Publisher not mentioned, 1895. Several such problems and occasionally their solutions are to be found in Hemendranath Palit, op cit. However, the book does not discuss the methods of solving.

Dineshchandra Sen gives the following list[59]:

1 = candra, śaśi (moon), mahi (earth), guru (mentor, God)
2 = pakkha (wings/fortnight), pākhā (wings), kar, bhuj (hand or arm)
3 = netra, locan (eye), agni (fire), Rama
4 = bed (the vedas), jug (yugas or epochs)
5 = bān, sar (arrow)
6 = mad (intoxicant), ritu (season)
7 = samudra (sea), asva (horse), muni (sage/seer)
8 = basu (a particular class of vedic gods), elephant
9 = graha [planets, i.e. the 5 traditional planets plus sun, moon, and rāhu and kétu (the two nodes where eclipses occur)], randhra (aperture)
10 = dik (the traditional Indian points of the compass, i.e. eight points on the horizon plus zenith and nadir).

Let us see the words in action.

Verse	Translation	Meaning
Muni muni bāme pākhā *Ḍāine bāro paṇ diyā sakhā* *Ṣolo diyā puriā ān* *Chāri Chārer janmasthān*[a]	Sage sage wings on the left Place twelve pans on the right Now multiply by sixteen The birthplace of four fours	Place 2 to the left of 77 Place 12 pans (i.e. ¾ on the right) Now, multiply by 16 You get four fours
The calculation		
16 pans make a kāhan, which stands for the number 1. Therefore, 12 pans indicate ¾. Therefore, we have $(277 + ¾) \times 16$ $= 4432 + 12$ $= 4444$		

Asiatic Society Manuscript no. 3749 A, p. 5

Now 4444 is a rather neat looking number. Further, the digits add up to 16, a number with immense significance in the *Śubhaṅkarī* tradition. Of course, it is also 4 times 1111.

It so happens that this number 1111, in its turn, has extremely interesting properties. The digits add up to four, the keystone number in the indigenous tradition. Moreover, it is the product of two very attractive primes, 11 (which is 10 + 1) and 101 ($10^2 + 1$). Now, 11 and 101 share a common feature—adding their digits yields the number 2 (the first prime). The prime succeeding 101 *that also adds up to 2* is 137.

What happens when you multiply 1111 with 137? You get the number 152207.

The number 15227 seems to have fascinated the *Śubhankarī* composers, for there are a large number of riddle-like verses that either yield the number as a result or includes the number in the problem.

[59] Dineshchandra Sen op. cit. p. 903.

In several verses this number has been described as *ashtakothā* (**eight**-places). This name is probably on account of the fact that the digits in this number add up to 8.

Multiply 152207 with 73, directs one verse. (73, of course, is also a prime.)

The product is 11111111.

This number also has been described as *ashtakotha* (the digits add up to 8).

Another Śubhaṅkarī verse asks us to multiply the number 12345679 with 9. The answer to this is 111111111. Here, the sum of the digits is 9. The verse has been described in the manuscript as *nabakothā*'s verse (the verse of **nine** places).[60]

Another verse proposes the multiplication of 5882353 with 17. Both these are prime numbers. Their product is the beautiful number 100000001. Yet another verse proposes the multiplication of 33346667 and 3, again two prime numbers, to yield the result 100040001. The numbers 100000001 and 100040001 share two common properties—they are both perfectly symmetrical about the digit in the middle and both of them are sums of squares [$100000001 = 10000^2 + 1^2$; and $100\,040001 = 10000^2 + 200^2 + 1^2$]. It is significant that both these verses occur on the same page of the manuscript.[61]

Thus, even in the rather pedantic multiplication exercises, there is a distinct awareness of the properties of numbers and a pleasure in exploring these properties.

1.9 Part VII: Śubhaṅkarī—Its Significance and Demise

Śubhaṅkarī was closely tied to the paṭhśālā, the main site of elementary mathematical instruction. Like the paṭhśālā, it was a completely indigenous development, growing in response to a social need. Although the tradition is linked to an eponymous Subhankar, several other authors also wrote similar verses that are an integral part of the tradition. This versification of mathematical instruction and algorithms is very much in the Indian tradition of gaṇita and jyotiṣa, as is the love of gargantuan numbers that we encounter in many Śubhaṅkarī arithmetical problems.

The social need for mathematical education was addressed spontaneously, organically, and non-uniformly. This is understandable in the absence of a central authority planning education (government, church organization, or organized citizen body) or even of a print culture that would quickly spread ideas and memes and help create cultural uniformity. Despite this, there is considerable and even surprising similarity among manuscripts from different areas and different authors, justifying their classification into a single broad tradition.

[60] From manuscript verses cited in Dineshchandra Sen, *Brihat Banga*, op. cit. Vol. II, p. 898.

[61] Asiatic Society Manuscript no. 3429 A, p. 8.

It would be absurd to judge the merits of the tradition in terms of modern mathematical pedagogy. The more important question is: *did the tradition serve the needs it was supposedly intended to serve.*

The objective was to impart mathematical instruction, primarily to schoolchildren. The main problem besetting the tradition was that, developing ground-up as it did, there was no possibility of central planning or quality control. Sources testify that schools were spread all over Bengal. However, in comparison to the scale of education, the number of manuscripts found is remarkably small. Adam himself has repeatedly reported the lack of appropriate manuscripts in paṭhśālās and finds the Elementary Persian schools much more thorough and systematic in that respect.[62] Of course, the Elementary Persian schools were much smaller in number and were usually attended by students from relatively well-off Hindu and Muslim families, aspiring for "respectable" and well-paying jobs.[63] This, of course, was due to the high status enjoyed by Persian and the fact it continued as official language of the Company Government until 1835 (and even thereafter continued to enjoy status, as the actual linguistic shift in the local bureaucracies took time[64]).

Adam also testifies that the quality of teachers and manuscripts varied widely across the schools he visited. As regards manuscripts, this is borne out by the ones that survive.

Rhymed verses, bound by the constraints of metre and rhyming, are good for memorizing but not always ideal for conveying principles. One also wonders how far the teachers succeeded in clearly explaining the rules contained in the verses and the reasoning behind them. Of course, similar questions regarding teacher quality and pedagogy could be posed of schools and colleges in any country and age.

Notwithstanding all limitations, people continued to find the paṭhśālā and the practical computational instruction it imparted useful even well into the nineteenth century, as testified by several sources, and most eloquently by Adam and Lal Behari Day. Day's account brings out the importance attached by a farmer to his son's learning of the accounting rules so as not be deceived by the officers of the Zamindar.[65] We have also witnessed something of the teaching of basic arithmetic and some algebra and of the attempts to introduce the student to the pleasure of playing around with numbers.

[62] William Adam, *Reports*, Anathnath Basu, op. cit., p. 151.

[63] For example, William Adam, *Reports*, Anathnath Basu, op. cit., pp. 111–12, 115, 155, 156, and 291.

[64] Amanda Lanzillo, *The Politics of Persian Language Education in Colonial India*, Ajam Media Collective, https://ajammc.com/2018/01/31/late-indo-persian/.

[65] Lal Behari Day, *Bengal Peasant Life*, op. cit., pp. 58–60.

1.9.1 The Initial Blows to the Paṭhśālā and the Śubhaṅkarī

Various currents resulted in a devaluation of living traditions of elementary education during the 1830s. There was Macaulays's brilliant and ill-informed rhetoric contributing to the shifting of education policy towards modern education in the English medium in new schools. There was also the love of the upper caste dominated bhadralok elite for English education and their apathy, if not hostility, towards paṭhśālā and popular education. The mesmerizing attraction of western learning and the gradual dying out of the cowries and their substitution by cheap copper currency were other factors. Thus, the paṭhśālā and the Śubhaṅkarī took a backseat for a few decades.

1.9.2 Renewed Interest in the Paṭhśālā and the Śubhaṅkarī

The story of renewed interest in the paṭhśālā and the Śubhaṅkarī cannot be told in detail. Suffice it to say that, as the decades rolled, the wisdom in the Adam reports gradually imposed itself on the education administrators. The efforts to widen the base of primary education through improved paṭhśālās had come in 1856.[66] However, the major impetus came in the early 1870s, with the egoistical, authoritarian, yet capable Lieutenant Governor George Campbell, who applauded the indigenous vernacular schools and their emphasis on writing and practical computation and saw in their strengthening the road to a broader-based primary education.[67] He alienated almost the entire bhadralok community, who saw in his education visions a mischievous effort to divert money away from higher education.[68]

The Śubhaṅkarī, however, had never really been out of the picture. For, although cowries had started becoming irrelevant by the 1830s (due to the new currency regulations and cheap coins), the new coins embodied the old ratios. Sixteen ānā (spelled 'anna') still made a rupee, 4 paisas made an ānā and, therefore, 64 paisa made a rupee. Moreover, the other units of weight and land measure were all organized on traditional lines—only more rigorously standardized. True, the verses pertaining to purchase of and by cowries no longer remained relevant. However, most other verses and rules were still exceedingly useful, as noted by a Śubhaṅkarī textbook writer in 1873 who expresses happiness in the new attitude

[66] *Report of the Indian Education Commission*, Calcutta: Government of India, 1883, pp. 36–37, pp. 96–97.

[67] *Report of the Indian Education Commission*, op. cit., p. 37, pp. 98–103.

[68] Bankimchandra was one of those rare bhadraloks who supported Campbell's primary education policy. See Bankimchandra Chattopadhyaya, 'Sir William Gray o Sir George Campbell', in Jogeshchandra Bagal (ed.), *Bankim Rachanabali*, Calcutta: Shishu Sahitya Samsad, 1361 BS, pp. 888–92.

of the Lieutenant Governor (Campbell).[69] During the late nineteenth century, there seems to have been a surge in the printing of Śubhaṅkarī textbooks.

1.9.3 *The Demise of the Śubhaṅkarī*

As the base of primary education widened and arithmetic textbooks used even in the paṭhśālās trained students in the mastery of fractional notation and the unitary method, students could now use the modern arithmetic notations and techniques to work with traditional units. The Śubhaṅkarī rules now started becoming redundant. However, Śubhaṅkarī computation continued to find some place in arithmetic textbooks as long as the old units held sway.

The dramatic shift to the decimal system came in the 1950s. Decimal coinage became legal tender in April 1957. The metric system in weights and measure was legally enacted in 1956 and gradually came into force during the 1960s. Although, bighās and kāṭhās continued to be in use until the 1980s (and still carry popular meaning), the die had been cast.

Thus, the shift to decimal metrication rang the real death knell for Śubhaṅkarī.

References

Poromesh Acharya, *Banglar Deshaja Shikkhadhara*, Calcutta: Anushtup Prakashani, 1989, pp. 147–92.

William Adam, *Reports on the State of Education in Bengal* (1835 & 1838), edited by Anathnath Basu, Calcutta: University of Calcutta, 1941.

Abū L-Faẓl Allāmī, *Ā-īn-i-Akbarī*, Vol-II, translated by H.S. Jarrett, corrected and annotated by Jadunath Sarkar, Calcutta: Royal Asiatic Society of Bengal, 1949.

Abū L-Faẓl Allāmī, *Ā-īn-i-Akbarī*, Vol. I, translated by H. Blochmann, edited by D.C. Phillott, Reprint, Delhi: Low Price Publication, 1989.

Francis Buchanan (Hamilton), *Geographical, Statistical and Historical Description of the District, or Zilla of Dinajpur in the Province, or Soubah. of Bengal*, Calcutta: Asiatic Society, 1833.

Santanu Chacraverti, *Śubhaṅkarī: An Indigenous Tradition of Elementary Mathematical Instruction*, Kolkata: Asiatic Society, 2007.

Santanu Chacraverti, 'The Re-inheritance of Heritage: The Construction of the Shubhankari Tradition in Colonial and Post-colonial Bengal', in Arun Bandopadhyay (ed.) *Science and Society in India, 1750–2000*, New Delhi: Manohar Publishers & Distributors, 2010.

Ranabir Chakravarti, *Exploring Early India*, Delhi: Primus Books, 2016.

Abhishek Chatterjee, *Copper coins of Subah-i-Bengal during Aurangzeb's reign*, Kolkata: The Numismatic Society of Calcutta, 2019.

Kirtichandra Choudhuri, *Manoganit*, Calcutta: 1873, Mahendranath Ghoshal.

[69] Choudhuri (1873), p. i.

Lal Behari Day, *Bengal Peasant Life*, London: Macmillan and Co., Limited, 1878.
John Deyell, 'Cowries and coins: The dual monetary system of the Bengal Sultanate', *The Indian Economic and Social History Review*, 47, No. 1, 2010, pp. 63–106.
John Deyell, 'Hard Money and 'Cashless' Economies Medieval Bengal and the Greater Asian World', in John Deyell and Rila Mukherjee (eds.) *From Mountain Fastness to Coastal Kingdoms: Hard Money and 'Cashless' Economies in the Medieval Bay of Bengal World*, London and New York: Routledge, 2020, p. 186.
Panchanan Ghosh, *Mānasāṅka sambalita Śubhaṅkarī*, Kolkata: Patrick Press, 1893.
Suchandra Ghosh, 'Use of Coin Terms in the Epigraphs of Early Medieval Bengal and Related Issues', *Numismatic Digest*, Vols. 36–7, 2012–13.
Alexander Hamilton, *A New Account of the East Indies*, Reprint, New Delhi: Asian Educational Services, 1995, Vol. II, Appendix.
Kusha Haraksingh, *The Revenue Administration of Sylhet District 1765–1792*, PhD Thesis, SOAS, 1973, https://doi.org/10.25501/SOAS.00034035
Jan Hogendorn and Marion Johnson, *The Shell Money of the Slave Trade*, Cambridge: Cambridge University Press, 1986.
Ashok Jhunjhunwala, *Indian Mathematics: An Introduction*, New Delhi: Wiley Eastern Limited, 1993.
Araj Ali Matubbar, "Saral Khetraphal O Kalikasha", manuscript composed in 1967. Printed in *Mijanur Rahamaner Traimashik Patrika*, *Ganit Sankhya*, Dhaka, January-March, 1995.
Hemendranath Palit (ed.), Ashok Kumar Palit re-edited, *Subhankari*: *Rarha Banger Ganit Padabali*, Bardhaman: Bardhaman University, 2001.
Tomé Pires and Francisco Rodrigues, *The Suma Oriental of Tomé Pires and The Book of Francisco Rodrigues*, translated and edited by Armando Cortesao, London: Haklyut Society, 1944.
James Prinsep, *Useful Tables, Forming an Appendix to the Journal of the Asiatic Society, Part I: Coins, Weights, and Measures of British India*, Second Edition, Calcutta: Bishop's College Press, 1840.
Sukumar Sen, *Bangala Sahityer Itihas*, Calcutta: Eastern Publishers, 1963 (1965), Vol I (Aparardha).
Dineshchandra Sen, *Brihat Banga*, Reprint, Calcuttta: Dey's Publication, 1993, Vol. II.
Andrew Sterling, *Orissa: Its Geography, Statistics, History, Religion, and Antiquities*, London: John Snow, 1846.
Jean Baptiste Tavernier, *Travels in India by Jean-Baptiste Tavernier Baron of Aubonne*, translated by V. Ball, Second Edition, edited by William Crooke, London: Oxford University Press, 1925, Vol. 1.
Lal C. Verman and Jainath Kaul (eds.), *Metric Change in India*, New Delhi: Indian Standards Institution, 1970.
J. Wilfrid Jackson, 'The Use of Cowry-Shells for the Purposes of Currency, Amulets, and Charms', *Manchester Memoirs*, Vol. Lx, 1916, No. 13, p. 4.
Bin Yang, "The Rise and Fall of Cowrie Shells: The Asian Story", *Journal of World History*, Vol. 22, No. 1, 2011, pp. 1–25.

Chapter 2
Vernacularity, Ruchi Ram Sahni, and the Pursuit of Science in the Colonial Punjab

Kamlesh Mohan

What is meant by vernacularisation? Its meaning and function depend upon historical context socio-religious and political urgencies, and engagements with the present as well as past heritage of a community or nation. It may be the vernacularization of science, religion, political economy, or any other discipline. For example, in modern Europe, the vernacularization of medieval texts in Latin and Greek into English for specialists in medicine enabled them to benefit from the already available knowledge and thus go ahead in the production of new theories and technologies. Thus, translation from classical languages into the currently used English language (vernacular) helped in building bridges with past knowledge in various disciplines. The essential tool in vernacularization and its diffusion was the printing press. Invented by Gutenberg, it played a crucial role in the Italian Renaissance and the Protestant Reformation. It enabled broader segments of society to access a wide variety of knowledge and information in science and other disciplines.

In colonial India, vernacularization has been used for multiple objectives: for religious propaganda by the various brands of Christian missionaries and political control as well as for scientific and technological domination by the East India Company. Bernard Cohn, the cultural anthropologist, has persuasively argued how the codification of South Asian languages in the late eighteenth and nineteenth centuries onwards had been used to convert 'indigenous' forms of textual knowledge into instruments of creating colonial difference and rule.[1] Translation was an integral part of this project. From the 1840s onwards, the great 'Translation venture' as Ulrike Stark

[1] Cohn (1996), pp. 15–56.

An earlier version of the chapter had originally been published in Kamlesh Mohan, *Science and Technology in Colonial India* (Aakar Books, 2014). Reproduced with the permission of the author and the publisher, Aakar Books, Delhi.

K. Mohan (✉)
Panjab University, Chandigarh, India
e-mail: kamleshmohan14@yahoo.com

S. Ahmed and S. Sarkar (eds.), *Decolonizing Science and Modernity in South Asia*,
https://doi.org/10.1007/978-981-97-1829-0_2

has defined it, then, expanded to the flourishing print culture.[2] The process had been initiated through mammoth educational projects including dictionaries, glossaries, and lexicographies. It had been brought forward by three groups: missionaries, scholar-administrators of the British East India Company, and Indian elites.

The Company's initiative in vernacular education was directed at the young British civil servants, who were to be trained in Oriental languages at Fort William College, established in 1806 for governing the Indian people to develop a standard language of governance, the local scribes had been appointed to translate Sanskrit, Persian, Hindi, Bengali, and other languages into English. The ultimate objective was to equip Urdu for a new role. It was to replace Persian as the language of administration. It was to serve as a clear and efficient language for administrators who had to communicate with their Indian subordinates and sometimes with the rural people when on tour. In other words, it was to function as a 'language of command' to use Bernard Cohn's meaningful expression.[3] The essential idea underlying the Company's educational policy was 'improvement'[4] of the 'native' mind, his social mores and habits through the spread of European knowledge in vernacular languages. It was to be implemented in collaboration with the Indian elite, who were regarded as the lynchpin in the entire project.

The educated Indians including scholars, translators, and entrepreneurs in printing press joined hands in facilitating the transfer of European knowledge in science, technology, medicine, literature, and arts through vernacular languages. Sir Syed Ahmad Khan (1817–98) and Munshi Naval Kishore (1836–95) played a critical role in circulating European knowledge. While Syed Ahmad Khan remained engaged in translation projects through Scientific Society (founded at Ghazipur in 1864),[5] Munshi Naval Kishore contributed by publishing 5000 books in Arabic, Persian, Sanskrit, Hindi, Bengali, Urdu and Marathi from 1858 to 1885. He has been rightly called 'Caxton of India'[6] who sought to make knowledge traditions in other languages available in Hindi and Urdu for readers.

Recent scholarship has uncovered a new dimension of scholarly enterprise by nationalist Indians. The regional intelligentsia, instead of being co-opted in the Campany's agenda—the pursuit of knowledge for hegemony—utilized vernacularization for their autonomous projects. These projects included the compilation of dictionaries and lexicographies. As advocates of vernacular languages, they began to forge technical terminologies, dictionaries, and lexicographies for facilitating the transmission of European scientific discourse among the non-English reading public. It may be called popularization of science among the masses and the potential young

---

[2] Stark (2007), p. 283.

[3] Bernard S. Cohn, *n.* 4, p. 38.

[4] For exposition of the idea of 'improvement' underlying Company's educational policy see Bayly (1996), pp. 223–24. Bayly has argued that 'improvement', which combined utilitarian and evangelical ideologies, reflected a wider change in the international division of labour and knowledge. The Indian elite engaged themselves in government institutions such as the Calcutta Book Society, in 1817. See also Hatcher (1996), pp. 44–68.

[5] Lelyveld (1996), pp. 77–81.

[6] Ulrike Stark, *An Empire of Books*, pp. 145–51, 283.

Indian scientists. Rajendralal Mitra (1822–1891), a Bengali polymath, made the earliest intervention in renegotiating unequal power–relations through his publication entitled *Scheme for the Rendering of European Scientific Terms into the Vernaculars of India (1877)*. Mitra and later lexicographers of Nagari Pracharani Sabha (established in 1893 by a group of students of Queen's College, Varanasi) used a multilingual perspective by drawing upon the knowledge of experts and philologists in other languages and regions. Thus, the *Sabha* compiled the *Hindi Scientific Glossary* (1906). The end–product of their hard labour was two-fold. It inspired such struggles for establishing the claims of vernacular languages as effective media for administration, education, and scientific knowledge at lower levels in Maharashtra, Andhra, and other regions. An inter-lingual space was opened for non-English readers.

Nonetheless, Indian scientists in the 1930s indeed chose to communicate scientific news and their views in serialized periodicals namely, *Current Science* (1932) and *Science and Culture* (1935). These two periodicals were modelled on *Nature* (started in 1869) and cost double the price of a Hindi monthly *Vigyan*. It seems as Charu Singh has commented, "The task of creating knowledgeable, science-literate vernacular publics remained to be carried out by other non-scientific' actors, with diffident language repertoires and other standards of cultural authority".[7] However, their contribution shouldn't be ignored as they had waged a parallel struggle along with the Indian practitioners of experimental and theoretical scientists in creating a space for their role in the global circulation of knowledge and scientific discourse. Hence, it was the beginning of the decolonization of science. Ruchi Ram Sahni (1863–1948) along with some prominent scientists from Punjab made a substantial contribution to the growth of scientific knowledge.

2.1 Exploring New Terrains and Contesting Imperialist Ideologies

Ruchi Ram Sahni represents that generation of Punjabis which laid not only the foundations of the middle classes in this region but also actively participated in the socio-cultural and intellectual fermentation, resulting from the East-West encounter. Interestingly, the small number of urban mercantile families, who were exposed to the initial blast of Western ideas and cultural innovations following the annexation of Punjab in March 1849 had also been obliged to reckon with the peasant ethos, oral tradition and popular culture brought by the sons (rarely daughters) of the ambitious village literate and rich peasants to the towns and cities of Lahore, Amritsar, Rawalpindi and Multan where they came for education and later employment.[8]

---

[7] For an analytical account regarding making of vernacular dictionaries, lexicographies, and glossaries see Singh (2022).

[8] For an interesting discussion of cultural fermentation in nineteenth century Punjab see Jones (1966).

Apart from experiencing the tensions and contradictions in this peculiar socio-cultural milieu, Ruchi Ram as an Indian had also to carry the grievous burden of colonial domination. Creativity and leadership roles of this generation in intellectual, social, political and cultural fields depended firstly on their perception and awareness of their altered circumstances, produced by the colonial states' transform artery programme and policies under the bannered development and modernization. Secondly, these depended on the extent of their dissatisfaction with the traditional patterns of culture and the intensity of resentment and apprehension aroused by colonial hegemonization through cultural and intellectual engineering.

These two points are equally relevant in any discussion on the cultivation of literature, arts and science under a foreign rule. Recent studies on the place of science in colonial projects have shown beyond doubt that science was inextricably woven into the whole fabric of colonialism. Both were locked into each other, sometimes in an inverse relationship. The state claimed superiority in terms of physical power, resources and race. Science, which claimed superiority in terms of knowledge, was ultimately co-opted by the colonial state and helped it to dismiss other epistemologies as primitive. Several colonial scientists felt uncomfortable yet they had to perform a dual role—to serve the colonial state and to pursue their intellectual calling, i.e., the cultivation of science.

As social control was regarded as the pre-condition for the perpetuation of the politico-economic control and authority of the raj, the colonial sociologist, anthropologist and ethnographer joined hands to produce scholarly times and empirical studies about Punjab—its ecological conditions, cadastral resources, its multi-religious society, institutions, manners, customs as well as religious and cultural traditions.[9] Under such compelling demands the national-minded scientists had achieved distinction in their profession and partial success in popularising science mainly on account of their personal commitment, and creative energy rather than through state patronage. In order to appreciate the magnitude of their enterprise in a predominantly rural society of Punjab, we need to understand not only the common cultural framework of this region but also the role of the new catalytic agents, introduced by the British.

2.2 Socio-cultural Milieu of Punjab

Ruchi Ram Sahni grew amidst awhirl pool of changes triggered by the installation of a new ruling elite, and the import of an equally new layer of experts, Indian but non-Punjabi However, eye intention in describing the socio-cultural universe of this region is to underline the shared aims of the movements for popularising science and for reforming religious beliefs, social practices and attitudes of the people, i.e., to create progressive society.[10] While the scientist used the discourse of rationality for

[9] Mohan (1996), pp. 171–8.

[10] I have borrowed this insight from Brooke (1991, 1993), pp. 82–117.

enhancing the common man's understanding about the working of natural phenomena and facilitate their productive control, the socio-religious activist and reformer made it an instrument for combatting polytheism, superstitious beliefs which had blocked the growth of individuality and reconstitution of contemporary society. In order to achieve their goal, both were required to understand the working of their societies in which they lived.

The socio-cultural universe which Ruchi Ram Sahni tried to challenge needs to be understood from the perspective of a total field of religion[11] as a common framework holding together its entire structure. The clientele of priestly religion despite its dependence on scriptures, often traversed the territory of popular religion, bowed to its presiding deities and sought their intercession in human affair. In contrast the believers in saint cults, nature-gods, spirits and demons inhabited the rural tracts and gave no space either to scriptures or religious codes in their daily lives. More concerned with solving real life problems rather than enquiring into their causes the simple-minded village returned to a variety of sacred resources, evil spirits and sorceresses to rescue them or grant their wishes.

It may be pointed out that Ruchi Ram Sahni's drive for popularisation of science in this region could not ignore the dominant presence of peasant culture even among the Western educated urban elite. À few empirical details about Punjab will Clarify the point. Punjab was predominantly an agrarian society in the nineteenth century. In 1881, 87% of its population lived in 34,000 villages and only 13% in urban centres.[12] The rulers and the ruled (except a small number of mercantile families) were dependent on income from agricultural production. The colonial rule in Punjab did not dramatically alter the situation. Denzel Ibbetson, who had been associated with Punjab in various official roles, recorded that this region had neither vast towns and cities like Calcutta and Bombay nor great factories nor varied mineral wealth.[13]

For much of the nineteenth century, the majority of the Punjabis continued to live in an 'enchanted universe' to use Max Weber's phrase.[14] Dotted with shrines of pirs or saints, *jatheras* or clan ancestors, temples of disease-goddesses and other local godlings,[15] the Punjabi countryside attracted devotees from towns and cities. Among their devotees were included a large number of Sikhs, Hindus and Muslims. The explanation of their fervent faith in the miraculous powers of Sakhi Sarar, Guzza Pir, Sitala Devi or Suraj Deva may be found in the popular perception of natural calamities, illness, disease and healing. In the pre-industrial society of Punjab, the people tended to attribute floods, drought, fire, famines and illness to the anger of

---

[11] Tambiah (1970), pp. 371–7. On the basis of his study of three villages in Thailand, Tambiah has argued that 'the idea of two levels (of religion) is an invention of the anthropologist'. His suggestion of looking at a 'total field of religion' gives a more integrated view of society.

[12] Ibbetson (1883), p. 17.

[13] Ibetson et al. (1919), p. 3.

[14] Weber (1962), p. 226.

[15] Description of saints, pirs, disease-goddesses, nature pantheon is based on various Census Reports of Punjab from 1881 to 1931; R. C. Temple, Legends of Punjab, 3 vols., Patiala, Language Department of Punjab 1962–63; first published 1884–86; Kinsley (1986) and Gilmartin (1984), pp. 221–40.

nature pantheon, malignant influence of a deity or to possession by an evil spirit or demon.

Let us take examples of physical afflictions. In a peasant society, crippling arthritis and chronic rheumatism of joints are highly prevalent. Working for long hours on agricultural chores and handling farming implements were major causes of physical deformities among the cultivators. The inventory of illnesses for which cures were sought included intestinal disorders, eye diseases particularly blindness, tetanus, leprosy, localised paralysis and psychosomatic disorders. As there was no well-developed public health system and medical science to treat diseases, psycho-social problems and to cure minor ailments, the unlettered peasants turned to supernatural forces, sorcerers, *ojhas, sianas* and astrologers.

Visits to the shrines of *pirs* (minor shrines) and *khankahs* (major shrines) formed part of standard curative practices. These shrines scattered not only to the spiritual needs of simple-minded rural folks but also provided cures to invalids from different denominations whose affliction had often unknown causes and an uncertain prognosis. In performing their roles as healers, the *mejaweri* (successors of pirs who managed the shrine) did not undertake to establish the aetiology of any disease but to administer cures. All diseases were not treated at the same shrine. There was a kind of specialization among different pirs and their shrines. For example, the main *khankahs* (shrine)of Sakhi Sarovar at Dera Ghazi Khan had specialised in dealing with problems of spirit possession among women. The shrine of Gugga Pir at Bikaner, whose supernatural powers were invoked to grant wishes, heal the sick, protect cattle and bestow a son upon the devotee, specialised in curing his worshippers of snake and scorpion bites. Another example is of Sitala Devi, the goddess of pustular diseases who was worshipped all over Punjab by Hindus and Sikhs alike. Denzil Ibbetson, the chief ethnographer of Punjab, has recorded the therapeutic rituals to appease the goddess in nineteenth century Punjab.

The observers of Punjabi society are struck by the absence of the confrontationist approach to nature among the rural (even urban) people in the nineteenth century.[16] One reason is the impact of cultural socialisation which inculcates a friendly and respectful attitude towards natural forces. The second reason is the availability of favourable ecological conditions on the whole as compared with the cold and harsh climates of European countries. Naturally, the Punjabi peasantry showed no inclination to control or dominate nature and their environment for individual and communal needs. That was why village gods were assigned important roles both in life cycle and agrarian cycle, particularly at the time of harvesting and sowing of crops. Apart from emotional bonds among their devotees, these godlings were instrumental in protecting them from diseases, threats from the outside world and evil spirits. For example, local godling Bhoomi, represented by a heap of stones and generally placed on the boundaries of a hamlet, was invoked to protect the land on which villages were settled.

Nature Worship was also an important constituent of popular religion. In its pantheon was included Sun or the Surai Deuta as the major planet to be propitiated,

[16] Oberoi (1994), p. 202.

the lesser ones being Mercury and Mars. Earth was widely venerated as Mother, i.e., Dharti Mata because the rural folks were heavily dependent on fertile soil for abundant agricultural crops. Another god in this pantheon was Khwaja Khizx, the water god, who rules rivers and streams. His favours were invoked both for protection from foods and adequate water for irrigating fields. Evidently, the relationship between the local gods and peasantry, for whom the village and its immediate environment represented the entire cosmos, was crucial in his day-to-day experiences.

It may be conceded that the popular religion and priestly religion had worked out a comfortable relationship over the centuries, implicating a process of mutual give and take. With the growth of the middle classes under colonial rule, the reciprocal influence of the cultures of the powerful and the subordinate acquired a new angle. The popular culture, especially the oral tradition of the numerically dominant peasant population, circulated freely in the small towns and the big cities through the English educated sons of the village literati who acted as a bridge between the priestly elite culture of the urban areas and the rural culture.

Interestingly, this generation of the new elite also represented the rationalist critique and offensive against the Brahminical domination and excesses, rigid social hierarchies based on caste as well as against superstitious faith in magic cures, saint cults and sacred sites associated with village religion. Influenced by the enlightenment ideology, they tended to regard science as the 'pioneer instrument' for all progress.[17] The socio-religious reformers of the period also shared faith in scientism and used it as a weapon to counter missionary attacks upon their religions be it Hinduism, Islam or Sikhism. Without going into the details of the reform process, I wish to emphasise that rationalists like Ruchi Ram Sahni and Mahendra Lal Sircar, who were trained as scientists, tried to understand the nature of their societies wherein they had sought to popularise science.

Ruchi Ram Sahni, well-known commentator on Punjabi society, colonial culture and national policies, informs us that the religious boundaries were not rigidly drawn in nineteenth century Punjab. Even in the urban areas, most of the people did see any clash between the tenets of Hinduism and Sikhism. Recalling his formative years, he observed, 'my religious beliefs were fragile and indefinite as that of my father, but in my case at least there was a distinct leaning towards the tenets of Sikhism'.[18] He worshipped idols and recited *Rehras* and *Sukhmani* with equal fervour. Ruchi Ram Sahni attributed the prevailing ambivalence towards religion to liberal socialisation in the family rather than scientific or rational thinking. The accommodative spirit of the Punjabis owed in large measure to the absence of Brahminical Dominance in a peasant society which valued the Jats for their utility in an agricultural economy, despite their low status in the religious hierarchy. It was reinforced by the egalitarian teachings of Sikhism and their large-scale conversion to the new creed.

---

[17] Worboys (1991).

[18] Sehgal and Mahanti (1994), p. 133.

2.3 Gearing Towards Change: The New Social Forces

The foregoing description of mid-nineteenth century Punjab has underlined the prevalence of polytheism, superstitions, supernaturalism, a diluted version of the caste system in the absence of socio-economic ascendancy of Brahmins and the other discriminatory social practices. This conservative social environment, which had slowed the pace of innovation in thought and practice including science and technology, was to experience tensions when exposed to European ideologies and institutions. Among the various catalytic agents, which affected the mindset and behaviour of the Punjabis, the major ones were the Western educational institutions, print culture and the Brahmo ideology.

The first major catalyst was the Western education system, introduced with a two-fold objective: utilitarian and ideological. Macaulay and many of his contemporaries believed that military conquest had only won them a precarious hegemony. The Most enduring and profitable conquest was over the mind. It acquired even more urgency in the case of Punjab which had to be controlled through conquest over the hearts and minds of its inhabitants, especially the social and economic elites. The colonial mission was an instant success as the first generation of Indian intelligentsia in nineteenth century Bengal began to believe that it was only through the medium of English education that the reservoir of European scientific knowledge would be open to them. Raja Ram Mohan Roy and many of his contemporaries, regarded Britain as the chosen instrument for leading India to the path of political and economic modernisation.[19]

The British rulers were keen that this ideological perspective should be effectively published in the recently annexed Punjab which had to be converted into a secure but also profitable possession. Obviously, a serious missionary effort for the spread of Western education, including literary and scientific, was made in order to colonise the consciousness of Punjabis.[20]

Lahore was chosen as the centre for administrative and educational institutions. By early 1870, it had developed into a typical colonial metropolis and by the 1880s became the hub of the cultural life of the north.[21] Possessing abundant resources, imperial aura, educational facilities and employment opportunities, it attracted young Punjabis thirsting for new knowledge in arts and science, eager to experiment with new gifts of the raj and willing to embark upon new careers. Many among them like Ruchi Ram Sahni, Lala Lajpat Rai, Gurmukh Singh and Bhagat Lakshman Singh were destined to become prominent intellectuals while others earned educational degrees in the hope of prosperous futures but lived in anonymity.[22]

---

[19] Panikkar (1986), p. 421.

[20] Gough and Inner (1984), p. 222.

[21] For contemporary impressions about socio-cultural configuration of Lahore see Mehta (1972), pp. 35–6; Ruchi Ram Sahni, 'Self-revelations of an Octogenarian' (Manuscript is lying with his grandson Professor Ashok Sahni, Panjab University, Chandigarh).

[22] Singh (1865), p. 133.

These new intellectual elites were the product of a network of Western educational institutions including anglo-vernacular schools of arts and science and engineering and medical colleges in Lahore, and other cities and towns of Punjab. The opening of the first Government school in 1851 at Amritsar was followed by a chain of such institutions at Rawalpindi, Gujrat, Shahpur, Multan, Jhelum and Jalandhar. By 1856, there were thirty-five such schools.[23]

The next major landmark in the development of Western education in Punjab was the opening of Government College in Lahore in 1864 which taught courses in Arithmetic, Algebra, Euclidean Trigonometry, Conic Sections, Logic, Economics, English, Persian, History, Physics and Chemistry. Under the principalship of GW. Leitner College emerged as a major educational institution. It was here that Ruchi Ram Sahni, first as a student and later as a faculty member, was drawn to the Western scientific ideas. His concerns and sensibilities, particularly a real love and admiration for the English language as a noble vehicle of thought and feeling,[24] matured in the company of his fellow students. As a student he also demonstrated his passion for Western scientific knowledge, technologies and his ability to improvise cheap alternatives for experiments and lantern slide demonstrations for public lectures.

The second Western innovation was the printing technology which had facilitated the exchange and circulation of books like any other commodity. Described as 'print culture' by McLuhan,[25] such a phenomenon transformed the Punjabi society which was based on oral and scribal knowledge, into a society where printing technology became crucial for the reproduction of ideas and knowledge. Thus, the printed word became not only a source of ideas and images but also a bridge between scholars, scientists, students, social reformers and rural gentry. Easy availability of books, tracts, journals and newspapers had excited the imagination of students who could go beyond the prescribed textbooks and familiarise themselves with European ideas on liberty, humanism and rationalism.

The eager students of Government College, Lahore, spent their leisure time in discussing, examining and scrutinising the implications of social theories and relevance of European ideas and scientific knowledge for shaping modern India. In his memoirs, Ruchi Ram Sahni recorded an illustration of one such stimulating reading session.[26]

The remarkable books, as they appeared both to Guru Dutt and myself, we also read together in our spare hours in the college veranda, or rather the vestibule. These were Mill's 'Utilitarianism' and Bentham's 'Theory of Legislation'. They were, of course, not included in the college 'Course', but that was of little consideration for both of us. We read and reread Mill's small book line by line, or paragraph by paragraph. Now and again, we could not do more than a sentence or two in the

---

[23] Education Commission Report, Calcutta: Government Printing Press, 1884, p. 8. Appendix contains the Report by the Punjab Provincial Committee with Evidence Taken Before Committee and Memorials addressed to the Commission.

[24] Ruchi Ram Sahni, 'Self-Revelations of an Octogenarian', op. cit., note 14, p. 128.

[25] McLuhan (1962).

[26] Ruchi Ram Sahni, 'Self-Revelations of an Octogenarian', op. cit., note 14, p. 139.

course of an hour, for either we could not agree as to what the author's real meaning was, or for some reason, the whole time was taken up with discussion about all the implications of the passage or how far we could accept his lead. Now and again, we would deliberately take up our stand on opposite sides, so as to be able to thrash our point as well as we could. As on such occasions, we would, as a rule, undertake to support the experimental view-point while Guru Dutt would become the exponent of the *Transcendental* school ... sometimes 'these discussions were attended by elderly persons from the city interested in one or other of the contestants. Now and again, the debates became quite animated and heated, each party pleading for his own thesis'.

Obviously, these eager students, who matured under the influence of the enlightenment ideology during the 1880s and 1890s, had evinced a great admiration for Europe's achievement in literature, political theory and practice of democracy, positivism, scientific knowledge and technology. Their intellectual growth acquired more flesh when they came into contact with the English educated Bengalis in Lahore who had already internalised the discourse of scientific rationality of the raj. The earliest evidence of the application of modern scientific rationality to reform Indian society is found in Raja Ram Mohan Roy's *Tuhfat-ul-Mowahidin* (1803) wherein he has tried to define a rational theology.[27]

The third important catalyst of the socio-cultural milieu in Punjab was the Bengali Brahmo community which had provided a large number of clerks, teachers, pleaders and doctors as the human underpinning of the raj. Having crystallised social force, it played a crucial role not only in altering the traditional social and economic hierarchy but also in changing habits of the mind among the literate classes. During the 1880s, Brahmo Samaj and its creed was the major ideological influence on Punjabis, particularly Hindu Consciousness. Ruchi Ram Sahni became a regular member of the Brahmo Samaj within a couple of years after his arrival at Lahore as a student in October, 1879. Among the other young men who turned to the Brahmo Samaj, included Devi Chand Gupta (better known as Shradha Prakash Deva), Bhai Sunder Singh, Bhai Harbhagwan, Bhai Chatter Singh, Pandit Girdhar Raj Biswasi and Chander Prakash Deva.

Ruchi Ram Sahni's decision to study science and assume the role of a crusader for scientific awareness was influenced to some extent by his close involvement in Brahmo Samaj activities in Lahore. To the Bengali Brahmo intelligentsia and their Punjabi brethren-in-faith like Ruchi Ram Sahni, science connoted certain values: unity over diversity, the compilation and successful application of useful knowledge about men, society and the universe, the search of natural laws, optimism about the vital role of science in progress.[28] Brahmos not only appropriated science and rationalism in a special way but also deified them. Brahmo scientists were the first modern scientists in contemporary India particularly in Bengal.

Most Indians, especially Punjabis, despite their resentment against the domination of the Bengali Brahmo elite in educational and administrative institutions of the raj and religious life for almost half a century, continued to regard Calcutta as an

[27] Cited in Panikkar (1975), Section III.

[28] Kopf (1988), p. 48.

intellectual centre. For many of them including Ruchi Ram Sahni, residence in a big city like Calcutta was in itself a 'continuous source of education'.[29] It proved to be useful in two ways. Firstly, his interest in chemistry grew into a passion under the inspiring guidance of Professor Sir Alexander Pedler in the classroom and laboratory of the local Presidency College where he was a student. While assisting his teacher in his work as an analyst to the Calcutta Municipality, Ruchi Ram acquired training in testing wines and methodical preparation of reports in a short time. As a student, he had not only formed a lifelong association with his aforesaid intellectual mentor but also with his class fellow Ashutosh Mookeriee who rose to be the Vice-Chancellor, Calcutta University later on.[30] Upendra Kishore Roy Choudhary, his class fellow in the M.A., strengthened his Brahmo sympathies. Secondly, the idea of his future role as a crusader for scientific awareness crystallised while listening to the public lectures of Professor Jagadish Chandra Bose, P. C. Ray and Father E. Lafont in the Indian Association for the Cultivation of Science, Calcuta.[31] This Institute's role inspired Ruchi Ram to set up the Punjab Science Institute in collaboration with Professor J. C. Oman, his colleague in Government College Lahore in the summer of 1885.

2.4 Ruchi Ram Sahni's Crusade for Scientific Awareness

Dissemination of scientific awareness in a colonial society implied theoretical understanding of the subject, ability for its practical application and demonstration and a talent for lucid transmission of factual knowledge to the people with an average intelligence in their own idiom. Even more serious was the lack of organisational/ institutional structure to aid the popularisation of science among the people. Establishment of as many as 170 colleges including several medical and engineering colleges and technical education institutions (affiliated to the five universities—Calcutta, Madras, Bombay, Lahore and Delhi) and ten scientific services had only trained and employed Indians as clerks, second-rank technicians, and engineers to run railways, shipping or canal construction ventures or to gather information about botanical, zoological and mineral resources.[32] Transfer and relocation of Western

[29] Ruchi Ram Sahni, Memoirs, op. cit., note 11, p. 1.

[30] Details about Sahni's associates in Calcutta are contained in his Memoirs, op. cit., note. 1, pp. 5–12.

[31] Mahendralal Sircar qualified in Medicine in 1860 from the Medical College Calcutta and received the MD degree in 1863. In 1876, he founded the Indian Association for the Cultivation of Science, the first institute for scientific research in India. He was a member of the Bengal Legislative Council from 1887 to 1893.

[32] McLeod (1975). Ten scientific services included the Meteorological Reporter, The Inspector General of the Civil Veterinary Department, the Director of-the Botanical Survey of India, The Reporter on Economic Products, The Inspector General of Agriculture, The Director General of Archaeology, The Chief Inspector of Mines, The Surveyor General, The Inspector General of Forests and The Director of Geological Survey.

scientific knowledge and technologies was only partly successful in terms of long-term benefits to the Indian people. It was so because these ventures were conceived of and remained as mere technological projects which did not target at diffusion of knowledge and skills. Indians were educated to a certain point and culture of technology was withheld from them.[33] As colonial subjects, non-Europeans including Indians were denied opportunities for enterprise, investment and experience. It is evident from Rai Bahadur Ganga Ram's bitter experience who was discouraged not only from making investment on lift-irrigation technology to harness water near Renala in Lower Bari Doab canal but also deprived of a contract to construct tube-wells on his plot in Upper-Chenab Colony.[34]

Painfully aware of the conflict between the colonial agenda of scientific development and needs as well as aspirations of the Indians, Ruchi Ram Sahni launched his mission for the transfer of scientific knowledge and grafting of technologies. It was difficult for any Indian to get state patronage or funding for projects concerned with dissemination of scientific awareness. Hence, paucity of financial resources made Ruchi Ram Sahni's task as a crusader for scientific awareness, indeed, daunting.

Despite all these handicaps, Ruchi Ram Sahni took the first step towards his goal by establishing the Punjab Science Institute in 1885.[35] The idea of this institute has originated with Professor J. C. Oman but it took a concrete shape after Sahni had seen the working of the IACS. Founded in mid-1885 with Professor Oman as Honorary Secretary and Sahni as Joint Secretary, the original aim and object of the Punjab Institute was the popularisation of all kinds of scientific knowledge throughout the province by means of lectures (in English and vernacular) illustrated with experiments and lantern slides, as well as the publication of tracts. These objects, according to Ruchi Ram Sahni, were expanded to include the encouragement of technical education, and, in particular, of chemical industries.

In pursuance of the original object of science popularisation Ruchi Ram Sahni managed to involve several Professors from local colleges in the activities of the Institute especially in its lecture-programme. For example, Professor Oman delivered several lectures on various aspects of 'Electricity' and 'Magnetism'. Dr. C. C. Caleb, who was on the faculty of Medical College, Lahore, gave a series of lectures on human anatomy. Caleb's two lectures, 'Man's Fear' and 'Smokes: Poisonous and Non-poisonous', became especially popular. Dr. Grant's Lectures on 'Soap Bubbles', 'Chemical History of the Candle' and 'Spinning Tops' were particularly well-received. Sahni himself gave as many as five hundred popular lectures in various towns of Punjab and repeated these in private gatherings on personal invitation from local elites. His lectures on weather with special reference to India were based on sound and practical knowledge which he had acquired in the course of his job as the second Assistant Meteorological Reporter to the Government of India.

[33] Inkster (1988).

[34] Ali (1989), pp. 218–22.

[35] My information regarding the establishment of the Punjab Science Institute, its objects and activities for the popularisation of science is based on Ruchi Ram Sahni, 'Self-Revelations', op. cit., note 14, pp. 256–67.

Public lectures arranged by the Punjab Science Institute Aroused tremendous interest among the residents of Lahore, especially among students who had enlisted their parents as active supporters of the Institute. As a result, Sahni received invitations for lectures from the rise of Patiala, Kapurthala, Mandi and Bahawalpur. Besides, he used to give a regular annual course of twenty lectures on the basic facts and principles of Physics and Chemistry to large audiences, consisting entirely of shopkeepers. Ten lectures were devoted to general topics such as 'How Does the Telegraph Wire Speak', 'The Common Flame', 'Electroplating', 'Electricity in the Service of Man' (a series of three or four lectures), 'Glass Making' and 'Puniab and its Rivers'. The last one was illustrated by a large relief map of India (Made in clay under Sahni's direction for an Educational Exhibition at Lahore). Several of these lectures, especially those giving advance information, were repeated in Lahore city itself and also in a few small towns. Ruchi Ram's tireless crusade for the popularisation of science created 'a real furore of enthusiasm in the province about scientific studies.'[36]

Indeed, the personal commitment and concerted efforts of Ruchi Ram Sahni and his associates, particularly Professor Oman and Dr. Caleb, had succeeded in arousing interest in popular science among the people. The Punjab Science Institute, which was substituted by the newly established Society for the Promotion of Scientific Knowledge, founded by local Medical College students in 1905, had been able to raise a substantial amount of funds owing to the good quality of lectures, illustrated by practical demonstrations and lantern slides.[37] It may be mentioned that Sahni had reinforced his mission by the establishment of the Punjab Science Institute Workshop. However, there is no evidence to suggest that enthusiasm and interest in popular science, which Ruchi Ram Sahni had generated so assiduously, matured into a science movement in this region.

Another aspect of Ruchi Ram Sahni's contribution was to underline the positive role of regional languages in the dissemination of scientific knowledge among the masses. Despite the fact that perceptions of the educated elites in Punjab during the 1880s had been influenced by the Brahmo leaders of the Bengal renaissance, they had increasingly turned to indigenous sources for inspiration and strength as the century drew to its close.[38] Hence, Punjabi was assigned a valuable role for social and intellectual regeneration. Ruchi Ram Sahni also preferred the use of Punjabi as a vehicle for scientific knowledge in his popular lectures. Judging by the response of his audience including average men and elites, Sahni concluded that the mother tongue was the best medium to communicate knowledge of modern science. It would enable the people to adapt scientific knowledge and technologies to their environment and finally contribute to the development of alternative technologies. That the local languages were best suited to the fruitful dissemination of scientific knowledge

[36] Ibid., p. 264.

[37] For details of Ruchi Ram Sahni's improvisations for making photographic lantern slides, see his *Self-Revelations*, pp. 259–61.

[38] The renaissance in Punjab shares this feature with the Delhi Renaissance. For a detailed discussion of the role of vernacular in the Delhi Renaissance, see Minault (1987), pp. 289–90.

among the masses was a deep-rooted conviction common among Indian intellectuals such as Raja Ram Mohan Roy, Master Ram Chandra and his contemporary Sir Syed Ahmed Khan. There was a definite political agenda underlying the translation of the best European writings into Indian languages. Sir Syed Ahmed Khan pleaded that.[39]

> Those bent on improving and bettering India must remember that the only way of compassing this is by having the whole of arts and sciences translated into their languages.

The strongest argument in favour of the translation project was that the constant use of English language by impressionable children would not only habituate them to express their thoughts in it but also 'denationalise' them. Both Master Ramchandra and Ruchi Ram Sahni had expressed their anxious concern for the alienation of the younger generation whose patriotic feelings ought to be nurtured. Both of them had actively contributed to the building of 'national character'.[40] While the former devoted his energy and talent to the translation of a number of European works on science in Urdu, the latter popularised information about the latest inventions and various branches of scientific knowledge through his lectures in Punjabi, the most widely spoken language in this region.

2.5 Conclusion

A review of Ruchi Ram Sahni's contribution towards the spread of Scientific awareness among the common people living in the urban areas (lecture-tours to villages being very few) draws our attention to the crucial issues (i) relationship between religion and science, (ii) tension between the practice of nationalism and cultivation of Western science and technology, and (iii) conflict between the agenda of colonial science and needs/aspirations of the subject people.

The first issue concerning the nature of relationship between science and religion, also addressed in Europe in the last decades of the eighteenth century and first half of the nineteenth century, does not necessarily imply conflict, but a half-way meeting of two human initiatives (seemingly opposed) to maximise social progress, welfare and happiness. Both the scientist and the social reformer in colonial India had sought to achieve this shared objective through the pursuit of knowledge and scientific rationality. While the scientists of Ruchi Ram Sahni's generation were engaged in acquiring 'superior' and modern scientific knowledge and technologies for nation building, the social reformers had chosen to mediate the recovery and return to Upanishadic or Vedic knowledge. In their critique of Indian cultural tradition, social institutions and practices, religious codes and knowledge-systems as well as in their formulation of models and strategies for recasting society, economy and polity,

---

[39] Mohammad (1972), pp. 231–2.

[40] Both Munshi Zakaullah, a student of Master Ramchandra in Delhi, and Akshay Kumar Dutt had shown acute awareness of and concern for the consequences of the English education, especially its role in alienating young students from their own national culture.

impact of the West, particularly Comte's positivism, bourgeois values and concept of democracy, liberty, progress and modernity was obvious. Very often their roles overlapped. Both of them had used rationalism as a major argument for denuding Hinduism of its excrescences: idol-worship, superstitions, blind faith in demons' spirits, caste distinctions and outdated religious rituals as well as for popularising Western scientific knowledge and technologies.

Ruchi Ram Sahni, whose thought process had largely been influenced by the Brahmo intellectuals particularly Jagadish Bose and Prafulla Chandra Ray, had sought to resolve the conflict between the Western scientific rationally and institutional-oriented Indian religions by turning to the Brahmo Samaj. Without rejecting the need and support of religion, he practised rational theology. After joining the Brahmo Samaj as a regular member he had discarded the sacred thread -the cherished mark of a high caste orthodox Hindu. His commitment to rationalism was demonstrated when he had parted with his Brahmo mentor Shiv Narain Agnihotri because the latter indulged in self-glorifications as godman of sorts. Thus, scientific rationality had been used as a tool for bringing about socio-cultural change not only by Ruchi Ram Sahni but by the Brahmo ideologies generally.

The second issue concerning tension between the practice of nationalism and cultivation of Western scientific knowledge was viewed by the Brahmos from a different angle. For example, Jagadish Bose who believed in Brahmo-inspired universalism, argued that knowledge was not the monopoly of any race and thus 'science is neither of the East nor of West but international'. Most of the Indian scientists sincerely believed that universalism and unity of knowledge were crucial to the growth of civilization and common heritage of mankind. However, their bitter experiences as colonial subjects obliged them to confront the cultural chauvinism of the British imperialists who claimed scientific genius and achievements as the prerogative of the Anglo-Saxon races and behaved arrogantly. For example, Ruchi Ram Sahni has recounted the blatant form of racial discrimination against Indians in daily life. His personal experience of supersessions in the Educational Service in addition to the frequent humiliations as an individual not only angered him but also enabled him to evolve his own code of conduct to live with dignity.

Despite wilful denial of opportunities of training, professional growth and promotion by the colonial state, the national-minded scientists neither boycotted scientific institutions and services nor honorific titles. They channelised their trust ration and anger into a more creative way like (i) creation of scientific awareness among adults and improving quality of science teaching in schools, (ii) opening of swadeshi chemical industries and scientific workshops and (iii) translation of Western scientific works into vernacular languages. These three forms highlight the common trend of thinking harnessing training and Western scientific knowledge for nation-building indirectly. It was an acknowledgement of the scientific achievements of the European nations and the keen desire of their colonial subjects to emulate them without abandoning their civilisation.

The third issue concerns the conflict between the agenda of colonial science and needs/ aspirations of the Indian people. The colonial rulers perceived three potential roles for science. Its first role was concerned with exploration and discovery, the second with solving environmental and disease problems and the third with providing ongoing advice and technical services. All these roles targeted at making the colonies reproductive and at exploiting their potential in terms of human and natural resources as a tribute exacted from the hapless colonial subjects. Neither the development of the intellectual potentialities of the subject people nor economic modernisation of their country were on the agenda of the British rulers who had increasingly harnessed science and technology to holster the empire and capitalism. It was not only scientific education which was underdeveloped but research as well. Factually speaking, training of Indians as qualified doctors, engineers and scientists was not a priority for the British rulers because these ranks were always available to metropolitan graduates in the biological and other sciences. Directed by the colonial-administrators-turned-ethnographers, the British officers in scientific institutions and services constantly used scientific means for social control with a view to perpetuating political hegemony. The statistical population surveys are a good example.

The foregoing discussion has shown that scientists like Ruchi Ram Sahni had been obliged to negotiate between the pulls and pressures of their intense desire to modernise their society through the instrumentality of Western scientific knowledge and the patriotic compulsion to flight for the defence of the Indian cultural tradition and for national freedom. Despite role-conflicts and limited resources and space as a colonial subject, Ruchi Ram Sahni had tried to lay the foundation of modern scientific tradition through his campaign for popularisation of science. However, his pioneering initiative failed to mature into a science movement. Firstly, Ruchi Ram Sahni failed to train a team of students to continue either this task of popularisation of science or of improving the quality of science teaching in schools. Those who were trained as scientists in this region were not farsighted enough to think beyond individual achievement. Secondly, the absence of large-scale industries also did not favour the growth of a science movement. Thirdly, the colonial state was not interested in the development of science in this region. But on the whole the examples set by Ruchi Ram Sahni in Punjab were in line with the current of 'limited modernisation' then passing in the local metropolis like Calcutta. Nevertheless, his crusade for the popularisation of science in Punjab had stimulated a great deal of interest in the pursuit of western knowledge without devaluing Indian cultural heritage and struggle for freedom.

References

Imran Ali, The Punjab Under Imperialism 1885–1947, Delhi: Oxford University Press, 1989.

Chris Bayly, *Empire and Information: Intelligence gathering and social communication in India, 1780–1870, Cambridge:* Cambridge University Press, 1996.

John Hedely Brooke, *Science and Religion: Some Historical Perspectives*, New York: Cambridge University Press, 1991, 1993.

Bernard S.Cohn, *Colonialism and Its Forms of Knowledge: The British in India*, Princeton N.J.: Princeton University Press, 1996.

Brian A. Hatcher, *Idioms of Improvement: Vidyasagar and Cultural Encounter in Bengal*, Delhi: Oxford University Press, 1996.
David Gilmartin, 'Shrines, Succession and Sources of Moral Authority', in B.D. Metcalf, ed., Moral Conduct and Authority, Berkeley: California University Press, 1984.
Charles Gough and Arthur D. Inner, The Sikhs and the Sikh Wars, First published 1897; Reprinted, Delhi: National Book Shop, 1984.
Denzil Ibbetson, *Census of Punjab, Report 1881*, Lahore; Government Printing Press, 1883
D. Ibetson, E.D. Maclagan, and H. A. Rose, A Glossary of the Tribes and Castes of the Punjab, Vol. I Lahore: Govt. Printing Press, 1919.
Ian Inkster, 'Prometheus Bound: Technology and Industrialisation in Japan, China and India Prior to 1914: A Political Economy Approach', Annals of Science, 45, 1988, pp. 422–3.
Kenneth W. Jones, 'The Bengali Elite in Post- Annexation Punjab: An Example of Inter-regional Influence in 19th Century India', *The Indian Economic and Social History Review*, III (4), December 1966, pp. 377–95.
David Kinsley, Hindu Goddesses: Visions of the Divine Feminine in Hindu Religious Tradition, Berkeley: University of California Press, 1986.
David Kopf, *The Brahmo Samaj and the Shaping of the Modern Indian Mind*, Princeton: Princeton University Press, 1979.
David Lelyveld, *Aligarh's First Generation: Muslim Solidarity in British India*, New Delhi: Oxford University Press, 1996.
R.M. McLeod, 'Scientific Advice for British India: Imperial Perception and Administrative Goals 1898-1923', Modern Asian Studies, 9(3), 1975, pp. 343–84.
M. McLuhan, *The Gutenbery Galaxy*, London: Routledge and Kegan Paul, 1962.
Ved Mehta, Daddyji, Delhi: Oxford University Press, 1972.
Gail Minault, 'Sayyid Ahmed Dehalvi and the Delhi Renaissance' in R. E. Frykenberg (ed.), *Delhi Through the Ages*, Delhi: Oxford University Press, 1987.
Shan Mohammad (ed.), *Writings and Speeches of Sir Syed Ahmed Khan*, Bombay: Asia Publishing House, 1972.
Kamlesh Mohan, 'Colonialism and Ethnic Identities: 19th to Early 20th Century', in A.J. Qaisar and S.P. Verma, eds., *Art and Culture: Endeavours in Interpretation*, Delhi: Abhinav Publications, (1996).
Harjot Oberoi, The Construction of Religious Boundaries: Culture, Identity and Diversity in the Sikh Tradition, Delhi: Oxford University Press, 1994.
K. N. Panikkar, 'Presidential Address' to the 36th Session of the Indian History Congress, 1975.
K.N. Panikkar, 'The Intellectual History of Colonial India: Some Historiographical and Conceptual Questions', in S. Bhattacharya and Romila Thapar, eds., Situating Indian History, Delhi: Oxford University Press, 1986.
Narender K. Sehgal and Subodh Mahanti (eds.), *Memoirs of Ruchi Ram Sahni: Pioneer of Science Popularisation in Punjab*, New Delhi: Vigyan Prasar, 1994.
Bhagat Lakshman Singh, Autobiography, ed. Ganda Singh, Calcutta: The Sikh Cultural Centre, 1865.
Charu Singh, 'Science in the Vernacular? Translation, terminology and lexicopraphy in the Hindi Scientific Glossary', *South Asian History and Culture*, Vol. 12, No. 1, 2022, pp. 63–86.
Ulrike Stark, *An Empire of Books: The Naval Kishore Press and the Diffusion of the Printed Word in Colonial India*, New Delhi: Permanent Black, 2007.
S.J. Tambiah, *Buddhism and the Spirit Cults in North-East Thailand*, Cambridge: Cambridge University Press, 1970.
Max Weber, The Religion of China: Confucianism, Lanism and Taoism, New York: The Free Press of Glenoe, 1962.
Michael Worboys, 'Science and the Colonial Empire 1849–1940', in Deepak Kumar (ed.), *Science and Empire*, Delhi: Anamika Prakashan, 1991.

Chapter 3
Characterizing the Debate on Constructing Scientific Terms in Hindi: A Study of Influences and Motivations

Sandipan Baksi

Abstract The early Twentieth Century witnessed a vigorous debate on the construction of words for glossaries of scientific terms in Hindi. While the main features of this debate have been captured in scholarship, this article attempts to highlight the underlying principles—linguistic puritanism on the one hand, and pragmatism on the other—that informed the debate. The article establishes that while the debate apparently played on linguistic grounds, it was essentially a reflection of a more fundamental political conflict around the nature of nationalism, and the corresponding role of science and language in shaping it. One side of the conflict was motivated by cultural and, especially, linguistic nationalism framed alongside or in conjunction with a revivalist perspective. The other side was driven by a feltneed for science popularization and economic nationalism.

Science writing in different Indian vernaculars began in the first half of the nineteenth century. The earliest initiatives in this direction came from the Christian missionaries and the colonial educational institutions. These initiatives were soon taken up by the burgeoning class of 'native' intelligentsia, who were at the forefront of the emerging print media in colonial India. The earliest attempts were dedicated towards preparing translations of books on science, and developing text books on various science subjects. By the second half of the century, science writing was to become a feature in the vernacular periodicals and magazines. There were some dedicated periodicals on various science subjects, but also many literary magazines that gave space to writings on and about science.

The 'native' attempts to write science in the vernacular inevitably encountered an impediment in terms of lack of a vocabulary that could accurately define and describe phenomena in the language of modern science. A lack that was well recognized and often articulated by the 'native' intellectuals as a barrier to the development of

S. Baksi (✉)
Foundation for Agrarian Studies, Bengaluru, India
e-mail: sandipanbaksi@gmail.com

S. Ahmed and S. Sarkar (eds.), *Decolonizing Science and Modernity in South Asia*,
https://doi.org/10.1007/978-981-97-1829-0_3

science in India.[1] An outcome of this recognition was efforts to develop dictionaries and glossaries of scientific terms in the different vernaculars, including Bengali, Gujarati, Marathi, and Urdu.[2]

This was a familiar trajectory for many vernaculars, and Hindi was certainly not an exception. The reality of inadequacy in the language in terms of vocabulary that could be used to define phenomena scientifically as well as describe scientific principles was strongly felt by the Hindi literati in the second half of the nineteenth century. The recognition of this handicap belied the tall claims of Hindi being the language most appropriate for development of science writing.[3] Consistent efforts were taken up by the Hindi intelligentsia since the last decade of the nineteenth century to develop and revise glossaries of scientific terms.[4] These efforts, however, were not just an exercise in linguistics. They were shaped by the peculiar nuances in the social and political context in which a particular form of Hindi evolved in the late nineteenth and early twentieth century. This chapter is an attempt to characterize the different principles that were employed in the construction of Hindi terms for developing glossaries of scientific terms in Hindi. It also highlights the motivations and influences underlying these principles—motivations that drew from important questions pertaining to the philosophy of science, as well as from the contemporary political background—shaped by nationalism on one hand, and a contentious movement for developing a 'modern' 'standard' Hindi on the other.

3.1 The Early Official Attempts

The first attempts to conceptualize vernacular words for scientific terms began in the mid-nineteenth century in a background that was largely defined by the victory of the 'Anglicists' among the colonial administrators over the 'Orientalists,' in a debate about the system of education to be promoted by the British colonial administration. The roots of this debate perhaps go back to the Charter Act of 1813. The Act, among other things, resolved to encourage 'native education,' with the intention to

[1] In a lecture that he delivered in the Madras university, Prafulla Chandra Ray, the famous chemist and historian of science, argued that the lack of scientific nomenclature in the vernacular was an important impediment to the development and popularization of science among the Indian population. The lecture was published in Hindi under the title *Pracheen Bharat Main Rasayan Vigyan Kee Khoj*, in a popular science periodical, *Vigyan*, in the year 1918, vol 7, See Mishra (2010), pp. 26–28. The lecture was a part of several lectures delivered by Ray at Madras University in 1918.

[2] See Singh (2015), pp. 107–111.

[3] See Baksi (2020), pp. 229–48. 2020 for a discussion around the claims made by the Hindi intelligentsia of late nineteenth, and particularly early twentieth century, that Hindi was the most suited Indian vernacular for developing scientific literature.

[4] The efforts towards, and the process of development of the glossary of scientific terms in Hindi has been described in great detail by Charu Singh, Science, *Hindi Print and Agricultural Development*, pp. 100–160.

"promote both Oriental culture and Western Science."[5] This apparent ambiguity laid open the ground for a debate among the colonial officials between those who wanted to promote indigenous education systems in the vernacular languages and investigate the culture of the 'native' people, and those who argued for a direct encouragement to knowledge essentially coming from the west in the English language. The debate continued unceasingly till the 1830s, before the utilitarians, represented by the likes of William Bentinck, James Mill, and Thomas Babington Macaulay, took control of the administrative power in India.[6] With the Minute compiled by Macaulay in 1835, the debate was seemingly settled in favour of science education, with English as the official language and medium of instruction.[7] The colonial state was to gradually withdraw support from all forms of 'native' education systems and establish new institutions for the propagation of modern knowledge.

Notwithstanding the apparently clear verdict favoring the Anglicists, different provinces continued to persist with local languages as the medium of instruction. The particular case of Northern India has been discussed in detail by Michael Dodson,[8] where Urdu was officially promoted as the medium of instruction for "useful knowledge," in the early 1840s, and was subsequently followed by a push for Sanskrit as the appropriate language for teaching "Western knowledge systems."[9] Dodson characterizes this as a "conciliatory educational policy whereby Western 'useful knowledge' would be introduced to Indians through the medium of Indian languages."[10] This approach led to efforts to develop translated texts in the vernaculars, in institutions such as the Agra School Book Society,[11] the Delhi College, and the Benares College. It is under the auspices of these institutions that the first systematic attempt to construct vernacular words for scientific terms took place in the mid nineteenth century in North India.

The Education Dispatch of 1854, and particularly that of 1859, pushed the administration of education policy further towards the provincial governments, which enjoyed relative freedom on questions related to pedagogy and medium of instruction.[12] Both English and the vernaculars were envisaged as mediums to propagate

[5] Whitehead (2005), pp. 315–329.

[6] See Baber (1996), pp. 186–205, for the details on the debate that finally established the Anglicist view of education by the mid-nineteenth century. Also see Whitehead, The Historiography of British Imperial Education.

[7] Ibid.

[8] Dodson (2005), pp. 809–835.

[9] Ibid.

[10] Dodson defines this approach as a hybrid of the liberal anglicist school and the orientalist school. He says, "The liberal concern to impart Western learning and science to affect India's civilizational progress continued to dominate the content of educational curricula, while earlier orientalist expertise in, and appreciation for, Indian languages and cultural forms were adapted for the purpose of more readily facilitating the introduction and acceptance of this body of knowledge among India's learned elites," Ibid.

[11] The school book societies played a pioneering role in translating books on modern science into different Indian languages. See Roy (2010), pp. 10–12, and Mishra, *The Beginnings*, p. 34.

[12] Whitehead, The Historiography of British Imperial Education.

science, or "European Knowledge," in general. Unsurprisingly, such a policy framework led to significant differences in spread of education across different provinces. It also resulted in many private endeavors in education, particularly at the primary level, that continued in the local languages. The language of instruction in the institutions of primary education in the North-Western Provinces (NWP), for instance, continued to be Urdu and Hindi.[13] This, in turn, created a demand for vernacular textbooks in various subjects including mathematics and science, and provided another thrust to science writing in these languages, and soon led to a felt need for construction of new vocabulary.

Unsurprisingly, the earliest attempts to construct local words for scientific terms were undertaken by colonial officials. Dodson, in his assessment of the early efforts of "government-sponsored translation," draws the contours of some fundamental debates that, as we will see shortly, continued to evolve and plague the question of construction of new words to describe scientific phenomena in Hindi in the late nineteenth and early twentieth century. Some of the main pointers of the debates include: the question of feeder language, the question pertaining to the process and principles of "engraftment," (in other words the principles to devise new terms), the question of the most appropriate vernacular language to translate western science into, and the question of need for improvement or development of Indian languages for making them capable of "reproducing European thought," and expressing modern science in particular.

3.2 Hindi Science Writing: A Background

Before we turn to the task of characterizing the debate on the construction of scientific terms in Hindi as it evolved in the early twentieth century, it is useful to briefly describe the specific background of development of science writing in Hindi, one that provides the context to the debate.

Science writing in Hindi began by the mid-nineteenth century, rather late an entrant when compared with other vernaculars such as Assamese, Bengali, Marathi, Tamil, and Urdu.[14] Dedicated attempts towards modern science writing in Hindi were to be witnessed after the establishment of the Agra School Book Society in 1833. The society published original books as well as translations in Hindi. The volume of Science text-books and translations in Hindi expanded in the second half of the century. This was also the period when some Hindi literary periodicals began covering topics related to science and scientific subjects. These periodicals provided a platform for science to be discussed along with literature and various other topics, thus

---

[13] See King (1994), p. 118, on the British Educational System in the United Provinces that, he claims, created a 'Hindi-speaking elite.' Also, see Mishra, *The Beginnings*, pp. 26–28 and 33–34, on the continued encouragement for producing textbooks in the vernacular in NWP.

[14] See Baksi (2014), and Mishra, *The Beginnings* for a survey on Hindi science writing. Also see Sehgal et al. (2000), for a discussion on some of the early efforts on science writing in different vernacular languages in colonial India.

allowing a space for debates, discussion, and construction of discourses, unlike the science textbooks which had more specific focus.[15] The beginning of the twentieth century greatly expanded the volume of science writing in Hindi.[16] Many more books on scientific topics were published, and a few literary periodicals regularly featured discussion on science and science related subjects. The first popular science periodical in the language, *Vigyan*, also began publication in 1915. Science writing in Hindi was to become an important endeavour for the Hindi literati.[17]

The most important development that was happening in parallel, and in the background, was the formalization and standardization of Hindi as a language. This process brought about some significant changes in the form of the language as well as the form and content of its literature, and gave rise to the literary Hindi—*Khari Boli* Hindi written in the *Nagari* script—as we know it today. The original thrust for the growth of modern Hindi came from the need to develop prose writing in the language.[18] However, this thrust in the case of Hindi was much more than a literary development. It was essentially an outcome of a contentious political process. The development of modern Hindi was shaped by an opposition to the hegemony of an older elite as well as the British colonial rule. On the other hand, it was also characterized by the politics, essentially communal in nature, of a Hindu national identity formation through a linkage that was sought to be constructed between language (Hindi), religion (Hindu), and region (Hindustan), which aided some form of linguistic and religio-cultural nationalism.[19]

In this context, the development of science writing in Hindi must be seen as having multiple dimensions, and driven by a multitude of forces.[20] Science writing in Hindi was certainly an act of popularizing science in a language that was arguably much more accessible. It was also an accessory to the objective of economic nationalism, defined broadly as utilizing modern science for the development of productive activity. Science writing in Hindi was, of course, also an instance of domestication or cultural appropriation of modern science, a dimension of cultural nationalism. At the same time, however, it was also an aspect of the ideology of linguistic nationalism, which equated the development of a particular form of Hindi as a hegemonic 'national' language with intellectual freedom.[21]

---

[15] See Baksi (2021).

[16] Ibid. Also Mishra, *The Beginnings.*

[17] Ibid.

[18] Trivedi (2003), p. 959.

[19] See Brass (1974), King (1994), Sharma (1977), Rawat (1998), Dalmia (1997), Rai (2001), Stark (2008), Orsini (2009), to appreciate the contours of the contentious trajectory of the development of Hindi..

[20] See Sandipan Baksi, Science journalism in Hindi, for a discussion on certain characteristic features peculiar to science writing in the Hindi language in the late nineteenth, and especially the early twentieth, century. Also, Baksi, Science Writing in Hindi in Colonial India, discusses the different motivations that encouraged Hindi Science Writing.

[21] Ibid.

3.3 A Lacuna to Address

The earliest attempts at scientific writing in Hindi depended on the Sanskrit vocabulary for constructing terms that could define scientific phenomenon in Hindi. This was evidently in continuation with the orientalist approach promoted by the colonial educators in Northern India, particularly the North-Western provinces. The dependency on Sanskrit, however, was not without disaffection among members of the emerging Hindi intelligentsia.[22] However, by the late nineteenth century, as the efforts towards science writing in Hindi picked up pace,[23] a lacuna in the language was beginning to be expressed by the Hindi literati that was gradually taking to science writing. This lacuna, they argued, was a great impediment to science writing and learning in Hindi.

For instance, Lakshmishankar Mishra's book on trigonometry, titled *Saral Trikonmity* (simple trigonometry), written in 1873, mentioned the acute need for fundamental as well as translated literature dealing with different science subjects in all vernacular languages, and especially in Hindi. Hindi, he felt, suffered from a terrible lack of scientific books as well as a proper scientific glossary.[24] The latter was asserted as a serious detriment to science writing in Hindi. Mishra encouraged the use of the "rich resources" of Sanskrit to strengthen the Hindi language. Sanskrit, according to him, provided the flexibility to create new words that could be utilized to develop scientific terminology in Hindi.[25] Noticeably, Lakshmishankar Mishra was not an ordinary Hindi commentator. He was a professor at the Benares College, and also served as a district school inspector.[26] He established the *Chandraprabha Press* in 1882, one of the printing presses that was engaged in, among other things, popularizing science in Hindi.[27] He was also associated with the *Nagari Pracharini Sabha* (Organization for the Promotion of the *Nagari* Script), an institution established in 1893, and dedicated to the development of a standard *Khari Boli* Hindi and the *Nagari* script.[28]

---

[22] Dodson shows that the "highly Sanskritized vocabulary advocated by the colonial educators like Ballyntyne, was largely ignored by the emerging class of Indian educators in their publication of Hindi school books in subsequent decades in North India, as these writers reverted instead to transliterated English vocabulary." See Translating Science.

[23] By the late nineteenth century, a literary culture that incorporated science writing, and appreciated its importance was beginning to take shape in Hindi. At the same time, the required organizational and infrastructural support was also now available for further publication of writings on science in Hindi. The ground for general science writing in Hindi as well as for Hindi science journalism was laid by the end of the century. See Mishra (2017), and Sandipan Baksi, Science journalism in Hindi, for some details.

[24] Mishra (2001).

[25] Ibid.

[26] See Pooja Mishra, Science Education and Science Writing.

[27] Ibid.

[28] King, *One Language, Two Scripts.*

Impressing on the same concern, Shyam Sundar Das, the founder-secretary of the *Nagari Pracharini Sabha,* and one of the editors of the first *Hindi Scientific Glossary* published by the *Sabha* in 1906, wrote in the preface of the glossary that "the situation was such that if someone was commissioned to write a scientific essay or book in Hindi, then he would accept it under the condition that somebody provides him the (scientific) definitional terms."[29] Das, once again, was not just any common commentator, but a renowned Hindi publicist and a leading representative of the Hindi intelligentsia of the period, who led several of the *Sabha's* projects to develop the language in a particular form. He was the general editor of the Hindi dictionary, *Hindi Shabdasagar*, prepared by the *Nagari Pracharini Sabha*, and was also involved in the campaign for official recognition of Hindi.[30] The lament about lack of scientific literature in different Indian vernaculars, particularly Hindi, continued to persist in the writings of the Hindi intelligentsia through the early twentieth century.[31] One of the main reasons for the paucity of science writing in Hindi, it was argued, was the continued official encouragement to English as the mode of instruction in the education system, which, in turn, prevented a demand for vernacular writings, and discouraged authors and publishers from taking efforts to developing scientific literature in local languages. The other major reason for the persistent lack in Hindi science writings, it was often articulated, was the lack of definitional terms in the vernacular, which "curbs the enthusiasm of even the self-motivated writers, who make effort to write in Hindi without seeking for any personal gains"[32] The poverty in terminology was consistently perceived as a serious barrier to Hindi science writing.

It is evident that the Hindi intelligentsia of the late nineteenth century had identified a lacuna, and spoke of a need for "improvement" in the language. In a sense they mirrored the Imperial discourse on 'improvement' of the 'native.' However, the felt-need was also an indication of the growing encouragement to science writing in Hindi, and the establishment of a literary culture that increasingly incorporated science writing and appreciated its importance.[33] The motivations driving the literary culture, in turn, were multi fold, from science popularization, to economic nationalism, to strengthening the language in order to ensure its acceptance as the "national"

[29] Cited by Satyaprakash (1930).

[30] See Charu Singh, *Science, Hindi Print*, p. 106. Orsini, *The Hindi Public Sphere 1920–1940*, Appendix, pp. 393–394.

[31] See for instance, the commentary by Sethi (1933). The author says, "while on one hand we witness the likes of Bankimchandra Chattopaddhaya, Dwijendralal Rai and the internationally renowned Ravindranath Tagore; on the other hand, in subjects such as the physical sciences, the chemical and botanical sciences etc., we do not have a single good book for even school children. Is this not a matter of embarrassment for our country?".

[32] Ibid.

[33] The *Nagari Pracharini Patrika*, a tri-monthly periodical published by the *Nagari Pracharini Sabha*, which also ventured into scientific topics, asserted "The dignity of any civilized community today is due to science, therefore the pursuit of science and scientific progress is the sole means to gain respect in these times." As quoted in Mishra (2012), p. 6.

language.[34] The outcome of this felt-need was the initiative by the *Nagari Pracharini Sabha* to develop a glossary of scientific terms in Hindi. In the words of Shyam Sundar Das,

> The literary public of India has now come to the *Nagari Pracharini Sabha* and has said make words we will use them to revive and enrich our moribund and poor vernacular literature and make it powerful for the service of the Indian people by translations, reproductions, and adaptation from the valuable works and ideas of the rising Western nations. This glossary is the result.[35]

The *Nagari Pracharini Sabha* took the first step to develop a glossary of scientific terms in 1898. It is noteworthy that this was one of the first major initiatives by the *Sabha*. The *Sabha* was established in 1893 and the idea to publish a glossary for science was already up for discussion in 1894.[36] The *Sabha* began publishing a tri-monthly literary periodical, *Nagari Pracharini Patrika*, in 1896. A sub-committee to conceptualize the glossary of science in Hindi followed in quick succession.

It was decided that to begin with the glossary will cover terms from Astronomy, Chemistry, Geography, Mathematics, Philosophy, Physics and Economics. Separate departments were formed to deal with each of the chosen subjects.[37] Webster's International Dictionary was used to select the words. The compilation of terms was completed by 1902. A committee was then formed with representatives from Bengal, Punjab, Madhya Pradesh and United Provinces, which held its first conference in the Central Hindu College, Banaras, in September 1903.

The policy to create definitional terms was decided as follows:

> 1. Give precedence to popular Hindi words, 2. When that is not possible: a) use appropriate words from other popular Indian languages like Marathi, Gujarati, Bengali and Urdu, b) in case even that is not possible, i. Use popular Sanskrit terms, ii. Use English terms, iii. construct new Sanskrit words. Once the glossary for different subjects was prepared, they were to be vetted and polished by some smaller sub-committees. The glossary was also to be referred to some scientists and scholars from Bombay and Pune for their opinion and suggestions.[38]

The final edition of the glossary with 356 pages, with a mention of 10,330 English words and 16,263 Hindi terms corresponding to them was published in 1906.[39] This glossary was of crucial importance to Hindi science writing, but a need to update it was felt in time. The criticisms of the glossary began finding voice in the Hindi periodicals by the second decade of the twentieth century. It was often seen as limited

---

[34] See Baksi, Science Writing in Hindi in Colonial India, for a detailed discussion on the motivations that drove Hindi science writing in the early twentieth century.

[35] Preface of the *Hindi Scientific Glossary, p. xv*, cited by Charu, *Science, Hindi Print and Agricultural Development*, p. 100.

[36] See Mishra, Science Education and Science Writing in Hindi, pp. 475–476.

[37] This brief account is based on Satyaprakash (1930). Further details on the process of preparing the glossary by the *Nagari Pracharini Sabha*, and the principles used for constructing new terms are to be found in Charu, *Science, Hindi Print and Agricultural Development*, pp. 112–140.

[38] Ibid.

[39] Ibid.

in scope, as it did not take into account many terms of current usage, as well as words that were found commonly in science books.[40] There were others who criticized the glossary's approach to translations. The principles of transliteration applied in the glossary were seen as impractical and difficult from the point of view of day-to day application.[41] Yet others found this whole exercise as distorting the natural growth of the language. It was denounced as an exercise that would possibly lead to Hindi becoming akin to a constructed language.[42]

The glossary was also seen as incomplete in its coverage. An essay published in 1933 argued that the glossary played an important role in promoting vernacular writings on science, not just in Hindi, but also in Bengali, Marathi, and Gujarati. Further, it claimed that the list of definitional terms published frequently by the *Banga Sahitya Parishad*, as well as those used in the English-Marathi dictionary, had borrowed a lot from the glossary of the *Nagari Pracharini Sabha*. The glossary, it therefore asserted, had greatly served [scientific] literature. Nonetheless, the essay concluded,

> the glossary was found absolutely wanting with regard to words from the Botanical and Zoological sciences. Same was true, but to a much lesser extent, with regard to words from the physical sciences and mathematics. It was found incomplete and unacceptable.[43]

Most importantly, the glossary was critiqued for its failure to impose standardization of scientific terms in Hindi. The lack of uniformity in the terms used to describe scientific phenomenon, and to develop scientific literature was criticized as a tendency that can lead to "literary anarchism."[44] The absence of a standard glossary, it was complained, meant that writers had to indulge in constructing words by themselves, devoid of any foundational and common principles or standards. This also meant having a multiplicity of words referring to the same phenomenon.[45]

3.4 The Endeavour to Construct Standard Terms

While the need for standardization was frequently articulated by the end of the second decade of the twentieth century, the question that remained to be resolved was, "what were to be the principles for arriving at the standards for constructing the Hindi scientific terminology?" The endeavour to develop such standards led to a prolonged debate which occupied ample space in the discourse on science writing in Hindi, particularly in the third decade of the twentieth century. There were sections that argued for a continued use of the popular scientific terms, as science, they claimed,

---

[40] See Charu, *Science, Hindi Print and Agricultural Development*, 129.

[41] Ibid., 132–136.

[42] Ibid., 127–128.

[43] Sethi (1933).

[44] See Charu, *Science, Hindi Print and Agricultural Development*, 129.

[45] Sethi (1933).

was international and beyond boundaries. There were others who argued for a more puritan approach and wished to utilize the Sanskrit vocabulary for the construction of definitional terms in Hindi. Some sections wanted to use both Sanskrit and Arabic sources, and there were yet others who claimed that tapping of classical sources could invoke a competition that could worsen the communal environment of the region and the nation. The different positions in this debate have already been captured by Charu Singh.[46] In the next few sections, we try to characterize the debate, and identify the broader influences that shaped the different positions in the debate.

While the debate seemed to have encompassed a significant section of the Hindi intelligentsia and the literary organizations, including the *Nagari Pracharini Sabha*, the *Hindi Sahitya Sammelan*, the *Vigyan Parishad*, the *Banaras Hindu University*, and the *Hindustani Academy*, it was played out most vigorously between two institutions that led the efforts to develop their respective versions of the glossary of scientific terms. The respective approaches propounded by these two institutions, as will see shortly, were almost diametrically opposed to each other. They reflected the two dominant strands that influenced the anti-colonial freedom movement since, at least, the 1920s, one that was modern and forward looking (perhaps defined most appropriately as economic nationalism), and the other, a form of religio-cultural nationalism shaped by a strong dimension of revivalism.

3.5 The Pragmatic Approach

The first of the two institutions that led the effort to develop a new glossary of scientific terms in Hindi was an association of teachers from the *Kashi Hindu Vishwavidyalya* (Banaras Hindu University), termed as the "Hindi Publication Board."[47] The Board was part of the Technical Terms Committee, constituted by the *Nagari Pracharini Sabha* to revise the glossary. The effort here was not to develop a new glossary, but to refine the one prepared by the *Sabha*.[48] It planned to prepare specific glossaries for different subjects such as Physics, Chemistry, and Geology.[49]

The Board boasted of many prominent names from the Banaras Hindu University who participated in the endeavour. The names included Professor Krishnakumar Mathur, who obtained his undergraduate degree from the University of London, with a specialization in geology and mining. He served as the Professor of Geology and Principal of the College of Science, Banaras Hindu University, and was responsible for the

[46] Charu, *Science, Hindi Print and Agricultural Development*, pp. 140–159.

[47] See Sethi (1933).

[48] Ibid.

[49] Charu, *Science, Hindi Print and Agricultural Development*, p. 104, reports that the revised editions of terms related to physics and chemistry were published in 1929 and 1930. Revisions for other subjects were also reported to be in progress. "In publishing these revised editions," argues Charu, the *Sabha* aimed at creating a common Scientific Vocabulary for the whole of India.".

preparation of a glossary of geological terms in Hindi.[50] He was a Fellow at Geological, Mining and Metallurgical Society of India, National Academy of Sciences, Allahabad, and Indian Academy of Sciences, Bangalore.[51] Dr. Mukundswaroop Verma, another member of the board, was the Chief Medical Officer at Banaras Hindu University, who consistently wrote on subjects pertaining to health and anatomy in Hindi.[52] Professor Fuldeo Sahai Verma, who was assistant editor of the Scientific Glossary of chemical terms being prepared by the board,[53] studied chemistry in the Presidency College, Calcutta University, and Indian Institute of Science, Bangalore. He then became a teacher at the chemistry department of the Banaras Hindu University. Verma wrote extensively on the subject in Hindi.[54] Dr. Nihal Karan Sethi was a teacher of Physics at the Banaras Hindu University. He was also elected Fellow of the Indian Academy of Sciences in 1935. He served in the Agra College both as a teacher and as its Principal. Sethi was involved in preparing the Hindi vocabulary for scientific terms in physics.[55]

Their careers reflect the fact that they were primarily men of science involved in the project of popularization of science in Hindi. Fuldeo Sahai Verma, and Nihalkaran Sethi, in particular, actively wrote popular science articles in Hindi periodicals. Both of them appear as spokespersons of the Hindi Publication Board in the debate on the approach to developing Hindi scientific terminology.

The policy propounded by the Hindi Publication Board to design and conceptualize scientific terms in Hindi was defined, by some of the members of the Board, as the "theory of middle path," or *Madhyavarthi Siddhant*.[56] As the name suggests this was a theory informed by pragmatism. Nihal Karan Seth summarized the policy as follows:

> 1) All name denoting words, which are used not just in English but also in other languages of the world shall be incorporated as it is. For instance, the name of chemical elements, electron, proton etc.; words from Modern Physical sciences, names of tools and machines like Motor and Dyanamo etc. 2) Symbolic names of chemical elements which are used to formulate chemical compounds and equations shall be incorporated as it is. 3) No change in the names of compound matter, carbonate, Nitrate etc. 4) Hindi words to be constructed for all other words, especially those that imply a meaning. The Hindi terms in such cases should always denote the meaning of the original word. It is not satisfactory to just change the pronunciation of the English term.[57]

The preface of the glossary of physical sciences published by the Hindi Publication Board elaborates further:

---

[50] See https://www.insaindia.res.in/detail.php?id=F00-0440, last accessed on January 22, 2022.

[51] Ibid.

[52] See Singh (2020).

[53] The revised edition of the chemistry section of the Glossary was published in 1930.

[54] Singh, Hindi Protagonists of Science and *Swadeshi*.

[55] Ibid.

[56] Sethi (1933).

[57] Ibid.

> Where a word has different meaning in different senses, an attempt has been made to see that the Hindi equivalent conveys the exact sense of the English word. A Sanskrit scholar has been associated with the committee to see that the Hindi word has been given its proper grammatical form. It is perhaps unnecessary to point out that no attempt has been made at puritanism. Words of foreign origin have been unhesitatingly accepted, and even English words have been taken with the nearest approach of their pronunciation in Hindi.[58]

Is clear that this approach was liberal in its use of English terms. Proponents of this approach, like Fuldeo Sahai Verma, argued for internationalism in the use of scientific terminology. He was against relying on the classical sources of Sanskrit and Arabic for formulating scientific terms. Words formulated from classical sources, he argued, would prove to be as incomprehensible for common people as would be the words of English or any other foreign language.[59] Thus, he argued, the import of such terms would not help the cause of science popularization in any sense. At the same time, Verma impressed upon the need to learn English for succeeding in international trade and commerce and in order to travel and communicate with the outside world. This need, he insisted, would continue even after India gained independence.[60] A liberal approach in developing a vocabulary of Hindi scientific terms, he argued, would also facilitate the study of scientific literature in English and other European languages. This would, in turn, also help in accelerating the rate of scientific literature in local languages.[61] Elsewhere, Verma says:

> Even when India is no longer under British rule, we will certainly need to retain the knowledge of English to maintain commercial ties with the world. English is the only language that can assist humans to navigate the whole world with ease, and express their emotions in every country. It is essential to have the knowledge of the English language in order to sustain our relations with the world even after the end of English rule. Also, the use of English words will facilitate the study of scientific books prepared in other European languages. This will surely lead to a swift expansion in the scientific literature in different native languages. The main reason behind the progress of Japan in the last few years is that while providing education in their local language, they have used the English scientific terms as it is in their language. This prevented them from the chaos around the use of scientific words that is prevalent in our country. This has resulted in a fast development of scientific literature [in Japan].[62]

Verma hoped that the "well-wishers of the Indian languages" would soon arrive at a similar conclusion.[63]

In a similar vein, Nihal Karan Sethi while making a point on the importance of directly using the English terms for scientific elements, stated that it was essential for any living language to accept words from different languages and broaden its own vocabulary.[64] He dismissed a puritan approach and advised against associating

---

[58] Cited by Satyaprakash (1930).

[59] This is based on an essay by Fuldev sahai Verma published in the periodical *Desh*, that was cited by Satyaprakash (1930).

[60] Ibid.

[61] Ibid.

[62] Verma (1925).

[63] Ibid.

[64] Sethi (1925). Also cited by Satyaprakash (1930).

the question of scientific terminology, and language in general, with the pride of community and the nation. His was clearly a moderate voice that consistently urged against any form of linguistic nationalism. Sethi argued:

> By linking this question [of language] to the nation and the pride of the community, and thus keeping the language unchanging, can never be the appropriate thing to do. Such love of *swadeshi*, and pride in community is false. As a result [of this pride], our progress is obstructed and we have difficulties at every step of the way.[65]

Sethi also asserted that scientific principles and discoveries were universal, and that no community exercised any special rights on them. He insisted that the terms that defined scientific principles could only be universal, and that their usage could not cause insult to any language.[66] He also argued that

> the students from India who will progress from primary science to the study of higher sciences will themselves be interested in participating in the process of development of science. At that point they will definitely have to learn these universal terms, as it would be impossible to read the periodicals from any country without their knowledge. Hence, ultimately every person will have to learn both kinds of words.[67]

In view of this he questioned the need for an elaborate exercise to construct separate Hindi terms for scientific terminology. Sethi was vocal in articulating the crucial need for expanding the knowledge of new scientific discoveries across geographies in a manner that was comprehensible to everyone who was pursuing science. Facilitating widespread dissemination of knowledge, he contended, was critical for the development of science, and should be the ultimate objective of developing scientific terminology in any language.

3.6 The Puritan Approach

The other institution that endeavored to develop its own glossary, in competition with the Hindi Publication Board of the Banaras Hindu University and the *Nagari Pracharini Sabha*, was the *Vigyan Parishad* of Prayag (Allahabad). The*Vigyan Parishad* was established in 1913 by four professors—Ganganath Jha, Hammiduddin, Ramdas Gour, and Shaligram Bhargava—at Muir College, Allahabad.[68] Its main purpose was to make science literature available in Hindi, in order to popularize science among the masses.[69] The other equally important purpose of the Parishad was to promote the development of Hindi.[70] The *Parishad* began publishing

---

[65] Ibid.

[66] Ibid.

[67] Ibid.

[68] See Prakash, *Another Reason*, 63. Also see is Mishra, "The Beginnings".

[69] The evolution of this institution is, in fact, a subject of an independent study. The only attempt in this direction is Mishra, "The Beginnings"; however, there is much scope for further study.

[70] See Baksi, Science Writing in Hindi in Colonial India.

a monthly periodical, *Vigyan*, in 1915. *Vigyan* was arguably one of the first popular science magazine in India that provided a platform for dedicated science writing in Hindi, although it cannot be regarded as a research journal in any sense. The debate around the construction of scientific terms in Hindi was carried on most expressly in the pages of *Vigyan*.

While many science teachers, such as Ramdas Gaur, Shaligram Bhargava, Nihalkaran Sethi, and Gorakh Prasad were closely associated with the *Vigyan Parishad*, its effort to develop scientific terms in Hindi was undoubtedly led by Swami Sathyaprakash, a follower of the *Arya Samaj*,[71] and a Chemistry teacher at Allahabad University. Satyaprakash was a science graduate from the Chemistry department of Allahabad University, who went on to become the head of the department. He took up the role of the editor of *Vigyan* in late 1920s and continued through the 1930s, and in this period was prolific in his writings on the subject of scientific terminologies in Hindi. While he was not the sole editor, his was a frequent voice heard increasingly over the years, for more than a decade and a half. He developed the glossary of terms for Chemistry, *Rasayanik Shabdavali*, and was the editor of glossary of scientific terms, which was published by the *Vigyan Parishad* in 1930–31, and contained 4821 terms. Interestingly, and perhaps unsurprisingly, Satyaprakash had a keen interest in Indian "traditional systems of scientific knowledge."[72]

Satyaprakash was the main representative of the approach taken by the *Vigyan Parishad* to develop the glossary of scientific terms in Hindi. He employed a unique method to publicize these words, even while they were being conceived. He periodically produced a list of scientific and technical terms in *Vigyan*, and then took effort to prepare books on scientific subjects using these terms.[73] While making an appeal to all editors and publication managers to promote the publication of scientific literature in various Hindi periodicals, Satyaprakash noted that the resolution of "the problems of literature demand a struggle, and the most appropriate ground for this struggle is laid by the papers and periodicals."[74] He further stated that the definitional terms that proved to be sustainable and popular through these periodicals will eventually be "victorious."[75] The editor was clear that the adoption of a particular form of Hindi, and Hindi terminologies, will ultimately be the consequence of an intellectual competition. He was in a competition with the Hindi Publication Board, and was trying to play it out in the public sphere. The public, according to him, would eventually decide the popularity and acceptance of either of the approaches.

The approach adopted by the *Vigyan Parishad* to develop a scientific glossary in Hindi was much more conservative, and, in essence, informed by puritanism. It

---

[71] *Arya Samaj* is a Hindu reform movement established in the late nineteenth century by Dayanand Saraswati. The movement had a strong dimension of revivalism, and maintained that India had a glorious past marked by social, cultural, and spiritual greatness. See Jafferlot (2007).

[72] He wrote a book titled *Vaigyanik Vikas kee Bharatiya Parampara* (The Indian tradition of scientific development), and another titled *Prachin Bharat Main Rasayan Kaa Vikas* (The development of chemistry in ancient India).

[73] Charu, *Science, Hindi Print and Agricultural Development*, 140–141.

[74] Satyaprakash (1930).

[75] Ibid.

argued for avoiding the direct usage of English words, or for that matter words from any "foreign" language, and instead prescribed translating them for the meaning of the "foreign" word, or constructing new terms using Sanskrit and Hindi. The use of foreign scientific terms, Satyaprakash felt, drained the melody out of the language, and "renders it cacophonous and rough."[76] In a sense, it was an argument for the boycott of all English or foreign words to the extent possible. The approach was polemically termed by Nihal Karan Sethi as "*Videshi Bahiskar Sidhaant*" or "*Angrezi Bahishkar Siddhant*," the doctrine of boycotting everything that was Foreign or English.[77]

In a series of articles published in *Vigyan* in the 1930s, Satyaprakash defended a glossary of scientific terms in Hindi that borrowed heavily from Sanskrit. He strongly argued for the use of such a glossary by all vernacular writings on science.

The basic principles devised by the *Vigyan Parsihad* to develop the glossary were as follows:

> 1) The first attempt is to look for a popular word in the language [Hindi] that can express the meaning of the foreign scientific word. 2) On not finding an appropriate popular word, such a word is sought after, which, although foreign, is popularly used in the factories in some distorted form. Such a word is considered as appropriate for use. 3) After this, easy Sanskrit words are constructed which can express the meaning of the foreign scientific term. 4) While constructing Sanskrit terms, it is kept in mind that the new word is not very big or difficult. Therefore, in cases where the foreign term appears to be small, easy, and fitting easily in our language, then it is used as it is or with some minor changes. 4) Incorporating all English terms, as it is without any change, is not in our principles, because it is known from experience that it is easier to find words from our language, words that have some relation to our language.[78]

It was ample clear to all those who were involved in this exercise that it was impossible to find local or even Sanskrit terms for all the scientific words, hence the need to construct new words was quite evident. The question was on the source for such constructions. Satyaprakash was convinced that Sanskrit will provide the ultimate source for this process. He was against the direct import of English or European terms as far as possible. The idea was to first look for colloquial Hindi terms (*Theth Shabd*), followed by a search for words from the family of Sanskrit (*Sanskritjaat Shabd*). Using English terms directly was only a last resort, to be used after exhausting the possibilities of using local terms from other Indian languages, or even popular Persian and Arabic words for that matter. Even in such a case, he was willing to use only such English terms that had gained a general popularity and were fitting in the pronunciation with some minor changes.[79] He resisted the arguments for assimilating Roman script even more vehemently and defended the capacity of the *Nagari* script to express science and scientific subjects.

---

[76] Satyaprakash (1930).

[77] Sethi (1933).

[78] Satyaprakash, Vaigyanik Paribhashik Shabd (2).

[79] Ibid. In this essay, the author gets into a detailed discussion on the principles to construct new words in Hindi.

Satyaprakash expressed a willingness to import the European words that represent patents in the industry and business.[80] However, even here he argued for the need to modify such terms wherever possible. In fact, he even argued for the use of Indian names for patented scientific and business products, when being sold in the Indian markets. Businesses, he asserted, would not hesitate to adopt "Indian" names of scientific products, if they were promoted with enough confidence.

> Once we accept the Indian scientific names (of products) then even the foreign businessmen/merchants will sell such products in India under Indian names. Their sole purpose is business. Hence it is not essential that a product with some patent name in Europe, has the same name in India as well. You will find two names, English and German, written on many bottles that are used in chemical labs. The foremost concern of Business people is the comfort of people. Since Indians like *Khaddar*, therefore thick cloth comes here from Lancashire and Japan under the name of *Khaddar*. Gandhi cigarettes and Gandhi Matches have come to our country after being produced in foreign land. Thus, if we agree to call scientific machines and chemical compounds by Indian names then there is no reason that the western businessmen will not agree to these names as well for our country. How difficult it is for them to send bottles with *Gandhkaaml* written on them instead of Sulphuric Acid, or *Panshuj Kashtet* instead of Potassium Oxalate. But the problem is that we lack self-confidence, we are not ready to accept our own words. When we are not concerned for ourselves, why should others care!![81]

Satyaprakash was quite direct in his criticism of the idea of internationalism, which he called an "illusory ghost", "under whose intoxicating influence many are eager to destroy their identities."[82] It is obvious that Satyaprakash was countering the arguments propounded by the members of the Hindi Publication Board for universality in the approach to assimilate science and formulate definitional terms. Satyaprakash tried to portray the futility of the notion of universality by giving the example of the persistence of separate terms for scientific concepts in different European languages such as English, German and French.[83] In doing so, he seemed to echo the logic used by orientalists such as J. R. Ballantyne, arguing about 75 years ago for coining Sanskrit equivalent terms for scientific nomenclature, as well as later day Indian translators such as Rajendra Lal Mitra.[84]

Satyaprakash asserted that while the idea of universalism of science might sound attractive, it had no relation to reality, and that "it was futile to sacrifice one's identity for something so far from reality.[85]

> I have already stated that the question of internationalism is just an illusion because when the English scientific terms are not popular even in Europe then the issue of assimilating

[80] Ibid.

[81] Ibid.

[82] Satyaprakash, *Vaigyanik Paribhashik Shabd* (1).

[83] Ibid.

[84] Singh (2021). Singh argues that the "European and Indian translators in South Asia were not only aware of the linguistic differences among European chemical communities but also deployed this linguistic diversity in their own arguments for the translation of European scientific knowledge into South Asian languages," p. 13.

[85] Satyaprakash, *Vaigyanik Paribhashik Shabd* (1).

> these words, across the seven seas, in India is laughable. Such an effort runs counter even to the laws of linguistics. It is correct that in today's situation we cannot limit and constrain our language inside the four walls of the nation. We can teach the idea of liberalism, but then there are limits to being liberal.[86]

He criticized the arguments for internationalism as a cover for a political imposition by a stronger colonial country on a weaker un-free nation. To make the point, he asks satirically

> Is it possible to get our terminologies of astronomy and mathematics established in Europe, by giving the reference of (our glorious) history? Would we not fail to turn the face of the stream of science towards Indian sources, despite great efforts? We might get equal recognition in the five nation novel conference, or in the famous League of Nations or even get the National Dominion Status. But to expect an international acceptance of our ancient, historically proven culture, even with such recognition, will just be a day-dream.[87]

Further, he gave the example of Japan that modified their script, "to render it with all qualities that are there in Roman script, rather than accepting the Roman script, in order to maintain the freedom of their community."[88]

It is important to appreciate that for Satyaprakash, the question of scientific literature in Hindi and a glossary of scientific terms is essentially part of a political objective, to establish Hindi as the national language against the dominance of English. The battle, of course, is primarily against English, but it is also equally against Urdu, and other vernaculars and local languages.[89] The nationalism that reflects in his discourse is closely entangled with the objective of establishing the hegemony of a particular form of Hindi. In this sense, his fight to popularize science in Hindi appears to build upon the legacy of Bharatendu Harischandra.

3.7 Points of Convergence

Notwithstanding the fundamental differences between the two approaches, both of them were against any idea of a wholesome usage of English terminology in Hindi science literature—an approach that was perceived by many of these commentators as being promoted by the official departments. Interestingly, there was also an evident perception that the approach of direct importation of English terms was preferred by men of business, and those with commercial interests. Satyaprakash's commentary on a book titled *Practical Photography*, by one Harigulamjee Thakur, makes the point; it reads:

---

86 Ibid.

87 Ibid.

88 Ibid.

89 See Baksi, 'Science Writing in Hindi in Colonial India'.

> although the language of the book is very polluted, the labour put in by the author is certainly praiseworthy. The author does not delve into the question of definitional terms at all. Words like camera, lens, slide stand, dry plate, printed frame, developer, toning, din, etc. have been used as it is. Similarly the [names of] chemical substances used in photography, such as Pyrogallic acid, Potassium Bromide, Metabisulphite, Liqour Ammonia, Metol, Hydroquinone, etc. have also been used as it is. In fact, the use of literary definitional terms in business or public works nowadays remain doubtful. Same is the case with medicine.[90]

Nihalkaran Sethi termed the approach of wholesome English usage as the *Swadesh-Bahishkaar* (Boycott Homeland) principle. He concluded that this principle could not provide an effective solution to the problem of definitional terms in the longer run. However, Sethi's criticism, notably, was not ideological. Rather, he asserted that usage of all English scientific terms without the respective grammatical rules being incorporated would make it extremely difficult to write in a comprehensible language. He argued,

> there is no doubt that this approach will first of all prevent the Hindi-Urdu squabble. At the same time, it will render redundant the extremely difficult task of constructing definitional terms. To a lay eye it also appears that this approach makes it very easy to write scientific books in vernacular languages. But from my own experience, I can claim that it is not at all easy to write books in the Hindi language by such wholesale use of foreign definitional terms. It could become somewhat convenient if we were to digest the foreign grammar along with foreign words. But we dare not call such scientific volumes as books in the Hindi language at all.[91]

Further, he elaborated that such an approach distorted the language to an extent that "it is impossible to comprehend which country it belongs to." Mockingly, he noted that it resembled the peculiar language spoken by "shopkeepers, washermen, barbers and chefs living in the cantonments of English platoons". "Such a language", he asserted, "can never be the language of our [Hindi] literature."[92]

Both the institutions were also united in opposing an approach that allowed different vernacular languages to continue with their own evolution, developing their own glossary of scientific terms in the process. Such an approach, it was argued, would prevent any standardization in scientific literature, and lead to confusion and lack of communication between different scientific writings.[93] It was also argued that such an approach would prevent an united effort to develop scientific literature in a "national language" in India.[94] It is striking that the appeal for such an united effort was, on one hand, motivated by a sentiment of nationalism, but on the other,

---

[90] Satyaprakash (1930). A similar example has been discussed in detail by Charu, *Science, Hindi Print and Agricultural Development*, (130–135), describing the criticism of the glossary developed by the *Nagari Pracharini Sabha* by one Mukhtar Singh, from the perspective of a "small manufacturer and scientific amateur." Mukhtar Singh, however, proposes more use of common and current Hindi names (*tijarati shabd*) of materials, instead of direct import of English terms.

[91] Sethi (1933).

[92] Ibid.

[93] Ibid.

[94] See Baksi, 'Science Writing in Hindi in Colonial India' for a detailed discussion on the arguments for the exclusive development of scientific literature in Hindi.

was a reflection of the attempt to establish a hegemony of Hindi over other vernacular languages, such as Marathi, Bengali, Gujarati, and Tamil, that were themselves trying to build scientific literature.[95]

It is thus evident, that despite the differences in the underlying influences, both the institutions were one in articulating the need for a single uniform glossary of scientific terms to facilitate scientific communication. Strikingly, while both the institutions aimed to develop a uniform scientific glossary in Hindi with a pan-India acceptance, such that other Indian languages can also inculcate the terms defined, they did not, however, consider Urdu as part of the effort toward standardization. This was perhaps indicative of the fact that while constructing new words both the approaches considered to base themselves, albeit in different degrees, on Sanskrit. There was an evident belief that all Indian languages, including the ones popular in the southern part of the country such as Kannada, Telugu, Tamil and Malayalam, were closely related to Sanskrit, and therefore would more readily accept Sanskritized definitional terms. Nihalkaran Sethi, for instance, while describing the need for a glossary that would be used by various Indian languages, stated "that constructing such words will not be a difficult task considering that Sanskrit is foundational to all these modern languages."[96] Satyaprakash, of course, was much more obvious when he stated that

> The question of Urdu, Arabic and Persian is out of context for the concern of a Hindi dictionary, because we want to promote one scientific glossary for all *Aryan* languages. And this scientific glossary will take help from Sanskrit Nouns, Prefixes, Suffixes to the extent possible.[97]

The exclusion of Urdu was also indicative of the larger political conflict between Urdu and Hindi, coloured by linguistic nationalism, as is reflected strikingly by the following statement by Satyaprakash, who while insisting that "for any national scientific glossary to be popular in India, it has to be strongly related to Sanskrit and very distantly to Persian and Arabic", stated that the "question of Urdu is neither an all-India question nor a concern for all *Musalmans*. It concerns only the language of some people from the United Provinces and Punjab." Reflecting a strong belief on the linkage between religion and language, he claimed that "Hindi is increasingly being understood by the Hindus in the United Provinces and Punjab, and in the next fifty years it will be a rarity to find Hindus who read and write in Urdu." Apart from making an argument on the numeric irrelevance of Urdu, he also made the grand claim that "the renunciation of Sanskrit words from the language will be a strong dilution of our nationalism. Thus, the more we accept Persian words, the more it will be a barrier to our nationalism, our Indianness."[98]

A majoritarian understanding of nationalism, that connected *Hindi*, with *Hindu*, and *Hindustan*, is clearly at witness in the discourse propagated by Satyaprakash.

---

[95] Ibid.

[96] "*Vaigyanik Paribhashik Shabdo kee Samasya*", Vigyan, March 1933, Vol. 36, No. 6, pp. 167–170.

[97] *Vaigyanik Paribhashik Shabd* (1), *Vigyan*, March 1931, vol. 30, No. 6, pp. 241–255.

[98] Satyaprakash (1931).

It shapes a critical motivation behind his imagination of the scientific glossary in Hindi, in particular, and Hindi science writing in general.[99]

It is remarkable to note that the original purpose of the *Vigyan Parishad* was to promote scientific literature in both Hindi and Urdu, but it eventually focused only on Hindi. Describing the scientific literature produced by the *Vigyan Parishad*, Satyaprakash unambiguously mentioned that although the general objective of the *Parishad* had been to produce scientific literature in Hindi and Urdu, "it is natural that the *Parishad* has been dominated by those interested in Hindi."[100] The effortless transition by the editor of *Vigyan* from Hindi to the "national language," and then to a "natural choice," was a clear instance of the attempt to establish the language, against English, as well as all against other Indian vernaculars, especially Urdu.

Satyaprakash concluded

> If the Urdu proponents of *Anjumann Tarakki* wish to expand the Urdu literature, they can do so merrily. We do not have any problem with their intention to construct scientific terms using the rules from Arabic, Persian and Turkish, because we are, anyways, not going to use those words. We can just say this much that neither will we be a barrier to your dictionary, nor should you oppose our scientific terms. Neither do you instruct us to use your words, nor will we insist upon you to use our Hindi words. Thus, there is neither a need for any cooperation between Hindi and Urdu nor the need for any opposition. If, possibly, these two languages unite in future, it will be good but even otherwise there is nothing to lose.[101]

The impact of Hindi nationalism, and particularly the conflict with Urdu, on the development of the Hindi writings on science, including the imagination of the Hindi science glossary is evident from the fact that even a liberal voice like Nihal Karan Sethi strongly expressed the inability to have a uniform glossary for the two languages. He noted that,

> although it is always intended that there should be only one set of definitional terminologies in the whole of India and also that there should not be any dispute between Hindi and Urdu, still at this point it seems almost impossible to resolve this conflict. That the definitional terms produced by the Usmaniya University will be acceptable to the language of any of the provinces of India is, even though not absolutely impossible, an unbelievable proposition, at least, at this juncture. Hence the effort to resolve the Hindi-Urdu conflict at this juncture is pointless.

3.8 Concluding Remarks

It is evident that the discourse on the construction of words for glossaries of scientific terms in Hindi in the late nineteenth century and, particularly, early twentieth century was informed by two opposing approaches, a principle of linguistic puritanism on one hand, and that of pragmatism on the other. The debate, though played on linguistic grounds, was in reality a reflection of a deeper political debate on the nature of nationalism, and the corresponding role of science and language in shaping it.

---

[99] See Baksi 2020.

[100] Satyaprakash (1930).

[101] Satyaprakash, *Vaigyanik Paribhashik Shabd* (1).

It is indeed revealing that the debate on the principles to develop scientific terms in Hindi did not appear to settle through the first half of the twentieth century, and lingered on even in independent India. The reason for the persistent vigour in the debate lies in the fact that the motivations and influences that shaped these two approaches were fundamentally opposed to each other in more than one ways. The essential conflict was between the underlying motivations—cultural and, especially, linguistic nationalism framed alongside or in conjunction with a revivalist perspective on one hand, and the drive for science popularization and economic nationalism on the other.

In light of the fact that English eventually became the international language of science, and was recognized thus in India as well, we know today that it was always a losing battle for all the vernaculars, including Hindi. It is noteworthy, however, that the political battle for Hindi was not just against English, as the symbol of colonialism, but also against other vernaculars to establish its hegemony as a "national language." The development of science writing in Hindi, including the preparation of the glossary, was essentially part of this political process.

References

Zaheer Baber, *The Science of Empire: Scientific Knowledge, Civilization and Colonial Rule in India* Albany: State University of New York Press, 1996.

Sandipan Baksi, *Attitudes To Modernizing Agriculture In The Colonial Era: A Study of Hindi Periodicals from 1876 to 1947*, Mumbai: M.Phil. Diss., School of Habitat Studies, Tata Institute of Social Sciences, 2014.

Sandipan Baksi, 'Science Writing in Hindi in Colonial India: A Critical View of the Motivations', in Rocio G. Sumillera, Jan Surman, and Katharina Kuhn (eds.), *Translations in Knowledge, Knowledge in Translation*, John Benjamins, 2020.

Sandipan Baksi, 'Science journalism in Hindi in pre-independence India: A study of Hindi periodicals', *The Indian Economic and Social History Review*, Vol. 59, Issue 1, 2021, pp. 1–23.

Paul Brass, *Language, Religion and Politics in North India,* Cambridge University Press, 1974.

Vasudha Dalmia, *The nationalization of Hindu traditions: Bhāratendu Hariśchandra and nineteenth century Banaras,* Delhi: Oxford University Press, 1997.

Michael S. Dodson, 'Translating Science, Translating Empire: The Power of Language in Colonial North India', *Comparative Studies in Society and History,* Vol. 47, No. 4, October 2005.

Christophe Jafferlot, *Hindu Nationalism: A Reader*, Princeton University Press, Princeton and Oxford, 2007.

Christopher R. King, *One Language, Two Scripts: The Hindi Movement in Nineteenth Century North India,* Bombay: Oxford University Press, 1994.

Pooja Mishra, *The Beginnings of Science Education and Science Popularisation in Hindi (1860–1935),* New Delhi: MPhil. Diss., Zakhir Hussain Center for Educational Studies, Jawaharlal Nehru University, 2010.

Pooja Mishra, 'Science Education and Science Writing in Hindi in the North West Provinces (1860–1900)', *Indian Journal of History of Science*, Vol. 52, No. 4, 2017, pp. 463–83.

Francesca Orsini, *The Hindi Public Sphere 1920–1940: Language and Literature in the Age of Nationalism*, Oxford University Press, 2009.

Alok Rai, *Hindi Nationalism (Tracts For The Times)*, Orient Blackswan, 2001.

Ramesh Rawat, '1857 and the 'Renaissance' in Hindi Literature', *Social Scientist,* Vol. 26, No. ¼, January–April 1998, pp. 95–112.

Arnab Roy, *Attitudes to Agricultural Science in Bengali Periodicals in the Years 1876–1947*, Mumbai: MPhil. Diss., School of Social Sciences, Tata Institute of Social Sciences, 2010.

Satyaprakash in, 'Vaigyanik Paribhashik Shabd (2)', *Vigyan*, Vol. 31, No. 1, April 1930, pp. 1–11.

Satyaprakash, 'Vaigyanik Paribhashik Shabd (4)', *Vigyan*, Vol. 31, No. 3, June 1930, pp. 89–95.

Satyaprakash, 'Vaigyanik Paribhashik Shabd (2)', *Vigyan*, Vol. 31, No. 2, May 1930, pp. 49–57.

Satyaprakash, "Vigyan Parishad aur Vaigyanik Sahitya," *Vigyan*, 32, No. 3, December 1930, pp. 108–116.

Satyaprakash*, Vaigyanik Paribhashik Shabd* (1), *Vigyan*, March 1931, vol. 30, No. 6, pp-241–255.

N. K. Sehgal, S. Sangwan and S. Mahanti (eds.), *Uncharted Terrain: Essays on Science Popularisation in Pre-Independence India*, New Delhi, Vigyan Prasar, 2000.

Nihal Karan Sethi, 'Tatvo Key Naam', *Vigyan*, Vol 22, No. 1 & 2, Oct–Nov 1925.

Nihal Karan Sethi 'Vaigyanik Paribhashik Shabdo kee Samasya', *Vigyan*, Vol. 36, No. 6, March 1933, pp. 167–170.

Ram Vilas Sharma, *Mahavir Prasad Dwivedi Aur Hindi Navjagaran,* Delhi: Rajkamal Prakashan, 1977.

Shivgopal Mishra, *Swantrata Purva Hindi Main Vigyan Lekhan*, Vaigyanik tatha Takninki Shabdavali Aayog, Human Resource Development Ministry, 2001.

Shivgopal Mishra, *Vigyan Lokpriyakaran tatha Patrakarita,* Prayag: Vigyan Parishad, 2012.

Dhrub Kumar Singh, 'Hindi Protagonists of Science and *Swadeshi* in the First Half of the Twentieth Century', *Indian Journal of History of Science*, Vol. 55, No. 3, 2020, pp. 235–246.

Charu Singh, *Science, Hindi Print and Agricultural Development in Colonial North India*, New Delhi: Ph.D. Diss, Centre for Historical Studies, Jawaharlal Nehru University, 2015.

Charu Singh, Science in the Vernacular? Translation, terminology and lexicography in the Hindi Scientific Glossary (1906), *South Asian History and Culture*, Vol. 13, Issue 1, 2021, pp. 63–86

Ulrike Stark, *Empire of Books: The Naval Kishore Press and the Diffusion of the Printed Word in Colonial India,* Ranikhet: Permanent Black, 2008.

Harish Trivedi, 'The Progress of Hindi, Part 2 Hindi and the Nation', in Sheldon Pollock (ed.) *Literary Cultures in History: Reconstructions from South Asia*, University of California Press, 2003.

Fuldeo Sahai Verma, 'Desi Bhasaon Main Vaigyanik Sahitya', *Vigyan*, Vol. 21, No. 1, April 1925, pp. 13–16.

Clive Whitehead, 'The Historiography of British Imperial Education Policy, Part I: India', *History of Education: Journal of the History of Education Society*, Vol. 34, Issue 3, 2005.

Part II
Plural Healing and Multiplicity

Chapter 4
Modernizing Ayurveda: Interactive Indigeneity and Medical Revivalism in Colonial India

Saurav Kumar Rai

The distinctive colonial context of the late nineteenth and early twentieth century in India led to the creation of various 'modernized traditions' around this time. This unique combination of oxymoronic traits was peculiar to colonial set-up manifesting 'the rule of colonial difference'. In the sphere of medical modernity, this peculiar phenomenon was substantially manifested through the emergence of 'modern' Ayurveda. In their revivalist quest the Ayurvedic practitioners and proponents of the time dramatically reshaped and redefined Ayurveda which was remarkably different from its pre-colonial version. It was very well reflected in Ayurvedic incorporation of the traits on the basis of which western medicine was claiming its superiority (such as scientificity, standardization, professionalization, institutionalization, etc.). Yet at the same time it attempted to contest the epistemological absolutism of western medicine as well in a very subtle way. The present essay tends to explore the aforesaid seemingly contradictory threads that went into the making of 'modern' Ayurveda as one of the 'modernized traditions' of India. In this regard, the present essay argues that even while the Ayurvedic revivalism was a product of (colonial) modernity, it does not necessarily symbolize the victory of modernism as the refurbished Ayurveda had been trying to derive its sanctity both from being 'traditional' as well as 'modern' at the same time.

4.1 Historical Backdrop

With the coming of modernity there appeared a broader shift in the history of medicine from being essentially 'clinical' to 'social' and 'political'. Commenting on this shift, Michel Foucault, in one of his lectures delivered at the Institute of Social Medicine

S. K. Rai (✉)
Gandhi Smriti and Darshan Samiti, New Delhi, India
e-mail: skrai.india@gmail.com

S. Ahmed and S. Sarkar (eds.), *Decolonizing Science and Modernity in South Asia*,
https://doi.org/10.1007/978-981-97-1829-0_4

(Biomedical Centre, State University of Rio de Janeiro, Brazil), in October 1974, delineated following characteristic pattern in the history of medicine in the case of eighteenth and nineteenth century Europe.[1] According to Foucault, since the eighteenth century, medicine had continually involved itself in what was 'not its business', that is, in matters other than patients and diseases. This, according to Foucault, was marked by four major shifts as follows.[2]

- Appearance of a medical authority, which was not restricted to the authority of knowledge, or of the erudite person who knew how to refer to the 'right authors'. Medical authority was by and large a social authority that could make decisions concerning a town, a district, an institution, or a regulation. It was the manifestation of what the Germans called *Staatsmedizin* viz. medicine of the State.
- Appearance of a medical field of intervention distinct from diseases: air, water, construction, terrains, sewerage, etc. In the eighteenth century all this became the object of medicine.
- Introduction of a site of collective medicalization: namely, the hospital. Before the eighteenth century, the hospital was not an institution of medicalization, but of aid to the poor awaiting death.
- Introduction of mechanisms of medical administration: recording of data, collection and comparison of statistics, etc.

Thus, as viewed by Foucault, medicine during this period became a 'social practice instead of an individual one' and was endowed with 'an authoritarian power with normalizing functions that went beyond the existence of diseases and the wishes of the patient'.[3] This broader shift in Europe allowed the use of medicine/medical discourse in colonial expansion and consolidation as well. Colonial medical authorities were integrally associated with colonialism which was manifested, in the case of colonial India, primarily through the rank and file of the Indian Medical Service (IMS). The cadre of the IMS in their respective capacities as civil surgeons of districts, professors and resident surgeons at the medical colleges, sanitary commissioners and deputy sanitary commissioners of a province performed the socio-political task of Foucauldian 'medical authority' and engaged throughout in consolidating colonial rule.[4] As a corollary, they not only undermined 'indigenous' healing systems of India,[5] but also thoroughly colonized the 'Indian land' and the 'body'.

---

[1] Foucault (2004).

[2] Foucault, 'The Crisis of Medicine', p. 13.

[3] Foucault, 'The Crisis of Medicine', pp. 13–14.

[4] For details on the multifarious functions performed by the cadre of the IMS see D.G. Crawford, *A History of the Indian Medical Service, 1600–1913*, 2 Vols., London: W. Thacker & Co., 1913–14 and Crawford (1930). Additionally, for the racial biases and predilections prevalent in the IMS and the consequent nationalist demand of its Indianisation and subsequent colonial reactions see Rai (2014).

[5] Keeping in mind the pluralistic society and history of India it is very difficult to determine which cultural artifact, system or tradition truly represents India's 'indigenous' culture, or system, or tradition. This was the case with 'indigenous' healing systems of India as well. India has traditionally been the abode of various healing systems ranging from Ayurveda to Unani to Siddha to other

The colonial medical authorities conceptualized the tropics in general and the Indian subcontinent in particular as the 'storehouse of disease and epidemics'. In fact, the colonial division of space was officially justified on the grounds of unhealthy practices of the 'natives' and their unhygienic surroundings. David Arnold, in his pioneering work on plague, has described the ways in which colonial medical authorities strived 'to form and give a seemingly scientific precision to abiding impressions of India as a land of dirt and disease, of lethargy and superstition, of backwardness and barbarity…and to contrast this Orientalized India with the cool headed rationality and science, the purposeful dynamism, and the paternalistic humanitarianism of the West.'[6] However, evidence shows that none of the diseases which the colonial sanitary authorities and western medical notions attributed specifically with the 'Indian land' was unknown to the Europeans.[7] Conversely, recent studies have revealed that colonial rule and its policies were in many ways responsible for the frequent outbreak of epidemics in colonial India.[8]

In a similar vein, attempts were made by the colonial medical authorities to thoroughly colonize the Indian body. This became particularly evident in the case of anti-plague measures adopted by the colonial regime towards the end of the nineteenth century.[9] As Anil Kumar remarks the anti-plague campaign 'was directed more against the natives than the plague bacillus'.[10] In fact, the nationalist leader Bal Gangadhar Tilak went on to claim that 'ever since the establishment of the British rule in India no measure was undertaken by Government which interfered so largely and in such a systematic way with the domestic, social and religious habits of the people'.[11] The Epidemic Diseases Act of 1897 which was introduced in the context of bubonic plague epidemic was the most blatant example of colonial pursuit

---

practices which do not fall under any 'system' (non-systemic healing practices). Nevertheless, colonial India witnessed the efforts both by the colonizers as well as the colonized to settle the issue of whatever is 'indigenous' to India. However, eventually both of them ended up making this category more messy, debatable and contentious than ever. Hence, I have put the terms 'indigenous' within quotes as one has to bear in mind the complexity of this category in the Indian context.

[6] Arnold (1993).

[7] As for instance while the colonial authorities tried to blame exclusively the Indian land for frequent outbreaks of cholera pandemics throughout the nineteenth century by naming it as 'Indian cholera', there is ample evidence which shows that the Europeans were already aware of this disease and also there are stances of its occurrence in Europe prior to 1817. For details on pre-colonial incidences of cholera outbreak in the West see Sen (2012).

[8] For example, Ira Klein shows that with the introduction of modern transport and railways the instances of cholera outbreak went on increasing in the Indian subcontinent in the second half of the nineteenth century. Explaining this crucial linkage between modern means of transport and spread of cholera Klein states: 'Modern transport multiplied domestic commerce and labour mobility and even social or religious travel, and people often staggered down dusty roads when rains failed, to seek succour from starvation in towns or on relief works. India's vast human mobility and commerce brought the disease bacillus in profusion to new locales, where it often found circumstances propitious for its acculturation' (see Klein 1994).

[9] See Arnold (1987); Arnold, *Colonising the Body.*

[10] Kumar (1998).

[11] *Mahratta*, 27 June 1897, Nehru Memorial Museum and Library (NMML), New Delhi.

of controlling the body of the colonized subject. This Act empowered the colonial authorities to detain the plague suspects, destroy or demolish infected property and dwellings, prohibit fairs and pilgrimages and examine the passengers at will.

At the same time, the colonial government was interested in promoting western medicine only as a 'tool' of colonial expansion and was never keen to expand the 'practical utilities' of western medicine in India. That is why the colonial government remained dispassionate when it came to develop the health infrastructure in India. The colonial government's adherence to Victorian notions of the minimalist state provided an ideological background for this. It went on visualizing health services as a 'philanthropic' and 'charitable' act, and nowhere perceived it as one of the fundamental tasks of the government. Also, the racial considerations undergirded colonial medical establishments. Despite showing its consistent disinterest in extending medical facilities to Indians, proper treatment even of destitute, sick and poor Europeans in India was high in the agenda of the colonial government's health policy. Several government hospitals were set up which were intended 'exclusively for the medical relief of civil European and Anglo-Indian government servants and European and Anglo-Indian paupers'.[12] As for instance, the European Civil Hospital at Allahabad, Eden Hospital or the European General Hospital in Calcutta, etc. Besides these exclusive 'health islands', there were also wards in other prominent hospitals reserved for the exclusive treatment of European and Anglo-Indian patients.

While doing all this, colonial medical authorities made frequent comparison/criticism of Indian systems of healing such as Ayurveda, Unani, etc. as 'primitive', 'prehistoric', 'stagnant', and 'non-scientific' methods of treatment, amounting to 'quackery'. This was part and parcel of creating the hegemony of colonial knowledge over the 'native' practices/knowledge system. These views at the same time were founded upon imperialist hauteur at its best. In fact, the insistence on English as the medium for all recognized forms of knowledge under the colonial rule in itself disadvantaged 'indigenous' medical knowledge.[13] Though the British, as Anil Kumar argues, were not averse to making use of the available 'indigenous' medical practitioners particularly during the situations of distress and duress created by frequent epidemic outbreaks, they were unwilling to spend any money for propagating these systems of medicine or in the opening of well-organized schools or colleges for their promotion.[14] In such a scenario, any claim of therapeutic efficacy of 'indigenous' healing systems were refuted and scoffed at by the colonial medical authorities.[15]

[12] 'Rules for the admission of patients to the European Hospital, Allahabad', F. No. 65, Box No. 31, 1919, Medical Department, Uttar Pradesh State Archives (UPSA), Lucknow.

[13] Bala (1991).

[14] Kumar, *Medicine and the Raj*, p. 72.

[15] In fact, by the late nineteenth century with the gradual acceptance of germ theory in western medicine, it moved away from all those systems of healing which were based upon the theory of humoral imbalance behind disease. The colonial disdain for systems like Ayurveda and Unani (both of which were based upon the humoral theory) was also a result of this broader therapeutic shift in western medicine, besides imperialist considerations.

4.2 The Idea of 'Interactive Indigeneity'

In the above historical backdrop, the rise and growth of nationalism and anti-colonial sentiments in the late nineteenth and early twentieth century India naturally created a favorable atmosphere for the revival of 'indigenous' healing systems such as Ayurveda. Nationalist leaders like Hakim Ajmal Khan writing in the Annual Report of the Ayurvedic and Unani Tibbia College in 1920 emphatically argued that '[I]f we want to take the administration of government into our own hands, we must right all national things, including the indigenous method of healing. Our real progress depends on these things. We fail in serving our country if we are dependent on outside things.'[16] Taking advantage of this situation, the Ayurvedic practitioners politicized the entire issue of health and healing and projected Ayurveda as the 'indigenous' alternative of western medicine. It aimed simultaneously to overthrow the colonial medical construct of the Indian land and body by reviving and reinstating a non-western 'indigenous' medical system.

Nevertheless, nationalist resistance and revivalism nowhere denied the possibility of 'interaction' and 'selective adoption' between the two distinctive systems of healing (viz. western medicine and 'indigenous' systems of healing). It was nowhere merely a question of complete 'resistance' or complete 'acceptance' of western medicine and its characteristic traits. In fact, two broad categories can easily be discerned while analyzing the 'indigenous' resistance against western medicine. The first category included those who advocated selective adoption from western medicine to enrich 'indigenous' systems of medicine, by keeping their 'indigenous' identity intact. They acknowledged the progress made by western medicine in past few centuries and did not reject it out rightly. The second category was composed of purists who made a call to restore the 'pristine purity' of 'indigenous' systems of medicine. They were not ready to acknowledge the superiority of western medicine in any respect.

However, within these categories there were multiple strands. For example, within purists there were those who denied the medical aspect of western medicine, but simultaneously adopted its institutional structure to further the cause of 'indigenous' medicine. This ambiguity of revivalist discourse on medicine was analyzed for the first time by Charles Leslie in his series of articles.[17] According to Leslie, even while talking of the revivalism of the classical systems of healing- viz. Ayurveda and Unani, both the pragmatic ideologues (also known as 'progressives', who advocated induction of new knowledge alongside classical learning) and the purists (also known as 'conservatives', who emphasized literal revocation of Ayurvedic and Unani texts) resorted to techniques which were alien to the classical system such as institutionalization of learning, professionalization of practice, manufacturing of standardized remedies in pharmaceutical factories, etc. all of which were traits of the 'modern' western system of medicine. The net outcome of this was the emergence of a 'modern' Ayurveda (as well as Unani). Interestingly, Leslie does not find such deviations in

[16] Cited in Metcalf (1985).

[17] Leslie (1973, 1974, 1976).

medical revivalism in India to be strange since, according to him, even in Europe, Renaissance and the ideas of Enlightenment while deriving sanctity from classical learning never literally revived them and altered classical conceptions significantly.[18]

Actually, in their revivalist quest, the Ayurvedic practitioners of the time dramatically reconstructed and redefined the Ayurvedic system of healing to incorporate the traits on the basis of which western medicine was claiming its superiority such as scientificity, standardization, pharmaceuticalization, professionalization and institutionalization. Thus, a whole lot of 'new' vaids came up during the period under consideration who were remarkably different from the vaids of the preceding era.

These western standards and traits were adopted by the 'indigenous' practitioners to such an extent that many 'bogus' medical institutions flourished in the early twentieth century evading government's eyes to provide certificates not only to unqualified allopathic practitioners but also to vaids. Here it is noticeable that earlier training of vaids and hakims mostly took place through the institution of family. '*Khandani*' (hereditary) vaids and hakims, in this regard, had more prestige in the pre-colonial society than those who had newly acquired such expertise. Contrastingly, by the early twentieth century, the mushrooming of such 'bogus' institutions for providing certificates in 'indigenous' medicine reveals the importance which the system of institutional training had gained over traditional forms of knowledge transmission even for the practitioners of 'indigenous' medicine. One such 'bogus' medical institution providing certificates in Unani, Ayurveda and Homeopathy was The Old Indian Medical College at Barnala in Patiala state. It was opened by Man Singh (son of Narain Singh Ahluwalia of Barnala, Patiala state). According to official records, he was 'a man of poor knowledge and ability; [who] opened this bogus institute and began to earn his livelihood by false advertisements and selling certificates of this bogus institution'.[19] Other examples of such institutions were The Royal College of Homeopathy, Gwalior; International Medical College, Calcutta, etc. In fact, throughout the early twentieth century, the Government of India was consistently engaged in restricting the growth of such 'bogus' medical institutions and degrees both in the field of western and 'indigenous' medicine.

Similarly, the standardization of drug manufacturing was another trait of western medicine which was internalized by the practitioners of 'indigenous' medicine by this time. Earlier vaids did not merely prescribe the medicines; rather they themselves prepared and gave medicines as well. The quantities in which various elements and herbs were to be mixed to prepare a particular medicine was highly personalized in nature and vaids used to determine this after looking at the severity of the disease. This skill of making medicine was so personalized in nature that, many a time, the methodology of preparation of a particular medicine disappeared forever after the death of the practitioner.

However, by the late nineteenth and early twentieth century, we see the emergence of some big pharmaceutical companies in different parts of India standardizing the

---

[18] Leslie, 'The Ambiguities of Medical Revivalism in Modern India', pp. 365–66.

[19] 'Bogus Medical Institution in India', F.No. 52–66, 34-H, Health Branch, Education, Health and Land Department, 1934, National Archives of India (NAI), New Delhi.

manufacturing of Ayurvedic drugs. For example, C.K. Sen & Company of Calcutta, established by Chandra Kishore Sen in 1878 which started large scale production of Ayurvedic medicines from 1898 onwards; N.N. Sen & Company of Bengal, which was established in 1898; Shakti Aushadhalaya of Dacca, established in 1901; and Arya Vaidyasala of Kottakkal (Kerala), established in 1902 by P.S. Variar, etc. Some of the nationalist leaders like P. C. Ray also pushed for the manufacturing of standardized Ayurvedic remedies. In fact, P. C. Ray himself hawked some of the Ayurvedic products, manufactured by his firm, in the streets of Calcutta.[20] Similarly, Madan Mohan Malviya opened an Ayurvedic Pharmacy in Banaras after consulting eminent vaids of the city. Interestingly, the establishment of the Ayurvedic Pharmacy preceded the establishment of the Ayurvedic College at the Banaras Hindu University.[21] At the same time one can also trace the origin of present day Ayurvedic pharmaceutical giants such as Dabur (1884), Zandu (1910), Baidyanath (1917) and Himalaya (1930) during this period only.[22]

Besides these giant pharmaceutical companies manufacturing Ayurvedic drugs, there were numerous small pharmacies in different cities operating at the local level. It shows the importance that the practitioners of 'indigenous' medicine were attaching to the standardization of manufacturing of 'indigenous' drugs. In fact, the rise and growth of patriotic fervour particularly following the partition of Bengal in 1905, wherein boycott of foreign goods was adopted as one of the tools of anti-colonial nationalist agitation, gave fillip to Ayurvedic pharmacies as well.[23] The contemporary newspapers often carried the advertisements of these big as well as small Ayurvedic pharmacies and medicines prepared by them (for illustration see Fig. 4.1). Study of such advertisements provides rich insights to understand how the modern visual culture and print media were tapped by the Ayurvedic practitioners and pharmaceutical companies to carve out a niche for themselves in the medical market.[24]

Thus, by the early twentieth century a kind of modern medical entrepreneurship came to develop even in the field of Ayurvedic medicine. Actually, the Ayurvedic practitioners of the time found the standardization of remedies as one of the factors behind the rapid growth of western medicine. This was largely because the standardization of remedies liberated the medical relief from the clutches of personalized knowledge system. One could buy medicines anywhere by producing the same prescription again and again and there was no need to visit the practitioner located at some distant place merely to get the renewed dose of medicine. In other words, it was more consumer friendly. Hence, 'new' vaids continuously laid emphasis on the standardization of manufacturing of 'indigenous' drugs and medicines. Besides

---

[20] Kumar (2001).

[21] Sharma (2012).

[22] For detail description of pharmaceuticalization of Ayurvedic drug manufacturing and consequent shifts see Banerjee (2009).

[23] Mukherjee (2015).

[24] For an interesting insight over it see Sharma (2009).

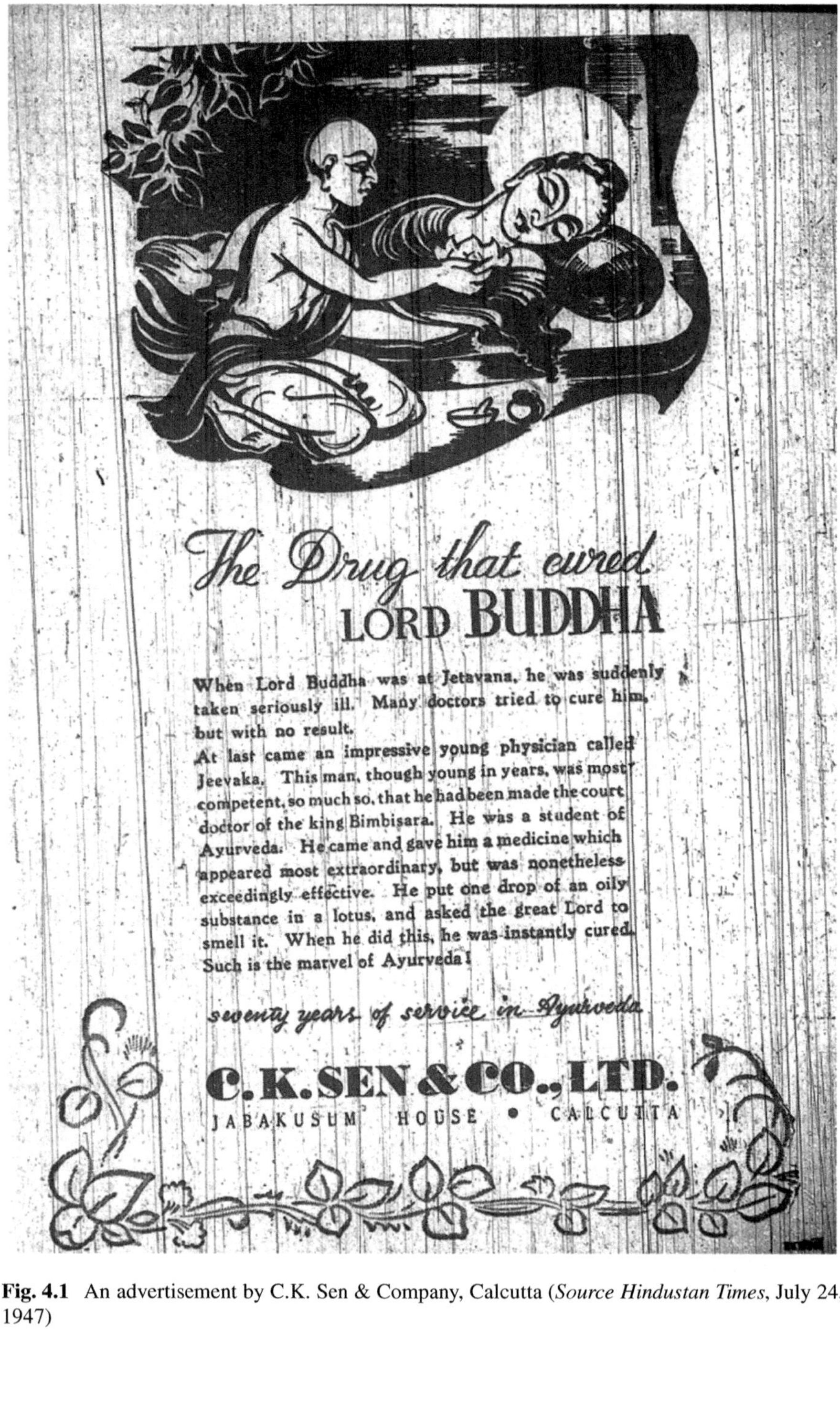

Fig. 4.1 An advertisement by C.K. Sen & Company, Calcutta (*Source Hindustan Times*, July 24, 1947)

this, many vaids saw the opportunity of handsome profit as well in the mass production of 'indigenous' medicine. Therefore, we see the mushrooming of Ayurvedic pharmacies in the early twentieth century where an established vaid often owned a pharmacy as well. Such pharmacies often gave postal guidance to the sufferer also and dispatched medicines using the postal services.

The intricacy of 'indigenous' resistance against western medicine can also be seen in the way vaids and hakims were trying to fight back the influence of western medicine. On the one hand they were thrashing the western system of medicine for its 'alien-ness' and 'colonial' motives, on the other hand they were trying to derive sanctity from the 'colonial' government to authenticate their newly discovered medicines. Consequently, the colonial government and the administrative authorities continued to be an indirect source of legitimacy even for the practitioners of 'indigenous' medicine at least for the first two decades of the twentieth century. One can find numerous petitions by vaids and hakims during this period praying that their medicine should be given a trial to cure the concerned disease. As for instance, petition submitted by Babu Satya Pado Banerji (1900) praying that his medicine for cholera may be given a trial in famine camps;[25] Hakim Sayid Muhammad Ibrahim (1900) submitting prescriptions for plague patients;[26] Kobiraj Jogendra Chandra Sen Mullick (1901) praying that certain medicines discovered by him may be experimentally treated and that he may be informed of result;[27] Messrs. Robindro Co. (1909) requesting that the malarial pills known as *Swarna Batika* invented by them may be given a trial.[28] Similarly, the advertisements of 'indigenous' medicine often carried testimonials of some men of authority. For example, one such advertisement by Babu Laxminarayan Kabiraj of Cuttack carried testimonials of a headmaster and a police inspector (both men of authority) to support its claim.[29]

Moreover, so far, the historiography on the emergence of 'modern' Ayurveda has focused particularly on the changes which were primarily institutional in nature or related to the outer sphere of Ayurvedic healing system. Seldom scholars have looked at therapeutic changes within Ayurveda initiated by modern technologies which brought it closer to global medical episteme thereby making it 'modern' in real sense of the term. Here Projit Bihari Mukharji in one of his recent works particularly focuses on day-to-day mundane small technologies introduction of which fundamentally altered the approach of an Ayurvedic practitioner (*Kobiraj* in the context of Bengal) towards his patient and pertaining disease. Pocket watches to check the pulse, thermometers to check the temperature, stethoscopes to hear the patient breath, microscopes through which germs and blood cells could be seen, injections of endocrinal extracts to rejuvenate the patients, and even glass phials (rather than carpet bags full of crushed herbs) in which medicines were stored are some of the mundane small technologies which Mukharji has taken up to show the therapeutic

[25] F. No. 2–3, Medical Branch, Home Department, September 1900, NAI.

[26] F. No. 138–39, Medical Branch, Home Department, May 1900, NAI.

[27] F. No. 42–43, Medical Branch, Home Department, April 1901, NAI.

[28] F. No. 193–96, Sanitary Branch, Home Department, December 1909, NAI.

[29] Pati (2001).

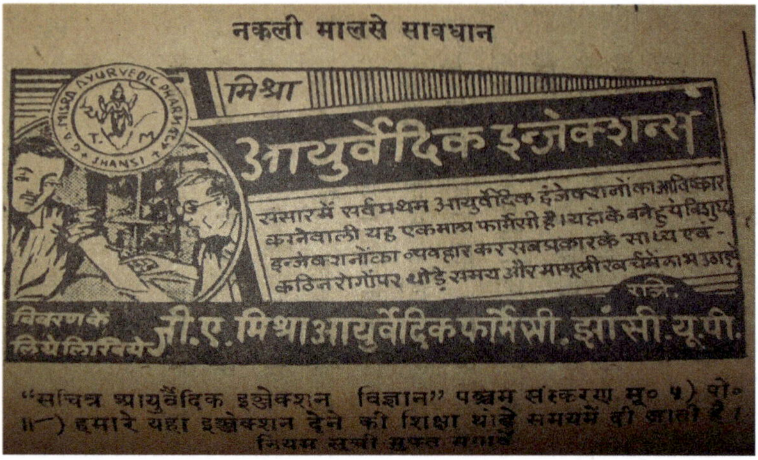

Fig. 4.2 Ayurvedic Injections [*Source Sudhanidhi*, Vol. 40, No. 10 (May 16, 1949)]

changes in Ayurveda in the late nineteenth and early twentieth century. Interestingly enough, around this time we see advertisement of an innovative 'Ayurvedic Injection' which explicitly traversed the boundaries of Ayurveda and western medicine and combined both the systems into one (see Fig. 4.2). According to Mukharji, these everyday small technologies were 'motors' or 'catalysts' of therapeutic change in modern Ayurveda through their constant presence.[30]

The above discussion explicitly shows that the Ayurveda which was revived in late nineteenth and early twentieth century was premised upon 'interactive indigeneity' and carried 'modern' traits and survival strategies thereby making it remarkably different from its pre-colonial version.

4.3 'Modernity' at the Service of 'Tradition'

Even though Ayurvedic healing system was modernized to a considerable extent during the period of our discussion, it should not be essentially viewed as epistemological victory of western medicine. As Neshat Quaiser has shown through his study of Ayurvedic and Unani Tibbiya College of Delhi that the changes which we witness in 'indigenous' healing systems during the late nineteenth and early twentieth century were, of course, modern, they were not devoid of cultural capital.[31] There was a constant attempt which was made by the practitioners of 'indigenous' medicine to maintain their distinctive identity despite modernizing themselves. This struggle "to fashion a 'modern' national culture that is nevertheless not Western" by

[30] Mukharji (2016).

[31] Quaiser (2011).

the nationalist reformist elites of the time was very much integral to anti-colonial nationalist milieu.[32] Hence, instead of succumbing to the hegemonic influence of western medicine, the 'new' vaids and hakims found their own ways to justify the changes, institutional as well as therapeutic, which they had brought in these age-long healing systems.

So far as institutional changes were concerned, such as standardization of training and pharmaceuticalization, these were seen primarily as ways to meet the requirements of the time through renewal and research so that the challenges offered by western medicine could be successfully met. Therefore, the institution like Ayurvedic and Unani Tibbiya College[33] was although outright manifestation of adoption of modern trait in 'indigenous' medicine, if we see the curriculum which was taught at this college and the kind of ideas propagated by it nowhere testifies the victory of market oriented and capital-intensive colonial modernity. According to Quaiser, 'the structuring principles of the Tibbiya College and its associated institutions arose primarily from the internal logic of the College's existence and from conscience based ethical values traditionally associated with the profession of *Tibb* (Unani medicine) against the capitalist, market oriented logic of profession-related ethics'.[34] Quaiser, in fact, goes on to argue that Tibbiya College was a significant intervention in the rampant colonial power that intended to destabilize the knowledge systems in the colony.[35]

Further, if we look at the standing rules passed by the Board of Trustees of the Ayurvedic and Unani Tibbiya College on 15 October 1915, it noted its objectives in following terms: 'To improve the Ayurvedic and Unani systems of medicine on modern scientific lines provided they *do not affect the essentials of these systems, or their importance* and with this object in view to cause compilations and translations of medical books and treatises'.[36] Thus, the idea was to secure the revival of 'indigenous' healing systems in its 'original' form through modern means. Here it is remarkable that Hindi and Urdu were used as the medium of instruction and expression at this College so that the students could conveniently learn the art of healing through their mother tongues.[37] In fact, it was probably the first time that Hindi and Urdu were used in such a systematic manner and scale to teach science and medicine which in

[32] This struggle can be best understood through the theoretical framework of 'inner' and 'outer' spheres of action of nationalist intelligentsia as delineated by Chatterjee in his influential work (1986).

[33] Aimed at harmonious promotion of both Ayurveda as well as Unani healing systems, an Ayurvedic and Unani Tibbiya College was established by Hakim Ajmal Khan in Delhi. Its foundation stone was laid by the then Viceroy of India, Lord Hardinge, in 1916, and was inaugurated eventually by Mahatma Gandhi in 1921 at the height of non-cooperation movement. In his inaugural speech at the opening of the Ayurvedic and Unani Tibbiya College, Mahatma Gandhi considered it a glaring symbol of Hindu-Muslim unity.

[34] Quaiser, 'Science, Institution, Colonialism', p. 524.

[35] Quaiser, 'Science, Institution, Colonialism', p. 524.

[36] Standing Rules of the Board of Trustees of the Ayurvedic and Unani Tibbiya College, Delhi (cited in Quaiser, p. 543, emphasis added).

[37] Nizami (1998).

itself was a serious challenge to colonial medical episteme dominated by preference of English language.

Similarly, the founding principle of the Unani and Ayurvedic Medicine Company Ltd. (later renamed as Hindustani Dawakhana), a pharmacy attached integrally with the College, was not to earn the profit through market, rather it was to make Ayurveda and Unani popular among the masses and to ensure easy availability of effective Ayurvedic and Unani drugs. Its income was spent for the purposes of the College to encourage further research and training in the field of 'indigenous' medicine.[38] Interestingly, as argued by Quaiser, Hindustani Dawakhana emerged as one of the most visible centres in the critical anti-colonial public sphere.[39] Another distinct unit of the College was Vedic Dawakhana and Ayurvedic Rasayan Pathshala which aimed at preserving the ancient Ayurvedic medicines. Statistics show that it incurred more expenses than income. However, Hakim Ajmal Khan ensured its sustained functioning and accorded to this unit an independent identity and status.[40]

Further, with regard to the therapeutic changes within Ayurveda, the Ayurvedic practitioners who were involved in deploying modern technologies which, in turn, led to fundamental shift in their approach towards the human body, nowhere saw them as 'alien' or 'new' to Ayurveda. Rather these small-scale technologies often acquired new genealogies or 'histories' in Ayurvedic discourse thereby making them integral to Ayurveda from 'time immemorial'. As for instance, Projit Bihari Mukharji shows that how the clinical uptake of a relatively new practice of diagnosing pulse using pocket watches by Ayurvedic practitioners was sanctified by invoking the concept of *nadipariksha* as the iconic and exemplary instance of Kobiraji knowledge. Not only this, while *Daktari* pulse diagnosis was shown to be superficial, quantitative, and incidental, Kobiraji *nadipariksha* was deemed as mystical, subtle, qualitative, and so on.[41]

In the same way, adoption of clinical thermometry using *tapman-yantra* (thermometer) was braided into Ayurveda by aligning Kobiraji concepts of *sontap* (body heat) and *jwor* (fever). Braiding measurement of body heat by thermometer with the concept of *jibatma* (life-soul) that was available in traditional *nadipariksha* texts, it was stated that '*jibatma* is happy at 98 degrees, any movement [of the mercury] upwards of it causes unhappiness'.[42] Thus, a relatively modern diagnostic practice of measuring precise body temperature was aligned with the philosophical idea of life and soul. In a similar vein, Mukharji shows the attempts on the part of the *kobirajes* to portray the enhanced vision of the *onubikshon-yantra* (microscope) as an inheritor of the supervision of ancient seers. The Indic *dibyodrishti* (cosmic vision)

---

[38] It is worth mentioning that the Unani and Ayurvedic Medicine Company Ltd. although began as a joint stock company, it was later on converted into an endowment vested in the College. Hakim Ajmal Khan gave to it all his family medical secrets which had never been leaked out to anybody for several generations (Nizami, *Hakim Ajmal Khan*, p. 43).

[39] Quaiser, 'Science, Institution, Colonialism', p. 546.

[40] Nizami, *Hakim Ajmal Khan*, p. 41.

[41] Mukharji, *Doctoring Traditions*, p. 78.

[42] Mukharji, *Doctoring Traditions*, p. 138.

was creatively braided with modern 'spiritual optics' or microscope. Interestingly, the *kobirajes* asserted that ancient seers (*rishis*) had had such microscopic visions or *dibyodrishti*, which they themselves did not possess. They argued that 'what separated the visually empowered ancients from the visually impaired moderns was nothing but historical time'.[43] In other words, they believed that microscope has just revived this long-forgotten art of ancient practitioners of India and there was nothing novel about it as such. Even the practice of administering injection by the Ayurvedic practitioners was sanctified by invoking Vedic authority and concepts like '*suchibhedan chikitsa pranali*' as delineated in ancient texts.[44]

The level of braiding was such that right from germ theory to organotherapy to neurology everything was traced back in Ayurvedic concepts and canonical texts. Ayurveda was portrayed as fountain of eternal wisdom and bearer of all seemingly 'modern' technologies. In their effort to prove this eternity of Ayurveda in the light of incorporation of modern therapeutic concepts and ideas, the *kobirajes* went on to the extent of altering the characteristic principles of Ayurveda. As for instance, the *Journal of Ayurveda*, a journal closely associated with the famous Ayurvedic practitioner of Bengal Gananath Sen, in one of the articles equated the Ayurvedic concept of '*doshas*' with 'hormones' instead of 'humors'.[45] Thus, the concept of '*tridosha'* or the three humors which is central to the Ayurvedic treatment underwent metamorphosis in the wake of hormonal interpretation of diseases and the earlier translation of '*doshas*' as 'humors' was rebuked as 'European misunderstanding'.

Therefore, the Ayurvedic practitioners defended the use of these apparent 'western' therapeutic/diagnostic technologies by arguing that these were clearly described even in the Vedas and other classical Indian texts but by the span of time these had been lost for want of research.[46] Further, adoption of many of the modern medical tools and technologies such as stethoscopes, thermometers, etc. by the Ayurvedic practitioners, as shown by Madhuri Sharma, allowed to maintain the *purdah* norms as it could be used to access women's bodies and to make an accurate diagnosis without touching her physically.[47] Similarly, these tools and technologies allowed in many ways to maintain the caste and class norms as well by allowing the physician to distance himself from 'poor' and 'dirty' patients with accurate diagnosis.

Thus, strings of broadly Ayurvedic and broadly 'western' scientific knowledge and practices were pulled together and 'braided' around the technological objects and institutions in the late nineteenth and early twentieth century thereby leading

---

[43] Mukharji, *Doctoring Traditions*, p. 160.

[44] Tripathi (1933).

[45] Mukharji, *Doctoring Traditions*, p. 193.

[46] For an interesting insight on how injection, stethoscope and other such things were adopted by the Ayurvedic practitioners of the United Provinces see Sharma (2013).

[47] Here Madhuri Sharma gives an interesting reference of Dr. Raghubir Sahaya Bhargav, Doctor of Homeopathic Medicines, who while explicating the usefulness of stethoscope argued: 'Major benefit of the stethoscope…those women who wanted to get their chest examined hold the one end of the tool which should be kept on the patient and doctor *behind the curtain* can put both leads of the tool into his ear and can hear the rhythm' (see Shamra, 'Knowing Health and Medicine', p. 169).

to the emergence of 'modern' Ayurveda which was considerably different from its predecessor, yet at the same time distinct from its western counterpart.

4.4 Conclusion

When India got independence in 1947, it had a number of Ayurvedic schools, colleges and hospitals functioning throughout the length and breadth of the country. Several provinces had already constituted Boards of Indian Medicine which were entrusted with the task of running 'indigenous' dispensaries or *aushadhalayas* and *dawakhanas* (dispensaries pertaining to Ayurvedic and Unani systems of healing respectively), registering practitioners of 'indigenous' medicine, and designing and regulating curricula for their training. Added to this were several Ayurvedic pharmaceutical companies having access to regional as well as national and occasionally international markets.[48] Besides, there were various associations of professional Ayurvedic practitioners sponsoring conferences, publishing journals and lobbying for privileges similar to that of practitioners trained in western medicine. All these eventually constituted the basic paraphernalia on which 'modern' Ayurveda thrived.

Here, as the above discussion explicitly reveals that modernization does not necessarily lead to end of tradition, instead it results into interaction. This interaction in the case of medical revivalism in India enhanced the ability of 'modern' Ayurveda to not only survive against the onslaught of hegemonic western medicine but also catch the global fetish in due course of time. It is true that many of the fundamental tenets of Ayurveda were compromised in this interactive journey, but it somehow managed to retain its distinctive identity thereby posing itself an alternative to western medicine.

References

Michel Foucault, 'The Crisis of Medicine or the Crisis of Antimedicine?,' trans. Edgar C. Knowlton Jr., William J. King and Clare O'Farrell, *Foucault Studies*, Vol. 1, December 2004, pp. 5–19, accessed November 18, 2020, https://rauli.cbs.dk/index.php/foucault-studies/article/view/562/607.

D.G. Crawford, *Roll of the Indian Medical Service, 1615-1930*, London: W. Thacker & Co., 1930

Saurav Kumar Rai, 'Indianisation of the Indian Medical Service, *c.* 1890s-1930s', *Proceedings of the Indian History Congress*, 75th session, New Delhi, 2014, pp. 826–32.

David Arnold, *Colonising the Body: State Medicine and Epidemic Disease in Nineteenth-Century India*, Berkeley: University of California Press, 1993, p. 292.

[48] In fact, Projit Bihari Mukharji begins his book *Doctoring Traditions* by mentioning his fortunate discovery of an old medicine bottle with an Indian script embossed on the glass dated from around the 1890s in Philadelphia. The bottle belonged to the well-known Ayurvedic pharmaceutical firm C.K. Sen & Sons. It testifies the transnational reach of 'modern' Ayurveda right from its inception (Mukharji, *Doctoring Traditions*, pp. 1–2).

Srabani Sen, 'Indian Cholera: A Myth', *Indian Journal of History of Science*, Vol. 47, No. 3, 2012, pp. 345-74.

Ira Klein, 'Imperialism, Ecology and Disease: Cholera in India, 1850-1950', *Indian Economic and Social History Review*, Vol. 31, No. 4, 1994, pp. 491-518

David Arnold, 'Touching the Body: Perspectives on Indian Plague, 1896-1900', in Ranajit Guha (ed.) *Subaltern Studies V*, Delhi: Oxford University Press, 1987, pp. 55-90

Anil Kumar, *Medicine and the Raj: British Medical Policy, 1835-1911*, New Delhi: Sage Publications, 1998, p. 197.

Poonam Bala, *Imperialism and Medicine in Bengal: A Socio-Historical Perspective*, New Delhi: Sage, 1991, p. 63.

Barbara D. Metcalf, 'Nationalist Muslims in British India: The Case of Hakim Ajmal Khan', *Modern Asian Studies*, Vol. 19, No. 1, 1985, p. 20.

Charles Leslie, 'The Professionalising Ideology of Medical Revivalism', in Milton Singer (ed.) *Modernisation of Occupational Cultures in South Asia*, Durham: Duke University Press, 1973, pp. 691-708

Charles Leslie, 'The Modernisation of Asian Medical Systems', in John Poggie and R. Lynch (eds.) *Rethinking Modernisation: Anthropological Perspectives*, Westport: Greenwood Press, 1974, pp. 377–94

Charles Leslie, 'The Ambiguities of Medical Revivalism in Modern India', in Charles Leslie (ed.) *Asian Medical Systems: A Comparative Study,* Berkeley: University of California Press, 1976, pp. 356-67.

Anil Kumar, 'The Indian Drug Industry under the Raj', in Biswamoy Pati and Mark Harrison (eds.) *Health, Medicine and Empire: Perspectives on Colonial India*, New Delhi: Orient Longman, 2001, pp. 356-85.

Madhuri Sharma, *Indigenous and Western Medicine in Colonial India*, Delhi: Foundation Books, 2012, p. 115.

Madhulika Banerjee, *Power, Knowledge, Medicine: Ayurvedic Pharmaceuticals at Home and in the World*, Hyderabad: Orient BlackSwan, 2009.

Sujata Mukherjee, 'Ayurvedic Medicine in Colonial Bengal: Challenge and Response', in Syed Ejaz Hussain and Mohit Saha (eds.) *India's Indigenous Medical Systems: A Cross-Disciplinary Approach*, New Delhi: Primus, 2015, p. 108.

Madhuri Sharma, 'Creating a Consumer: Exploring Medical Advertisements in Colonial India', in Biswamoy Pati and Mark Harrison (eds.) *The Social History of Health and Medicine in Colonial India*, London and New York: Routledge, 2009, pp. 213–228.

Biswamoy Pati, *Situating Social History: Orissa 1800–1997*, New Delhi: Orient Longman, 2001, p. 18.

Projit Bihari Mukharji, *Doctoring Traditions: Ayurveda, Small Technologies, and Braided Sciences*, Chicago: University of Chicago Press, 2016, p. 11.

Neshat Quaiser, 'Science, Institution, Colonialism: Tibbiya College of Delhi; 1889–1947', in Uma Das Gupta (ed.) *Science and Modern India: An Institutional History, c. 1784–1947*, Delhi: Pearson Longman, 2011, pp. 523–562.

Partha Chatterjee *Nationalist Thought and the Colonial World: A Derivative Discourse*, London: Zed Books, 1986.

Zafar Ahmad Nizami, *Hakim Ajmal Khan*, New Delhi: Publications Division, 1998, p. 39.

Ramakant Tripathi, *Sachitra Injection Chikitsa Arthat Sui kee Pichkari Dwara Rogon kee Chikitsa*, Mathura: Sukhsancharak Company, 1933.

Madhuri Sharma, 'Knowing Health and Medicine: A Case Study of Benares, *c.* 1900-1950', in Deepak Kumar and Raj Sekhar Basu (eds.) *Medical Encounters in British India*, Delhi: Oxford University Press, 2013, p. 170

Deepak Kumar and Raj Sekhar Basu (eds.) *Medical Encounters in British India*, Delhi: Oxford University Press, 2013, p. 170.

Chapter 5
Bazaari-Subaltern Unani Medical Public Sphere: Blanket Binaries, Disjunctions, Implications and the Making of Alternative Texts and Archives Convivially

Neshat Quaiser

5.1 Introduction

The essay, in addition to an introduction is divided into five sections with subsections. Section 5.2 explains the preliminary theoretical considerations with regard to the relationship between the *bazaari*-subaltern Unani practices and textual Unani; and the ashrāfiya[1] casteist predispositions and moral economy; Sect. 5.3 deals with the questions of blanket binaries, disjunctions and impedimental implications in the light of the recent endeavour to recover "the realm of subaltern therapeutics"; Sect. 5.4 deals with bazaar-bazaari in social history of South Asian medicine; bazaar-bazaari medicine and the colonial state; bazaar, bazaari and subaltern: concepts in context; nuances of bazaar, bazaari, 'market' and Unani; caste-class, economic dimensions

[1] The term ashrāfiya herein includes Muslim nobles of foreign origin and high caste middle class Muslims.

This essay heavily draws on Quaiser 'Medical Public Sphere in Colonial India: Competing Meaning Systems of Western Medicine and *Bazaar* Unani', paper presented in the Conference on Imperialism, Medicine and South Asia: A Socio-Political Perspective, 1800–1950, Cambridge, 14–16 June 2001; Quaiser (2010a) 'Subalternising the Unani—Sub-continental Experiences', at the workshop on 'Situating Subalternity in South Asian Medicine', under the auspices of the Warwick University and Centre for the Study of Developing Societies, New Delhi on 18-19th February 2010; Quaiser (2016). 'Subalternising the Unani—"Nationalist" Appropriation, Bazaar and the Identity Politics', at the International Seminar on Indian Medicine: Between State and Village, organised by the International Institute for Asian Studies, the Netherlands, at Leiden, the Netherlands—23 to 24 June, 2016.

N. Quaiser (✉)
Centre de Sciences Humaines (CSH), Delhi, India
e-mail: neshatquaiser@yahoo.com

S. Ahmed and S. Sarkar (eds.), *Decolonizing Science and Modernity in South Asia*,
https://doi.org/10.1007/978-981-97-1829-0_5

and bazaari Unani; post-colonial turn and implications for bazaari Unani; and institutionalised explanations; Sect. 5.5 examines the framework and types of bazaari-subaltern Unani medical public sphere as a convivial dialogic terrain within general medical public sphere; eleven cases to illustrate the bazaari Unani sphere; and the question of how Unani is bazaari-Unani and how different it is from non-Unani bazaar medicine; and Sect. 5.6 is an epilogue summing up the critical issues with a focus on the making of bazaari Unani texts and archives at the end.

Illustrative cases are drawn from the provinces of Bihar, Bengal, Tamil Nadu, Karnataka, Telangana, Delhi, and United Provinces of both colonial and post-colonial periods.

5.2 Preliminary Theoretical Considerations

The essay deals with the completely neglected sphere of bazaari-subaltern Unani medical practices demonstrating, how contrary to the predominant view within the studies of South Asian medical traditions, the terrain of Unani was marked by simultaneous presence of the two broad spheres—paradoxical in theory and practice—namely textual Unani, and bazaari-subaltern Unani. Significantly, contrary to the dominant view, even the textual Unani terrain, on several counts, has been a dispersed, and democratic terrain marked by complementarity and intersections. The essay critically examines the recent endeavour to recover "the realm of subaltern therapeutics" which remained eclipsed by a persistent focus on "both textual, erudite Unani and Biomedicine"[2] (see Sect. 5.3). In spite of erudite Unani's opposition through its textual authority to bazaari-subaltern Unani sphere, we would, however, argue that the relationship between the two cannot be comprehended through blanket binary frames such as elite[3]-subaltern; state-regulated-unregulated; and one-dimensional colonial-nationalist binaries. This is particularly for the following three factors that allow us a more nuanced and layered frame to propose a different approach:

1. In spite of certain post-colonial self-contradictory distortions through contested statist regulations and capitalist market compulsions and enticements, Unani's structuring features, in spite of its textual fidelity, also include it being a dispersed, discursive, and relatively relaxed terrain marked by complementarity and intersections; *critical anti-colonial Unani medical public sphere*, distinct from one-dimensional nationalist anti-colonialism,[4] that strongly resisted the western medical absolutism in defence of *Desi* (indigenous) medicine against *ghair-mulki*

---

[2] Hardiman and Mukharji (2012), particularly Attewell, Guy et.al. *Agendas*, pp. 1–5; Hardiman, David and Mukharji, Projit Bihari, *Introduction*, pp. 6–38.

[3] The term elite herein includes ashrāfiya.

[4] Questioning the idea that the relationship between South Asia's indigenous medical traditions and western medicine during the colonial period best be comprehended by the idea of encounter, Projit Bihari Mukharji has argued that most of the studies of the two are guided by the polemically charged "nationalist framework of anti-colonialism" Mukharji (2011a).

(foreign) western medicine; its practitioners' concerns for moral economy[5] with *informed* conscience based medico-ethical values; its opposition to capitalist-market-oriented profit making logic of western medicine by establishing manufactories such as Hindustāni Dawakhāna to produce affordable Unani medicine; and "medical communalism"[6] are crucial, and which, in a way, also went even in favour of bazaari Unani.[7]

2. Much of even textual Unani medicine and practice remained outside the "state's sanction and regulation".
3. Moreover, the concept of subaltern itself is to be re-examined contextually, particularly to account for the ways in which even a textual tradition, such as Unani, itself could be rendered to a state of subalternity, for example, due to "medical communalism".

However, the textual Unani based on long tradition of exclusive disciplinary expertise and practical experiences with rich written texts is regarded the authentic body of Unani medical knowledge. It officially always disapproved of the bazaari 'untrained', 'ignorant' hakims. Abu Bakr Muhammad bin Zakariyya al-Razi (Rhazes) (850–925 AD) had warned that:

> Never consider any illiterate and common person who has no experience in reasoning by analogy and excogitation as an expert in *tibb*. In case, sometime he gets something right, it should be considered as sheer luck chance.[8]

This disapproval of bazaari hakims and even of '*attars*[9] continued during colonial and contemporary periods, as they brought "bad name to…*sharīf pesha* (in the sense of engagement with a cause to serve people) of *tibb*",[10] thus the popular saying—neem hakim Khatra-e Jān ((ill-trained hakim poses danger to life) reflected textual authority. However, khāndāni hakims controlling the textual Unani establishments could not do much except lamenting about it for several reasons,[11] and bazaari Unani continued to co-exist and catered vast masses at all social locales. Thus, the relationship between the textual and bazaari Unani was not very antagonistic at the ground level.

This opposition to bazaari hakims was not essentially propelled by western notion of 'reform' but emanated from the retentionist logic of 'renewal' (*tajdeed*) as the

---

[5] For the concept of moral economy see: Thompson (1991) and Tripp (2006).

[6] Quaiser (2012a).

[7] Quaiser (2001, 2010, 2012b).

Moreover, the Internalist-Externalist debate on the philosophy and history of science may be of help to comprehend the Unani's opposition to the hegemony of western medical absolutism, see, Kuhn (1977); and, in spite of positivist leanings, Lakatos (1978). Moreover, for the role of social roots of the origins of science and the role of specific cultural values for the development of scientific thinking, see Joseph Needham *The Grand Titration: Science and Society in East and West*. London: Allen & Unwin. Also see Latour (1993).

[8] al-Razi (1994).

[9] One involved in Unani apothecary who also acted as hakims.

[10] Quaiser (2001). Politics, Culture and Colonialism, (2010) Science, Institution and Colonialism.

[11] See Quaiser (2010)—Science, Institution and Colonialism.

core of Unani in the given context, with conscious democratisation without fear of contamination. This democratization was not just because a large number of aspiring hakims were imparted training "through institutionalisation" of Unani at the beginning of the twentieth century as some scholars have argued, but it was self-constituted democratic logic for long, as we shall see in the pages to come.

5.2.1 Ashrāfiya Casteist Predispositions and Moral Economy-Dominance and Kaffara

In spite of discursive nature textual Unani and associated ethical values of moral economy, ashrāfiya hakims were not free from primary predispositions of 'epistemic ashrāfiya-morality'[12] and associated casteist predispositions in their everyday life, which also reflected in their medical practice. For example, Mohammad Ahmad Khan (1885–1975), a high-born khāndāni hakim of Delhi, would come out of his matab (clinic) in the porch/veranda to see patients of low-castes such as dhuniya (carder of cotton), julāha (weaver), bhangi,[13] chamār,[14] saqqa (water-carrier, water-vendor), bhatyāra (innkeeper, considered very low and immoral in Muslim caste hierarchy) and others of the similar caste category as they were not allowed to enter the matab. He would disdainfully ignore as a trifle their *salutation with the head and body bowing down*; maintaining a physical distance he would listen their *kaifiyat* (condition of illness), and would whisper to his prescription writers, who would then tell them about medicines.[15]

A more blatant casteist prejudice and discrimination is reflected from the following episode:

> Once a chamār came to hakim Mohammad Ahmad Khan and said, hakim sahib I have headache. Hakim sahib instructed his people: Hit him on his head with shoe. Headache is the disease of shurafa (nobles/gentlemen). I would have medically treated you if your armpit was louse-infested, if you had stinking abscess/canker or had gāgan. Headache and chamār! What does that mean? He does not seem to be from the sperm of his socially known father. Take him out and hit him on his head with shoe.[16]

Mohammad Ahmad Khan (1885–1975) was a prominent high-born khāndāni hakim of Delhi. Like others of his social location, he too was brought up with epistemic ashrāfiya casteist or class prejudices, which reflected in medical practice as well as is evident from the above narratives. Guided by his primary ashrāfiya predispositions, he was ready to treat a Chamār for other diseases which were caused by

[12] For the concept of 'epistemic ashrāfiya-morality' see Quaiser (2021).

[13] Street sweeper; traditionally restricted to cleaning service latrines and handling dead bodies.

[14] Cobbler. A very low caste even among scheduled castes; also engaged in disposing off carcasses of dead animals and skinning them.

[15] Dehlvi (1991).

[16] Dehlvi (1991). Op. cit. P. 47.

their dirty occupation and dirty living conditions, but not headache which he associated only with shurafa. As a result most low caste-class people would seek for bazaari medicine. He was not only practicing the ideology of casteist physical segregation but was also creating a hierarchy of diseases based on ascriptive casteist occupation, while even by textual Unani medical principles anybody of any occupation and social status can have headache as a CCRUM official trained at Nizāmia Tibbia College affirmed.

However, at the same time, a system of moral economy was practiced driven not only by the religious ethics, piety and patronage but perhaps also as an act of *kaffarah* (expiation, atonement) for their ashrāfiya caste-class prejudices that was against their strict religious injunctions. For example, poor and low-caste patients were "prescribed medicine costing not more than one or two paisa,[17] and if with the medicine costly diet is required such as fruits, chicken curry, which those poor patients could not afford, ten or five rupees were distributed among them".[18] But this system of moral economy also served as a system of ideological domination, which further reinforced the ashrāfiya epistemic casteist-class morality. And yet there were grey areas, intrinsically linked with the complex issues of Unani practitioners' medico-ethical thinking and occupational values engendering the system of moral economy. It is for this reason that the available evidences suggest that kakims generally did not ostensibly allow casteist-class prejudices to interfere in their practice of *tibb* to the extent possible and the occultating system of moral economy largely based on the feudal patron-client relations became expedient.

The textual and bazaar-bazaari-subaltern Unani practices represented bodies of knowledge with different meaning systems about body, health and medicine, despite certain elements of complementarity in terms medicinal ingredients at the ground level. This is to suggest that the bazaari-subaltern Unani practices on their own account, in a sense constitute a fluid parallel 'tradition' producing their own 'texts' and making their own 'archive'. The act of archiving is not antithetical to bazaari-subaltern realm.

5.3 Blanket Binaries, Disjunctions and Implications

This section situates the division between bazaari-subaltern Unani practices and textual Unani, within the recent concerns for the 'recovery of subaltern medicine and therapeutics', and argues that the issue cannot be comprehended through blanket binary frames such as elite-subaltern, state-regulated-unregulated and one-dimensional colonialis-nationalist.

A group of historians of South Asian medical history intended to explore the "non-elite forms of medical and therapeutic practices in South Asia" in order to recover "the realm of subaltern therapeutics" which remained eclipsed by a persistent

[17] Paisa—a unit of money in Indian subcontinent worth one hundredth of a rupee.

[18] Dehlvi. Op. cit. p. 46.

focus on "both textual, erudite Unani and Biomedicine". Hardiman and Mukharji[19] have employed the concept of medical marginality to focus on the relationships of power and hierarchy within the field of medicine, therapeutics and governance, by differentiating between medical practices that are regulated by the state, which includes both biomedicine and AYUSH and those which are not regulated by the state. According to them during the colonial period biomedicine and erudite Unani "served only a relatively small elite"; and that both biomedicine and Ayurveda and Unani even in the postcolonial period "have been largely inaccessible for the mass of the people". As a result, "a range of healing practices have flourished" including local bonesetters, midwives, spiritual healers etc. and even "unqualified or so-called 'quacks'" too thrived."[20]

This shift in focus is sure to initiate a fresh debate and discussion. However, this 'project' represents a particular historically qualified view of the relationship of South Asia's indigenous medical traditions with the western medicine and its postcolonial manifestations. 'Agendas' of these historians seems to be a judgement of perception. Agendas intend to replace the internal properties of object, though not immutable self-regulatory mechanism, with the perception of fact formed outside the fact as often desirably perceived and attribute it to the constitution of fact. Thereby, Agendas intend not to uncover the concept of cause of perception beneath the surface and that is how validity of a fallacy is validated. Thus, their arguments remind of what Charles Leslie had argued in 1970s. Leslie had highlighted the existence of medical pluralism, but ultimately his argument was to rationalise the 'modernisation and its ambiguities' argument for the 'revival' and 'reform' of Unani in modern India in the mirror image of western medicine.[21] Leslie's argument crucially continues to influence the scholars of Unani even today. Moreover, Agendas focussed more on indigenous/Unani or put western medicine and Unani on equal plane for eclipsing the subaltern medicine. While, it is the western medicine that exercised power and authority to set the standard for what is scientific, denigrating not only subaltern practices but also textual Unani as *wahshi* (irrational). This has serious implications in the context of semi-colonial and neo-liberal Indian subcontinental states dominated by western global power structures of which western medicine is a crucial component.

The whole argument of this group of historians is situated within the background historical evidences, thus we briefly reflect on some of the key issues accordingly.

David Arnold had argued that under colonialism indigenous medicines projected themselves as a single, homogenous 'system' of medicine "not to make a judgment about their relative efficacy or to interrogate their intellectual, physiological and therapeutic claims in the harsh light of Western scientific medicine; but to recognize that many millions of people across Asia subscribed to these systems....".[22] Arnold

---

[19] Hardiman, David and Mukharji, Projit Bihari, *Introduction*, in David Hardiman and Projit Bihari Mukharji edited *Medical Marginality in South Asia: Situating subaltern therapeutics*. Routledge pp. 6–38.

[20] Attewell et al. (2012).

[21] Leslie (1974, 1976).

[22] Arnold (2003).

identified three underlying assumptions which characterise and form a 'system' argument—continuity, coherence, and completeness. Following Arnold's conceptual thread, Attewell argued that 'the use of the word "system"...consolidates the impression of continuity, connoting internal coherence, discreteness, completeness and homogeneity'.[23] The arguments suffer from several disjunctions. For example:

1. Arnold has no problem with the use of the word 'system' for the deeply canonised biomedicine and characterises it as 'Western scientific medicine', but use of the word 'system' for indigenous medicines is arrogantly questioned. However, it has conclusively been shown that projection of Unani as a system of medicine was not in the mirror image of western medicine but on account of its own disciplinary logic, and endeavours to reform it was based on the *retentionist* logic of *tajdeed* (renewal) with the spirit of learning from others without losing its distinct identity[24] which is contrary to the argument that that the pressure for reform "came from within an expanding and fractured Unani profession"[25] to establish authority over "the unqualified Tabibs".[26] Moreover, Attewell deployed the colonial capitalist notions of 'reform' and 'profession' to Unani, and argued, while Unani's idea of '*pesha*' and reform were distinctly different from those.[27] It was not a 'profession' or pesha as we understand in the sense of capitalist marketised ideology but in the sense of engagement with a cause to serve people in need.
2. This definition of 'system' should prompt them to 'recover' subaltern medicine in the western worlds to replace canonised biomedicine.
3. The argument that that both textual Unani and western medicine eclipsed bazaari-subaltern medicines, strikingly placed both on the same ideological plane, ignoring the paramount significance of Unani's self-reflexive agency manifested through *literate critical anticolonial Unani medical public sphere* challenging western medical absolutism,[28] which had denigrated and eclipsed both textual and bazaari Unain. Unani responded to the western medical absolutism of the colonial period and the communal machinations of the post-colonial period also by tactically essentialising the body.[29]
4. They put biomedicine and textual Unani under the same category for being inaccessible to the mass of the people, and also questioned Unani's efficacy. These claims would not stand the test of historical records. Arnold himself has shown the very limited reach of western medicine and that the majority of Indians accessed indigenous medicine during colonial period.[30] Efficacy of Unani and its wider reach was also evident from the evidences provided by the scholars

[23] Attewell (2007).

[24] Quaiser (2010, 525, 2001, 2012b). For more on science and colonialism see, Kumar (2006).

[25] Attewell (2005).

[26] Attewell (2007, p. 31).

[27] Quaiser (2010, 525–526).

[28] Quaiser (2010, 2012b).

[29] Quaiser (2013).

[30] See Arnold (1993).

such as Mridula Ramanna from the vast Bombay Presidency,[31] and Hume who has shown that Unani practitioners had very wide reach and they contributed significantly in epidemics control in the vast Punjab province in the second half of the nineteenth century[32] among others.
5. They strongly favour medical practices which are not "regulated by the state". Would the pure tricksters and quacks also be considered practitioners of subaltern medicine as they too manage to get their share of clients?
6. They ignored democratic, secularising and discursive character of Unani's medical knowledge, practice and moral economy concerns. This democratization was not because a large number of aspiring hakims were trained "through institutionalisation" of Unani at the beginning of the twentieth century under pressure as some scholars have argued.[33] Instead it was self-constituted democratic medico-ethical logic for long, and later the idea of *renewal* (tajdeed) instead of 'reform'.[34] For instance, Unani openly incorporated Indian local medicines and some methods of treatment into its body of knowledge since its arrival in India.[35] Similarly, in 1835–36 a glossary of Arabic-Persian Unani terms into Urdu was published for the benefit of 'attars and common people[36]; in 1861 a book was published in Urdu with ingredients and method of preparing Unani medicine for a large number of ailments for the benefit of common people.[37] This process continued into colonial and post-colonial periods.[38] The prominent historian Barbara Metcalf was not off the mark when she characterised Unani tibb as 'inherently open'[39] though it was not completely free from contradictions in its everyday practice by the ashrāfiya hakims.
7. Hardiman and Mukharji[40] have argued that today 'traditional' medicines are available within the matrix of state legitimacy. However, the matrix of legitimacy is Unani's textual principles, and state regulations principally relates to education of Unani for granting degree by a recognised university, for which

---

[31] Ramanna (2006).

[32] Hume (1977); see also, Quaiser (2010)—Science, Institution and Colonialism and (2001)—Politics, Culture and Colonialism.

[33] Liebeskind (2002) and Speziale (2010).

[34] Quaiser (2010)—Science, Institution and Colonialism, (2001)—Politics, Culture and Colonialism and (2012b)—Unani Medical Culture.

[35] Such as Ali b. Rabban's, Firdausul-Hikmat (A.D. 850) and Bahwa bin Khwas Khan's Ma'dan al-Shifa-e Sikandar Shahi written in the period of Sikander Lodhi (1489–1517) are important early texts among others.

[36] Copy on record. Manuscript.

[37] Ilaj-ul Ghuraba (1861).

[38] For example, votaries of Unani openly shared their prescriptions for the benefit of common public; they opposed the capitalist profit-making motives of western medicine by establishing Hindustani Dwakhana to provide affordable Unani medicine Quaiser (2010); Hamdard Dawakhana, the manufacturer of textual Unani medicines, at the same time published a two volume book in Urdu for rural population in in late 1950s.

[39] Metcalf (1985).

[40] Hardiman and Mukharji (2012).

curricula is designed by a group of Unani experts. In other words, research, development, therapeutics and treatment do not come under state control.[41] Thus, India state's Drugs and Cosmetic Act, licences, registration, and Good Manufacturing Practices guidelines etc. for standardisation have been continuously resisted by the textual Unani practitioners highlighting specificities of Unani manufacturing methods.[42] Thus, it remains textual Unani not biomedicine, despite statist agendas, and the 'post-colonial medical communalism' which aims to impede its growth.

5.4 Bazaar, Bazaari, 'Market': Concepts and Practices

5.4.1 Bazaar-Bazaari in Social History of Medicine

Indian subcontinental Bazaar itself has not received much scholarly attention. Some of the works on bazaar include Bayly (1983) where merchant dominates; Sen (1998) written with a focus on East India Company; Bakhshi (2009) is an account of contemporary corporate financial market and contemporary bazaar mostly relating to Western world and draws largely on Western sources; and Rezavi (2015) provides an account of state's planning to construct markets in medieval cities of Delhi, Agra and Fatehpur Sikri.[43]

Within the domain of the scholarship on South Asian medical traditions, bazaar medicine have not received adequate attention; and South Asian bazaari unani medicine is left almost unstudied. Recently, some scholars have paid attention to various dimensions of bazaar and medicine. Pratik Chakrabarti,[44] outlined the economic dimensions of eighteenth century bazaar as meeting point between local and western medicines where European merchants bought local raw medicinal ingredients to be exported to overseas. Projit Bihari Mukharji[45] has interrogated the role of the market, in the late 19th and early twentieth centuries, as place of interaction between distinct medical traditions and the crucial role in vernacularizing 'western' medicine, which this author has elsewhere called "localization and 'Unani-ization'

[41] Interview with a former Director of Central Council of Research in Unani Medicine (Hyderabad) who is and actively involved with CCURM and a trained hakim-October, 2022; and author's field notes.

[42] See Quaiser (2012a)—Tension, Placation, Complaint.

[43] Bayly (1983a, b), Sen (1998), Bakhshi (2009) and Rezavi (2015).

[44] Chakrabarti (2007).

[45] Mukharji (2009).

of western medicine".[46] Maarten Bode[47] has perceptively studied India traditional medicine's dynamic contact with modern market with a focus on shifting image. Nandini Bhattacharya,[48] has explored why adulteration assumed significance in the Indian drug market in early twentieth century. Some scholars such as Hayes, Sharma, Gaudilliere among others have dealt with the questions of restorative medicine and aphrodisiac market with a focus on advertisements in the first half of the twentieth century.[49]

However, these works have predominantly dealt with 'market' not with the pre-'market' bazaar, and with a focus on Ayurveda, and mostly in the period of entrenched colonialism.

5.4.2 *Bazaar-Bazaari Medicine and the Colonial State*

With the advent of colonial rule in South Asia both textual and the widespread bazaar-bazaari Unani faced institutionalised denigration by the western medicine, as spurious, quackery, and inauthentic. And yet, the colonial state also deployed bazaar medicine not only as cost cutting strategy but also and more importantly due its inability to provide health facilities to the vast majority of the Indians particularly in remote qasbas (towns) and villages.

Therefore, it was handy to take recourse to easily available local bazaar medicine. Several editions of Waring's book[50] on Bazaar medicine indicated its popularity. All the editions were brought out in English for European or English knowing people. It included those recipes also ingredients of which included English wine.[51] By the time the 5th edition came out in 1897 Waring included diagrams of a clinical thermometer, with a few notes to guide the non-professional in its use',[52] we can see a process of *bazaarisation* of English medicine taking place. Gradually *Allopathic* (western medicine) also made inroads into this sphere. Importantly, this domain defied also the terror of western medicine in several ways.

---

[46] Quaiser (2010). 'Science, Institution and Colonialism. However, despite such tendencies, opposition to the ideological hegemony of western medicine remained a dominant trend on the part of Unani even on disciplinary grounds. The relationship between the two knowledge systems during the colonial period was characterized by both encounter and exchange (see Quaiser 2001—Politics, Culture and Colonialism; 2010—Science, Institution and Colonialism; 2012b—Unani Medical Culture).

[47] Bode (2008).

[48] Bhattacharya (2016).

[49] Haynes (2012). "Selling Masculinity: Advertisements for Sex Tonics and the Making of Modern. Conjugality in Western India, 1900–1945," *South Asia: Journal of South Asian Studies* 35, no. 4: 787–831; Sharma (2012).
Gaudilliere (2010).

[50] Waring (1897).

[51] Ibid. pp. 183–85.

[52] Ibid. p. v.

5.4.3 *Bazaar, Bazaari and Subaltern: Concepts in Context*

The relationship between textual Unani, and bazaari-subaltern Unani—paradoxical in theory and practice though—cannot be grasped comprehensively through a blanket binary opposition between subaltern and elite. Thus, moving away from Subaltern Studies Group's understanding (Ludden 2003) the term subaltern is employed herein not strictly through and against the coloniser and colonized; instead it includes also the non-colonial local caste-class forces of domination and forms of knowledge which may often colluded with the colonial state and ideology: the disenfranchised people of low castes; how certain practices rendered low and inauthentic on caste or religious grounds; or how even certain high tradition, such as textual Unani, itself can be subjected to aspects of 'subalternity' and othered through colonial or communal technologies of domination.

Etymologically the word bazaari is derived from the word bazaar in abstraction meaning those people who earned their living from bazaar; either they have businesses, shops in bazaar, or who earned their living through engaging in different sorts of manual work, brokering, open-air sellers of assortment of locally made goods, medicine seller and healers etc. in bazaar. For example, in a 1904 personal Diary it is stated that after the plague epidemic in a Mouz'a in the province of Bihar—"Ab bazaari bhi Āte-Jāte hain" (Now bazaar people also have begun to come to bazaar).[53]

Bazaar was marked by a deep sense of conviviality, camaraderie, ease, bonhomie, respect and concerns for inter-personal relationship. A particular Calcutta bazaar in 1830s under colonial-western influence was described as "a great resort to Epicureans", but bazaars in non-urban areas or in semi-urban urban centres (qasbas) not under western influence were of different nature. It was different from the 'market' as we shall see below. However, there exists a synonymity between bazaar and bazaari sharing certain derogatory connotations. For example, bazaar medical practices were/are considered bazaari i.e. inauthentic, quackery and fraud for being outside the framework of textual Unani or western medicine. Thus, herein we have at times employed the two terms having synonymity.

However, the term bazaari acquired an independent conceptual status with specific caste-class-cultural, power of knowledge trappings, denoting not being of khawās (people or things of high culture distinct from plebeian and vulgar).

Thus, bazaari in elite-ashrāfiyia parlance is referred with disdain as it denotes to ordinary, low, vulgar, thus inauthentic—like a *bazaari 'aurat* (a woman of lax morals) who is easily available on demand therefore not be trustworthy. As a virtuous woman is pure and authentic when protected within the safety of home, similarly textual purity of the body of medical knowledge is protected and authenticity maintained within the four walls of khāndāni or trained hakims (or practitioners of canonised western medicine). And if it goes to bazaar, it will be adulterated at the hands of bazaari hakims as bazaari woman loses purity of her body. Thus, *bazaari* is a derogatory expression employed by the textual Unani tradition for subaltern Unani therapeutics, because it is beyond its textual imagination, therefore it is declared

[53] Hussainābādi, Hāshim. *Sakrāt*. Bihar Foundation. Edited by Jabir Husain.

ordinary, unprofessional, and inauthentic. Colonial state and western medicine too denigrated bazaari medicine as unscientific, spurious and quackery. And yet, even the colonial state also deployed bazaar medicine not only as cost cutting strategy in case of British officials and other functionaries but also and more importantly due its inability to reach majority of the masses as discussed above and shall further see in Sect. 5.5.

However, bazaari Unani medical public sphere does not denote to inaccessibility of specialised textual knowledge system to bazaar people—the commoners—but essentially to a different notion of body, health, medicine located in caste-class status, local natural and cultural surroundings, even if some time certain textual elements are part of it.

Patients' idea of something which is within their reach and relaxed relationship and negotiability between the practitioners and patients are crucial aspects of bazaari-subaltern practices. Bazaari-subaltern practitioners' demonstrable message that they are there with them whatever happens, is an important component. The bazaari-subaltern medical practitioners do not consider themselves inauthentic as it a knowledge handed to them by the society drawing strength from everyday practical experiences, and is ever-evolving. Thus, sources of legitimacy for bazaari-subaltern and textual traditions are different. Ultimately, these dispersed bazaari-subaltern healing practices indicate the ways in which people negotiate with themselves in differential situations of existence individually or collectively.

Bazaari-subaltern medical sphere were/are characterised by the elements of domination and denigration by the textual western medicine and Unani during colonial period and their post-colonial implications, with caste-class trappings. It contested two textual traditions by its existence, sustenance and continuation with ease in spite of opposition. However, there were marked difference between the modes of dominations of western medicine and Unani as western medicine denigrated even the high textual Unani, and Unani's vigorous opposition to it, and defence of *Desi* (indigenous) medicines. Thus, the issue under consideration cannot be comprehended through and against a neat and blanket elite-subaltern binary.

Bazaari medical practices did not exist literally only in physical bazaar, but as a parallel practice it could be found even in homes, busses, trains and other geographical and social locales such as in print medium as well.

5.5 Nuances of Bazaar, Bazaari, 'Market' and Unani

This bazaari-subaltern Unani medical sphere was different from 'medical market'. Thus, in many cases physical appearance of bazaar may necessarily not be radically different from the 'market' but it is the operational logic of the two that made the crucial difference. Bazaar was marked by a deep sense of conviviality, respect and concerns for inter-personal relationship. Exchange was not strictly driven solely by profit-oriented normative 'market' motives and rules. Exchange was convivial where familiarity between the buyer and seller played a crucial role. For example, one crucial element of bazaar was/is the practice *phā'o*—a small quantity given by the shopkeeper over and above the quantity purchased, not only to establish rapport with the customer to boost sale but also and importantly due to familiarity with the local buyers and the economic status of certain customers. Bargaining too signified the conviviality of a bazaar as opposed to printed 'fixed-price' of the 'modern' market, Thus, an early nineteenth century Urdu text bazaar is depicted as: "bazaar har eik be-āzār"[54] (bazaar is where everybody is without distress and complaint). Thus, this bazaar was/is not the Bayly Bazaar, where the merchant was bazaar and bazaar was merchant.[55] However, it was/is not completely free from money-making interests including the practice of dandi mārna (to give short weight); but it was/is predominantly relaxed with elements of conviviality and casualness and did not operate solely or primarily for commercial interests.

This conviviality of bazaar had serious implications for Unani as it formed a crucial constitutive element of bazaari Unani medical sphere. The patients and bazaari hakims, pansāri (one who sells ingredients of medicine, spices), all acted in consultation with each other in a dialogic relationship with humane and ethical concerns. Thus, bazaari Unani medicine and its practice also represented actor's view of her/his body, health, and medicine. Here actor intervenes—gives her/his opinion about the disease and medicine beyond the fixed boundaries of the written textual medical authority.[56] Organising principle of bazaari Unani sphere was/is not based on a regimented, strict normative approach to body, health and medicine, but marked by conviviality, where negotiations, familiarity, respect and concerns for others, consultations, gaining experience and learning from others without discrimination etc. were the structuring principles. Bazaari Unani is practiced in bazaar and in all other social locales including homes.

With the advent of colonialism there emerged capitalist market and marketization of medicine as well having serious implications for textual Unani during colonial period. Unani had critically opposed the capitalist market oriented ideology of medicine and health, through the establishment of Hindustani Dawakhāna in 1903 to promote authentic Unani, projected as *Desi* (indigenous) medicine at affordable price.[57]

5.5.1 Caste-Class, Economic Dimensions and Bazaari Unani

There is a definite economic, class and caste dimensions of the bazaari Unani medical practices that catered to the health needs of vast poor, illiterate, politically disenfranchised masses who constituted majority of the population in villages, qasbas (towns) and semi-urban areas. Majority of them constituted "*kameen Jāt* (low-caste mean)

[54] Copy on record.

[55] Bayly (1983a, b).

[56] For 'patient's view' see *also Roy Porter's* following pioneering work: Porter (1985).

[57] Quaiser (2010, 546–49); Dehlvi op. cit. P. 46.

peasants", and other low caste people with ascriptive occupations.[58] They did not have access to high textual Unani particularly from villages and qasbas. The colonial state with western medicine too was unable reach this vast population.[59] Thus, it was the bazaari medicine that the majority of the population related to and could afford. Bazaari Unani was/is practiced largely by illiterate or semi-literate with low socio-economic status, predominantly men in villages, towns or in some semi-urban areas and at homes. Most of them claim to have traditional knowledge of disease and local medicine transmitted from one generation to another. Some self-trained literate too practiced bazaari medicine. It is sold in bazaar or accessible easily and affordable; it is, procured, prepared and administered personally without mediation of any trained practitioner.

Recently, Waltraud Ernst[60] has done terminological and conceptual reappraisal of Medical Plurality, Medical Pluralism and Plural Medicine, which have been debated in the context of the role of Complementary and Alternative Medicine (CAM). However, there is a need to make a necessary distinction between bazaar-bazaari-subaltern healing practices and medical pluralism-plural medicine. This distinction is necessary to make as medical pluralism-plural medicine are generally used in terms of accepting/accommodating different types of already institutionalised medical practices, such as AYUSH, as matter of state policy, and CAM have the potential for being mistaken in public perceptions for bazaari-subaltern medical practices.

5.5.2 Post-colonial Turn and Bazaari Unani

In the post-colonial India the situation is further compounded with a distinct shift from the relaxed *bazaar* to 'market'. Colonially introduced profit oriented logic of medical market is further consolidated in newer forms. As a result even textual Unani too is forced to follow the logic of increasingly competitive medical market to cope with the contemporary market demands and also the damaging effects of intense medical communalism.[61]

With the increasing competitive marketization in post-colonial India textual Unani medicines have become costly; more attention is paid to packaging but marked by Unani identity; hakims are charging higher amount of fees; young hakims would like call themselves as doctor; and cross prescribing is found to some extent; and yet

[58] Mitra (2010); Dehlvi, op. cit. p, 46; author's field notes.

[59] Arnold (1993), Mridula Ramanna, Hume (1977), Quaiser (2001)—Politics, Culture and Colonialism; author's field notes.

[60] Ernst (2013).

[61] Quaiser (2012a).

moral economy dimension has not grossly diminished even today[62] particularly in the vast non-urban and semi-urban areas.[63]

However, the post-colonial India is marked by two distinct features in this regard: firstly, post-colonial Indian state's medical thinking tilted towards biomedicine; secondly, Hindutava propelled cultural nationalism and brazen communally driven well-orchestrated patronage to Ayurveda projecting it as "the Indian Medicine" has rendered even the textual Unani 'subaltern', as against hegemonic state and Ayurveda.[64] Thus, the market has been influenced by non-market factors. This signifies a situation of duality between a relaxed, democratic and secularising *bazaar* and the modern capitalist market with certain critical non-market trappings. Numerous reports and studies have provided sufficient evidence to show how communal riots aimed at destroying economic base of Muslim across sections.[65] 'Medical communalism' resulted in a crucial need to construct an essentialised body of Unani's self in the mirror of new Muslim-Unani identity on the one hand and the construction of Ayurveda-Indian state—the 'other'—in Unani's new victimised image, on the other.[66]

Consequently 'medical communalism' adversely affected even the bazaari-subalterns Unani due to general post-colonial 'everyday communalism', as in everyday situations one is almost forced to conceal one's religious or linguistic identity, where even open identification with Unani could be harmful. In 2001 author of the present essay interviewed the nonagenarian hakim Chetan Ānand who had to conceal his Unani identity when he migrated from west Punjab. His experience in the North Indian province of Haryana in 1960s of concealing his Unani identity is a telling case of the impact and extent of immediate post-colonial 'medical communalism'. He said: *Log achanak samajhne lage ki iska (Unani's) taa'lluq Musalmanon se hai—garche Tibbiya College mein ayurved section bhi tha, magar maine to uUnani section mein hi daakhila liya tha* (People suddenly thought Unani was associated with Muslims, although, Tibbiya College of Delhi had ayurved section as well, but I had taken admission only in Unani section in 1930s, he agonisingly said.

As result, in post-colonial India bazaari-subaltern Unani is faced with a situation that has produced a process of triple-subalternisation by:

1. The state's medical thinking in favour of biomedicine and Ayurveda;
2. The 'post-colonial medical communalism';
3. And the institutionalised Unani and biomedicine.

62 Intizar Hussain. 1999. P. 57; Hussāmi, hakim Qudratullah 1994. *Waqt ka Mua'lij*. Fazil Cademy, Hyderabad.

63 Author's field notes.

64 For a full-length discussion on post-colonial Medical Communalism see Quaiser (2012a).

65 Ibid.

66 Ibid.

5.5.3 *Institutionalised Opposition to Bazaari Unani*

Khāndāni, or institutionally trained and CCRUM related hakims continue to be critical of bazaari hakims in spite of state's patronage to biomedicine and Ayurveda. They considered bazaari hakims to be unreliable, inauthentic as they are not formally trained. Hakim Mohammad Ismail trained in Nizāmia Tibbia College and associated with CCRUM, asserted that "bazaari hakims know only of prescriptions and already tested medicines; they are incapable of diagnosis but only of treatment; they are either somewhat educated or illiterate)". He complained that "every dawasāz (apothecary, druggist) became a hakim, and even a pansāri (grocer) dispensed medicine."

Hakim Wahīd also a Nizāmia Tibbia College product and also associated with CCRUM talked about to other set of problems with regard to bazaari or folk medicine practitioners. He said: "Government wants to pay attention to folk medicines and AYUSH in which CCRUM are also involved. Sources of knowledge of these practitioners is based on experiences and traditional transfer of knowledge. But government wants these medicines to be tested. But traditional bazaari hakims and healers are not ready to reveal their medical knowledge and nushkha (prescriptions), which are transferred lineally. They do not want to come openly in public domain".

However, these views need to be critically examined as bazaari Unani continues to flourish in different forms as we shall further see below.

5.6 Bazaari-Subaltern Unani Medical Public Sphere: A Convivial[67] Dialogic Terrain

The domain of medicine and health has been a crucial constitutive element of social life since the very beginning of society as it is ontologically linked with life and death conditioned by class, caste, race relations, religion and the character of state and nature of society at a given point in history. Thus, in due course of time there emerged multiple medical public spheres related to the cognition of body, health and medicine.

Within the general medical public sphere in South Asia, Unani medical public spheres has been a crucial component since pre-colonial days. With the advent of colonial rule the domain of medicine and health emerged as the most potent site where sentiments against the claims of superiority of Western medical knowledge as a technology of domination were articulated.[68] And within this domain, literate Unani's opposition to western medicine was not only much before nationalist anti-colonialism[69] but also much before Ayurveda's opposition.[70]

---

[67] For the concept of conviviality see Quaiser (2019).

[68] Quaiser (2010, p. 527, 2012b).

[69] Quaiser (2010, p. 527).

[70] Quaiser (2012a, 139–40).

It is beyond the scope of this paper to go into the over-all medical public sphere constituted by multiple textual and non-textual medical knowledges and practices. This paper deals only with the neglected bazaari Unani medical public sphere. As discussed in Sect. 5.4, we have herein at times employed the terms bazaar and bazaari having synonymity.

5.6.1 Bazaari Unani Medical Public Sphere

The bazaari-subaltern Unani medical public sphere is broadly constituted by five types of practices indicating their physical areas of operation and logistics, which also signified social and economic status of their practitioners and users. This sphere, thus, represented a critical caste-class medical consciousness and a distinct perception of body and health:

1. Bazaar sphere refers to a sphere where medicines/ingredients of medicine are openly available which can be obtained by individuals on their own and practiced. Such medicines and practices have existed simultaneously alongside textual medical traditions. These are known individually or domestically or from public/neighbours or acquired lineally and which are prepared, administered and even tested by them independent of trained practitioners; these are largely self-diagnostic by untrained individuals, family members or by extended family members of neighbourhood, friends etc.; or at times people buy ingredients on the suggestions of a trained or untrained hakims but prepare and administered on their own without strictly following textual methods.
2. Bazaari hakims who sold self-prepared medicine may at times examine patients and administer medicine literally in bazaar. They call out patients loudly, describing elaborately in ornate language the uniqueness, importance, and effectiveness of their medicine; describe rarity and extreme difficulties in accessing the ingredients from far off mountains and jungles. They are generally illiterate or somewhat literate. Their methods of reaching out to their clients in substance, may be regarded as the precursor of the modern techniques of advertisements. Defining characteristic of these bazaari hakims is that they have a direct connect with people through a shared conception of body, health and medicine.
3. Knowledge of the bazaari medicine and therapeutics which are practiced at home are transmitted lineally, and also freshly acquired through various non-family sources. Some of these medicines may also have some elements of textual Unani but they are obtained, prepared and administered by untrained people including illiterate, literate and educated women.
4. Even those having some amount of non-medical formal education and practicing Unani with a matab are also termed as bazaari hakims. They are mostly from non-elite, humble family background. Their knowledge of hakimi is transferred from family tradition or gained individually. They attract large number of patients.

5. Bazaari Unani sphere includes healers such as bone-setters, *Dais* (midwife), *jarrahs* (barber surgeons), bazaari *Dandān-sāz* (dentist) etc.

In terms of expertise bazaari unani practitioners are broadly divided into two groups: (1) those who treat common ailments; and (2) those who claim to be specialist of certain ailments such as stone in liver and gall bladder, jaundice, asthma, dentistry, *jarrahi* (simple surgeries) etc.

In the following, with the help of eleven specific cases, we illustrate specificities of practices literally in bazaar and at other locales outside the bazaar covering the above four bazaari Unani practices.

As discussed in Sect. 5.4 subaltern Unani was *bazaari* because it is considered by high textual Unani with disdain as common, ordinary, low, vulgar thus inauthentic and *jaa'lsaazi* (fraud) like a *bazaari au'rat* (a woman of lax moral, therefore not be trusted). It were the illiterate or semi-literate, socio-economically and politically disenfranchised bazaari hakims and other healers such as bone-setters, midwifes etc. in villages, qasbas (towns) or in some semi-urban, or people individually obtained medicine/ingredients as home remedies, along with the formally non-trained lower/middle-range somewhat educated hakims in rural, urban and semi-urban areas who catered the medical needs of large masses. These practitioners obtained ingredients easily available in bazaar or from jungle/mountains. Bazaari Unani was/is practiced in almost all social locales such as home, neighbourhood, circles of friend, and circles of women, busses and trains. Some literate/educated people too were involved in subaltern healing practices independent of professional hakims and state. Situation seems to be quite relaxed and democratic where practitioners and patients were in a convivial dialogic relationship.

5.6.2 Case I

The following account from a Persian text by a high Muslim Indian noble of foreign origin, written in the first half of the eighteenth century, describes a scene from a bazaar in the Indian capital city of Delhi that typifies the literal bazaari Unani medical public sphere. Original Unani names in Urdu of all the diseases and medicines are omitted in this English translation:

> Those frauds who practice the profession of medicine have spread colourful sheets on the open field of the Chowk. In the shop colourful bags are arranged properly. In reality, these bags must be containing dust from the roads but not medicines. They sit in their expensive dresses and turbans. They make colourful speeches in such a manner describing the qualities of medicine that foolish people try to precede others to get those medicines to the extent that not even a speck of dust is left in the shop. Among the medicines offered you find everything—dry powdered medicine, emollient medicine, syrup, electuary medicine, medicine made up as wafer, ointment. Whatever the customer demands is available, especially the medicines to treat the adverse effects of masturbation, strengthening of penis, medicine to prolong sexual pleasure, the venereal disease—syphilis, and gonorrhoea. People of low income even pawn their clothes and even their body to buy liniment used for impotency and ointment/plaster. This master medicine seller on the strength of his stylish speech, having received money,

hands over a prescription for donkey like penis to one and the ingredients for elephant like penis to the other and these parasites who live off their wives income acquired through illicit means go home happily. Whatever the calcined metal/calx having hot effects and efficacy made of different metals you desire are available. At several places you see the fire is burning and smoke of these nine metals is reaching up to the nine skies. A crab and sāndās,[71] which are principal ingredients of ointment/plaster, tied with strings are available at several places for those who desire to have them.[72]

The above narration described bazaari hakims as j'ālsāz (frauds) for being untrained and bazaari, thus, inauthentic and unreliable. The account also revealed the aspects of choice of medicine, self-control, inaccessibility to high medical establishments. It also provided anonymity as far as aphrodisiacs are concerned and also covered poverty of respectable people. It challenged the authority of text, khādāni hakims and their families.

5.6.3 Case II

The memoirs of a Mughal prince written between 1796 and 1806 narrated the pathetic condition of 'pure Unani' at the hands of 'jāhil (ignorant) hakims' in the Nizam-Karnataka region. He states:

Hakims of Karnataka make great blunders…they know little bit of tibb (Unani) but they do not at all pay attention to pure Unani medical treatment…Listing some of their blunders, he said that, for example, they never pay attention to the blood-letting therapy, no matter even if a patient reaches near death due to plethora; Some of them think that blood-letting therapy is not suitable because in these regions little blood is produced in the body of living beings, although no authentic book would support such a view. Few hakims who want to follow the pure Unani are silenced by these jāhil people…Majority of the people—respectable and lower class—have got used to these jāhil hakims…[73]

Azfari's account outlines how away from the centre the peripheral Unani practices did not strictly follow the textual Unani and had their own discursive local flavour and experiences. For example, local hakims reluctance to use the blood-letting therapy for the reasons given is quite understandable as in the region people were ravaged by a series of catastrophic famine caused mainly by severe droughts for long including eighteen and nineteenth centuries causing huge loss of life due acute lack of food, as the historical records and Government Reports sufficiently showed.

[71] Name of an animal resembling a lizard, the oil of which is supposed to have restorative powers for gout and impotence.

[72] Khān (2014).

[73] Azfari (1957).

5.6.4 Case III

The following narrative of a village wandering kakim relates to the second half of the nineteenth century rural Bengal.[74] The narrative presents a complex web of colonially propelled agrarian structure and relations, western medicine, plots, disease, epidemics, public health and a wandering bazaari hakim. The story revolved around three key figures—the wandering hakim of the Daryaganj village, and the two talukdars—Wali Khan of village Daryaganj and Pashupati Poddar of Plashpur village, both under one zamindar.

Although the narrative is constructed from a play and locales and characters may be fictional but details and events relating to pathetic health conditions of the period in the region can be corroborated with historical evidences.[75]

The Unani in this narrative does not seem to be the kind of non-textual medical practices of Bengali Muslim peasants which were 'institutionalised' even by non-elites in early twentieth century Bengal as suggested by Projit Bihari Mukharji.[76] Unani herein is discursive Unani; and hakim seems to be self-trained. His medicinal ingredients were locally obtained and he did not follow the normative manufacturing practices of textual Unani as an illiterate housewife Gangamoni acted as hakim's assistant for preparing medicines. Hakim's most of the patients were *kamīn Jāt* (low-caste, mean) peasants.

Hakim prepared mainly two types of medicine namely Sharbat-e Humma (syrup for fever) which was used for fever, headache, nausea, vomiting, loose-motion, etc. using certain ingredients like cucumber seeds, jhau-pata (Tamarix dioica leaves), Basak-pata (Justicia Adhatoda leaves), Thankuni phool (Centella flowers), and blood-red rose petals. It was Sharbat-e Humma which was hakim's primarily medicine; and secondly, Hubb-e Jālīnūs (Galen's Pills) for improving blood circulation and general energy, but its preparation and use by the hakim is not mentioned in the text.

Names of his medicines suggest that he was somewhat aware of textual Unani. However, Sharbat-e Humma is not found in any standard qarābādīn (Unani pharmacopeia) and surely was hakim's own multi-purpose concoction, but Hubb-e Jālīnūs is.

As a wandering hakim he visited patients on his lame donkey in remote locations. During the time of malaria and kala-azar epidemics hakim on his lame donkey beating a dhol (drum) would roam around villages calling out people for medicine for fever, cough, eyesore, body ache assuring them "you will get medicine for all diseases". This was "unlike the doctor", as a character commented "today we search for doctors but earlier hakim would look for patients". Moreover, a doctor charged fees[77] that common people could not afford. Hakim got vegetable, milk, fruits, rice, pulses etc. instead of cash from the patients.

---

[74] Constructed from the Bengali play of Mitra, Op. Cit. passim.

[75] See Arnold (1993, passim) and Bose (2006).

[76] Mukharji (2011b).

[77] On the fees question see Quaiser (2001)—Politics, Culture and Colonialism and (2000)

Once when Gangamoni was preparing Sharbat-e Humma hakim appeared and asked about the ingredients used and she named them; on the amount of water used, she replied: seven and half pitcher; on the time for which the ingredients were boiled, she said: 'for the time one takes to dry hair in sun light after bathing'. Suddenly, the hakim yelled—oh! Petals of twenty blood-red roses you did not put. Gangamoni said, yes I did, to which hakim said do not tell lies—if you had put, the smell and colour would not be what it is now. She began crying and admitted her mistake, to which hakim lamented—'how will I prepare medicine without roses'! It is obvious that the hakim in order to negotiate with the pathetic health condition had to establish his authority by using Unani names for his medicines and even made false claims that rose petals are essential part of all Unani medicines.

As a result of hakim's treatment, number of deaths due to measles, pox, cholera, fever, dysentery etc. were far lower in Dariyaganj compared to Palashpur. Thus, the Palashpur talukdar would do anything to lure the hakim of Daryaganj, therefore, he planted his courtesan Mohor Jaan of Calcutta at Daryaganj to prevent hakim from the use of roses so that he will be forced to go to Plashpur for roses. Underlying factor was that because of lower death-rate, Dariyaganj had higher number of tax payers compared to Palashpur, and more tax meant that the zamindar and the colonial state were pleased with Dariyaganj talukdar.

The above narrative tells about modes of Unani therapeutics, nature of medicine, ingredients used and method of its preparation, all of which were outside the state regulation and textual Unani. Extent of bazaari-subaltern Unani's reach and its efficacy and the limited reach of western medicine. It also tells that even an illiterate Gangamoni can gain knowledge through practice. Significantly, even the bazaari hakims carried clinical trial on human as in case of Mohor Bai. The narrative poignantly brings out the ethical values and moral economy associated even with bazaari Unani practices, against the colonial capitalist market norms of profit making; and also the colonial administration's failure to provide health care to the vast masses of India even at the time of epidemics.

5.6.5 Case IV

A seventeenth to eighteenth century renowned Persian poet Mirza Abdul Qadir Bedil A'zīmābādi was also a Tabīb (physician). His "Mā'jūns (electuary medicines) and Golian (pills) that he prepared from herbs (were) famous in whole of Delhi. And his nephew earned his living by selling them".[78]

[78] Khān, Dargāh Quli. Op. cit. p. 123.

5.6.6 Case V

A personal diary in Urdu has the following entry for the year 1901 from a remote qasba in the Indian province of Bihar.

Kāzim's urinary tract and liver problems persisted. All the prominent hakims and allopathy doctors failed to treat him. I decided to consult Lucknow hakims. However, before that I met moulvi Sayed Ahfadul Hussain sahib in the qasba mosque. He told to postpone the trip to Lucknow, and suggested the following treatment.

Apply five dry earthen cups on each side of his vertebral column on the site of pain in the back for few days. And also keep giving him cooling refreshing drink. After few days Kazim was fully cured.[79]

This method of treatment is called Hijāmah bila-surt (cupping without scarification) used by trained hakims as well and also formed a part of Prophet's medicine. What is important is that the treatment was suggested by a local moulvi with no training in Unani tibb, and was administered by the untrained family members, both of them guided by local experiences and wisdom.

5.6.7 Case VI

The above diary has the following entry for the year 1907.

"Ā'zam's wife was suffering from zīqun-nafs (asthma) since her childhood. Last year it was aggravated. All treatments of Unani and doctory disappointed us. We gave up hopes of her survival. I decided to resort to my own hand-written notebook containing many old hakimi prescriptions and selected one for asthma. Having obtained all the required ingredients I personally prepared the medicine. She is finally fully cured and for the last one year she has no problem".[80] In this case as well the diary writer was not a training hakim or 'attār to prepare and administer the medicine, nor this was certified by the state.

5.6.8 Case VII

The above diary has another entry for the year 1907.

"I was very scared of the wound that developed in the parallel side of this girl. This girl's mother too had the same problem, and a doctorny (allopathic lady doctor) had scarified the wound but that did not heal, and in the third year she died of shock. This girl remembered the whole thing, and pleaded not to hand her over to a doctorny in Patna. I was greatly disturbed. At this place there was no lady physician or any competent jarrah (one who scarifies and dresses wound, a barber surgeon)

---

[79] Hussainābādi. Op. cit. P. 136.

[80] Ibid. p. 253.

and the wound was of serious nature. Relying on God's kindness, I looked into my personal hand-written note-book and selected a prescription of *poltis* (a bundle of Unani medicine wrapped in a finely woven piece of cloth) and placed on the wound and tied. By the same night the wound opened up and the pus that filled the two large flat dish came out; and using the same ingredients prepared a medicine to clean the wound; and then prepared a medicine for drying the wound. In a month's time the wound was fully healed".[81] This too was done by the same diary writer with no training in tibb.

5.6.9 Case VIII

There exist a realm of domestic bazaari unani and therapeutics which are prepared and administered by medically untrained people including women domestically.

For example, One eighty year old professor Ansari told in 2012 that his Nani (maternal grandmother) who was not educated and but was familiar and comfortable with Unani medicine and methods of treatment. She would use magnifying glass for examining motijhara khasra (measles) its texture, colour, size of pox; she diagnosed, and treated with the medicine that she learned from her family and which was prepared by herself. She used sufoof (powder) made of some agricultural product (Kashtkari safaida) and some other herb's powder to treat boils.

This sphere also includes totkas, hakim Luqman's bazaari unani medicine among others.

5.6.10 Case IX

An 80 year old Tamil Muslim man of a town in deep down south Tamil Nadu, a self-trained hakim has been practicing what he called Unani—in fact a local non-codified Unani that he learned from his family members and other locals. It was known as Unani also in his immediate locale constituted by Tamil-speaking Muslims who were drawn to pan-India Muslim politics in early twentieth century, and many of them had even learned Urdu.

5.6.11 Case X

In a regular bazaar in old Patna City (Bihar), I met a bazaari hakim Mohammad Luqmān who specialised in dentistry. Criticising greedy English-type dentists for

---

[81] Ibid. p. 256.

looting innocent people with expensive treatment leaving loopholes to force patients to come again and again.

He was selling medicine for pyorrhea, cavity and for shining white teeth. He kept the medicine ready in a puḌyā (packed in a piece of paper). First he explained the ingredients, methods of preparing the medicine and how to administer it. He said: Take tīn-chār dāne (three or four grains) of Haleelah Zard (yellow myrobalan), powder it—put this powder in one glass of water—put on fire till it becomes half of the glass—sieve it—put this water in the mouth for 5 to 10 s—rinse.

He called out people loudly explaining the efficacy of his medicine and attracted a fairly big crowd and was able to sell his medicine to several people who either endorsed him or were convinced by his explanation. In other words, those who do not want to buy the medicine can benefit from his prescription with details provided. It provided an example of how acquired knowledge is shared by passing on to others as a plea against the capitalist monopoly of knowledge.

5.6.12 Case XI

Similarly is the case of Hakim Chetan Ānand whom I interviewed in 2001 when he was in his 90s. He was originally from a village in undivided Punjab. He got himself admitted as a student in Delhi's famous Tibbiya College. But he could not complete his education but had enough knowledge to practice Unani medicine; but he "made a mixture" of textual and "*muqāmi Unani adviyat*" (local Unani medicines), *jo keh muqāmi 'ilm aur tajurbe ke hisāb se istemāl ki jāti hain* (which are used according to local knowledge and experiences). Hakim Chetan Ānand told about one Udho Das, who was a clerk-hakim in his village in undivided Punjab. Udho Das in his leisure time acted as a hakim. He prepared medicine from Unani jari-buti (Unani herbs) himself and dispensed it free of cost. He had no training from anywhere but learnt it from his village people.

Bazaari-Unani and other bazaari medicines. These cases logically lead us to address the question of how Unani is bazaari-Unani and how different it is from other forms of bazaari medicine.

Bazaari Unani is an integral part of subcontinental medical public sphere and continues to be so in spite of statist and textual Unani regulatory mechanisms since the time of the pre-colonial days. Textual Unani practices percolated down to bazaari hakims in dispersed and discursive forms. However, its conceptions of body, health and disease, and prescribed methods of preparation and administration of medicine were mediated through evolving local experience based wisdom acquired through generations; and local requirements of health and medicine and easily accessible affordable medicinal ingredients played a crucial role in the making of bazaari Unani. Although bazaari hakims were not formally trained but they belonged to a popular discursive tradition of Unani medical cultural.

For example, at the textual level, there have been many commonly used ingredients and medicines used by both textual Ayurveda and Unani and yet Unani is different from Ayurveda. This difference is because of Unani's conceptions of disease, medicine and body differ from Ayurveda, which is concretised through method of preparation; method of administration; and method of employing, meaning which disease requires which medicine and how to administer it, which also includes the knowledge of the nature of medicinal ingredients and medicines; and certain specific Unani methods such as cupping therapy (Hijāmah) and blood-letting (fasad kholna) among several others.

Thus, commonly used medicinal ingredients and medicines across the systems of medicines were determined according to the Unani principles and made them part of Unani pharmacopoeia and codex and accordingly used them, thus, they were Unanised. For example, Ajwāin Khurāsāni (carom seeds of Khurāsān[82]) and Ajwāin Desi (local-Indian Ajwāin), or Ustukhuddūs (Arabian or French lavender-lavandula stoechas) used to come from Iran and its Indian variety was a bit different, but they were adopted as per Unani principles. Thus, Unani methods and occasions of using medicinal ingredients are different from other systems. For example, Asgand (Nāgori—originally from Persia, and local Indian) is used in Unani for Waja-ul-Mafāsil (pain in joints) which was part of Unani pharmacopoeia before it came to India, but in Ayurveda it is used to modulate the immune system.

Because of wide popularity textual Unani, a Unani public sphere came into existence which percolated down to bazaari hakims and transmitted lineally or socially in discursive and dispersed forms mediated by the local health requirements and easy availability of medicinal ingredients and peoples' perceptions, which shaped bazaari Unani conception of body, health, disease and medicine, as Hakim Chetan Ānand's account in case XI and in several other illustrations discussed in Sect. 5.5 suggest. It gained legitimacy as many ingredients used in textual Unani were also used by bazaari Unani but not according to the prescribed textual methods, making it distinct according to locales, practitioners and users providing alternative options to the users.

In addition, with the advent of print culture bazaari Unani was made accessible to reading public through numerous low-cost Urdu medical and other popular journals and manuals both in colonial and post-colonial periods. Significantly, this sphere also included textual Unani made-simple to reach vast reading public with methods to prepare and administer Unani medicines, which further democratised it to serve people, which is the motto of fann-e sharīf (noble science of Unani). Thus, with the advent of print medium, the questions authenticity and legitimacy acquired new meanings as it is in print so there is a proof and answerability, thus, bazaari Unani in a way gained legitimacy.

[82] **Khurāsān**—a vast historical region comprising now lying in north-eastern Iran, southern Turkmenistan, and northern Afghanistan.

5.7 Epilogue: Making of Bazaari Unani Texts and Archives

Medical and healing practices outside the textual traditions cannot be grasped onto-epistemologically in terms of neatly drawn binary oppositions as there are grey areas that cut across various divisions even within the textual traditions. Thus, we have attempted to show that the relationship between the bazaari-subaltern Unani sphere and textual Unani cannot be comprehended through blanket binary frames as they do not facilitate but distort our understanding of medical multiplicity in the Indian subcontinent, and probably elsewhere too. Following which we have in the above outlined the significant critical dimensions constituting an epistemology of bazaari Unani sphere, which can be summed up as follows.

- The relationship between the bazaari-subaltern Unani sphere and textual Unani cannot be comprehended adequately through blanket binary frames such as elite-subaltern; state-regulated-unregulated; and one-dimensional colonial-nationalist anti-colonialism binaries, as such a framework hides more than it reveals.
- Bazaari-subaltern Unani practices as parallel terrain of medical/healing tradition outside the boundary of 'high history' have a life of their own, and did not need to be legitimised by textual Unani. These practices represent bazaari Unani way of cognising the body, health, disease, medicine and healing outside the textual traditions.
- The practitioners of Bazaari-subaltern Unani practitioners were/are described as frauds and bazaari for being untrained in the textual Unani traditions; and they did not practice in a clinic/respectable homes. Therefore, they are declared unprofessional, inauthentic, unreliable and unethical despite some complementarity and intersections. However, these bazaari-subaltern Unani practitioners do not consider themselves inauthentic as their knowledge is rooted in historical experience and draws strength from everyday practical experiences. Thus, sources of legitimacy for bazaari-subaltern and textual traditions are different. Ultimately, bazaari-subaltern healing practices indicate the ways in which people negotiate with themselves in differential situations of existence individually or collectively.
- Patients' idea of her body and disease, relaxed relationship with and negotiability between the practitioners and patients are crucial aspects of bazaari-subaltern Unani healing practices. Bazaari-subaltern practitioners' demonstrable message that whatever happens they are there with their patients, is very crucial component of this sphere.
- Bazaari-subaltern Unani sphere strikingly offer the wide range of medicines and therapeutics for wide range of ailments with non-textual experiences and knowledge. They are part of everyday life of people who are comfortable, familiar with them and which materially and spiritually 'suits' them. The bazaari Unani sphere constitute an integral part of the collective memory of common people.
- The Bazaari-subaltern Unani sphere reveals that the diagnosis of disease, preparation and administration of medicine and treatment were not the preserve only of the khādāni textual hakims and their trained disciples and those who were later

trained at Tibbiya colleges. Bazaari Unani challenged the authority of textual traditions.

- The language and strategies employed by the bazaari Unani practitioners to communicate with their audience/patients/potential buyers/users instilled confidence in them and instantly established a kind of personal rapport with them appealing their collective memory.
- The bazaari Unani sphere also revealed the aspects of patients' choice of medicine, self-control, and inaccessibility to institutionalised medical establishments.
- Bazaari sphere also provided an atmosphere of anonymity as far as aphrodisiacs are concerned and also covered the low economic status of respectable people.
- Bazaari Unani is practiced largely by illiterate or semi-literate, socio-politically disenfranchised people, men and women at home, in villages, town or in some semi-urban areas; people with traditional knowledge of disease and local medicine, and some time by literate public as well; it is sold in bazaar or available easily and procured, prepared and administered personally.
- In the post-colonial India the situation is further compounded with a distinct shift from the relaxed *bazaar* to 'market', as the colonially introduced capitalist profit oriented logic of medical market is further consolidated in newer forms. However, the post-colonial India is marked by two distinct features in this regard: firstly, post-colonial Indian state's medical thinking is tilted towards biomedicine and Ayurveda; and secondly, the intense medical communalism.[83] Thus, textual Unani too is forced to follow the logic of increasingly competitive medical market, and also to cope with the damaging effects of intense medical communalism, which consequently, adversely affected even the *bazaari* Unani.

  As result, in the post-colonial India bazaari-subaltern Unani is faced with triple-subalternisation by: The state's medical imagination and policies favouring biomedicine and Ayurveda; the 'post-colonial medical communalism'; and the institutionalised Unani and biomedicine.
- 'Epistemic ashrāfiya-morality' and associated primary casteist predispositions of ashrāfiya hakims restricted the accessibility of low-caste patients to high Unani establishments having implications for the popularity of the bazaari Unani sphere.

Making of alternative texts and archives: The bazaari-subaltern medical sphere, in a sense assume the status of a fluid parallel 'tradition' outside the domains of textual Unani and western medicines producing their own 'text' without being immanently well-ordered, fixed and complete but ever evolving.[84] Studying bazaar Unani medical-healing practices actually is a convivial act of archiving sites of alternative knowledges—and an act of developing an archive of subaltern healing practices and their 'text'. Archive and subalternity may not necessarily be antithetical. Archiving here means bringing them within the zones of recognition—recognition of authenticity, relevance, arts (and science) of negotiating with one's own self and world in the given structured situations and not dismissing them for not being part of 'high fact'

[83] Quaiser (2012a).

[84] Quaiser (2010).

and 'research'. It is not only that these practices have not been chronicled, archived and lie outside the zones of recognition but what is perhaps more important to note is that they refuse to get chronicled in a particular way. This refusal is important as they refuse to conform to the normative structures of the canonised/institutionalised traditions—in a sense that they tend to refuse to adopt the processes of *medical ashrafisation.* This actually means they provide a different, a third way, to look at the body, disease, health, medicine and the social world. '

Acknowledgements With thanks to Maarten Bode for his incisive comments on the present draft, and to David Hardiman for his valuable suggestions on the 2010a draft; thanks to hakims Mohammad Ismail, M. A. Waheed and Tanweer Ahmad for certain clarifications; and thanks to Dilip Kumar Basu, Nabanipa Bhattacharya and Uthkarsh for help.

References

Hardiman, David and Mukharji, Projit Bihari edited. 2012. *Medical Marginality in South Asia: Situating subaltern therapeutics.* Routledge

Mukharji, Projit Bihari. 2011a. Symptoms of Dis-Ease: New Trends in the Histories of 'Indigenous' South Asian Medicines. *History Compass* 9/12 (2011): 887–899.

Thompson, Edward P. (1991). The Moral Economy of the Crowd in Eighteenth Century. *Customs in Common.* New York: New Press

Tripp, C. 2006. Islam and the Moral Economy: The Challenge of Capitalism. Cambridge: Cambridge University Press.

Quaiser, Neshat. 2012a. 'Tension, Placation, Complaint: Unani and Post-Colonial Medical Communalism', in *Medical Pluralism in Contemporary India.* Hyderabad, Orient BlackSwan

Quaiser, Neshat. 2001. 'Politics, Culture and Colonialism*: Unani's* Debate with *Doctory'.* In Pati, B and Harrison, Mark (ed) *Health, Medicine and Empire: Perspectives on Colonial India*, Orient Longman

Quaiser, Neshat. 2010. Science, Institution and Colonialism: Tibbiya College of Delhi –1889–1947 in Uma Das Gupta Ed. *Science and Modern India*: *An Institutional History, c.1784–1947.* Pearson Longman

Quaiser, Neshat. 2012b. Unani Medical Culture: Memory, Representation, and the Literate Critical Anticolonial Public Sphere, in Poonam Bala ed. *Contesting' Colonial Authority: Medicine and Indigenous Responses in 19th and 20th-century India.* Lexington Books, Lanham, Maryland.

Thomas Kuhn, *The Essential Tension*, University of Chicago Press, 1977

Imre Lakatos, *The Making of Scientific Research Programmes*, Cambridge University Press, 1978, vol. 3

Bruno Latour. 1993. *We Have Never Been Modern.* Cambridge: Harvard University Press.

al-Razi, Zakariyya. 1994. *Kitab al-Murshid*. Tarraqi-e Urdu Bureau, Delhi. Urdu trns. R. I. Nadvi. P.115.

Quaiser, Neshat. 2021. 'Epistemic *Ashrāfiya*-Morality and Urdu Theatre Public Sphere in Nineteenth-Century Bihar: Muslim *Internal*-Decoloniality', *Explorations*, ISS e-journal, Vol. 5 (2).

Dehlvi, Akhlaq Ahmad. 1991. *Yaadon ka Safar.* Maktaba Aalia. Lhore. P. 46

Attewell, Guy et. al. 2012.*Agendas*, in Hardiman and Mukharji, edited *Medical Marginality.* Ironically, the first introductory essay in the volume is called Agendas, p. p 1–5.

Leslie, Charles. 1974. 'The Modernization of Asian Medical Systems'. In John J. Poggie & Richard N. Lynch (eds.), *Rethinking Modernization: Anthropological Perspectives*, 69-108. London: Greenwood Press

Leslie, Charles. 1976. 'The Ambiguities of Medical Revivalism in Modern India;. In Charles Leslie (ed.), *Asian Medical Systems: A Comparative Study*, 356-367. Berkeley: University of California Press.

Arnold, David. 2003. 'Plurality and Transition: Knowledge Systems in Nineteenth Century India', Princeton History of Science Seminar, October.

Attewell, Guy. 2007. *Refiguring Unani Tibb: Plural Healing in Late Colonial India.* Orient Longman, Hyderabad. P. 24.

Quaiser, 2012b. Unani Medical Culture. For more on science and colonialism

Kumar, Deepak. 2006. *Science and the Raj.* Oxford University Press, Delhi.

Attewell, Guy. 2005. 'The End of Line? The Fracturing of Authoritative Tibbi Knowledge in Twentieth- Century india'. *Asian Medicine: Tradition and Modernity.* Vol. 1, no. 2. 387-419.

Quaiser, Neshat. 2013. 'Essentialising the Body: Locating Medical Domination and Resistance'. *The IIAS' The Newsletter*. Autumn.

Arnold, David. 1993. Colonizing the Body: State Medicine and Epidemic Disease in Nineteenth-Century India,Berkeley and Los Angeles, University of California Press. P. 3.

Ramanna, Mridula, 2006, 'Systems of Medicine: Issues and Responses in Bombay Presidency', *Economic and Political Weekly*, July 22,

Hume, J.C, 1977, 'Rival Traditions: Western Medicine and Yunan-I Tibb in the Punjab, 1849–1889' *Bulletin of the History of Medicine,* Vol 51, pp 214–231

Liebeskind, Claudia. 2002. 'Arguing Science: Unani *Tibb,* Hakims, and Biomedicine in India, 1900-50' in Waltraud Ernst ed. *Plural Medicine, Tradition and Modernity, 1800-2000,* Routledge, London, pp 58-75

Speziale, Fabrizio. 2010. 'Muslim or Greek? Past and Present of YUnani Medicine in the Deccan'. In Rais Akhtar and Nilofar Izhar (eds.) Global Medical Geography: In Honour of Prof. Yola Verhasselt, 305–332. Jaipur. Rawat Publications.

Metcalf, Barbara. 1985. 'Nationalist Muslims in British India: The Case of Hakim Ajmal Kham' *Modern Asian Studies*, Vol. 19, No. 1, pp. 5; 3–9.

Hardiman, David and Mukharji, Projit Bihari. 2012. 'Introduction', *Medical Marginality in South Asia: Situating subaltern therapeutics.* Routledge. P. 11.

Bayly, Christopher A. 1983. *Rulers, Townsmen and Bazaars: North India Society in the Age of British Expansion. 1770–1870* New Delhi

Sen, Sudipta. 1998. Empire of Free Trade: the east India Company and the Making of the Colonial Marketplace. Philadelphia

Bakhshi, Rajni. 2009. *Bazaars, Conversations & Freedom: for a market culture beyond greed and fear*. Penguin India. Delhi

Rezavi, Ali Nadeem. 2015. 'Bazars and markets in medieval India', *Studies in People's History*, 2, 1: 61–70.

Chakrabarti, P. 2007. 'Medical Marketplaces Beyond the West: Bazaar Medicine, Trade and the English Establishment in Eighteenth-Century India', in M. S. R. Jenner and P. Wallis (eds.), *Medicine and the Market in England and Its Colonies, c. 1450–c. 1850* (Basingstoke: Palgrave Macmillan, 216–37).

Mukharji, P. B. 2009. *Nationalizing the Body: The Medical Market, Print and Daktari Medicine* (London: Anthem Press.

Bode, Maarten. 2008. *Taking Traditional Knowledge to the Market: The Modern Image of the Ayuredic and Unani Industry -1980–2000.* Orient Longman, Hyderabad.

Bhattacharya, Nandini. 2016. Between the Bazaar and the Bench: Making of the Drugs Trade in Colonial India, ca. 1900–1930. *Bulletin of the History of Medicine,* vol. 90, no. 1.

Haynes, Douglas. 2012. "Selling Masculinity: Advertisements for Sex Tonics and the Making of Modern

Sharma, Madhuri. 2012. *Indigenous and Western Medicine in Colonial India* (New Delhi: Foundation Books);

Gaudilliere, Jean-Paul. 2010. "The Visible Industrialist: Standards and the Manufacture of Sex Hormones," in *Evaluating and Standardizing Therapeutic Agents, 1890–1950*, ed. Jonathan Simon and Christoph Gradmann (Basingstoke: Palgrave Macmillan, 174–20.

Waring, EdwardJohn. 1897. Remarks on the Uses of Some of the Bazaar Medicines and Common Medical Plants of India: With a Full Index of Diseases, Indicating Their Treatment by These and Other Agents Procurable Throughout India. J & A Churchill. London.

Bayly, C. A (1983) Rulers, Townsmen and Bazaars

Porter, Roy. 1985. 'The Patient's View: Doing Medical History from below', *Theory and Society*, Vol. 14, No. 2, pp. 175-198.

Mitra, Manoj. 2010 [1994]. *Galpo Hekimsaheb* (Tale of Hakimsaheb). Kalabhrit Publishers, Kolkata

Ernst, Waltraud. 2013. Medical Plurality, Medical Pluralism and Plural Medicine, *Medizin, Gesellschaft und Geschichte*, Supplement. 46: 11-23.

Quaiser, Neshat. 2019. 'Shari'a Politics, 'ulamā and Laity Ijtihād: Fields of Normativity and Conviviality', *South Asian History and Culture,* Volume 10. Issue 2.

Khān, Dargāh Quli. 2014(1926). *Safar Nāma Dargāh Quli Khān.* Tr. Khaja Abdu Hameed Yazadaani, Translated as Murraqa-e Dehli. Sang-e Meel. Lahore. 26–27.

Azfari, Mirza Zahirud-Din Ali Bakht. 1957. *Waqi'at-i Azfari.* Edited by T. Chandrasekharan and Syed Hamza Hussain Omari (Madras: Government Oriental Manuscripts Library, 1957). Persian, pp. 214–16, passim

Bose, P. K. 2006. Health and Society in Bengal, A selection from late 19th Century Bengali Periodicals, Sage, New Delhi, 2006, Introduction, 10-46

Mukharji, Projit. 2011b. Lokman, Chholeman and Manik Pir: Multiple Frames of Institutionalising Islamic Medicine in Modern Bengal. *Social History of Medicine.* Pp 1–19.

Quaiser. N. 2000. Colonial Politics of Medicine and Popular *Unani* Resistance, *Indian Horizons*, Vol. XXXXVII, No. 2

Chapter 6
Medicinal Practices of a Downtrodden Castes: Reflections on the Folk Medicinal Practices of the Rajbanshis of Sub-Himalayan Bengal

Rup Kumar Barman

The practice of 'folk medicine' is a global phenomenon. 'Folk medicine', often termed 'traditional medicine' is widely used as health-seeking behaviour all across the world especially where mainstay health care is not available. It is often underestimated as a part of health care. However, traditional medicines of proven quality, safety, and efficacy have drawn much attention in the domain of health care. From a historical perspective, 'folk medicine incorporates healing practices and ideas on health care which are confined to a particular group in a culture, and are usually transmitted orally or word of mouth.'

Most of the primitive communities with their interactions with nature had gradually developed techniques and methods for healing with the available plants, animals, and other naturally occurring substances. The ancient civilizations of China, India, Greece, the Arabian World, Egypt, South East Asia, and many other countries also contributed to the growth of institutionalized classical medicine (like *Ayurveda*, *Unani*, *Siddha*, Naturopathy, *Yoga*, and Chinese classical medicine). Side by side, little-known and un-institutionalized practices particularly, folk streams of traditions, continued to survive among the indigenous communities. In this context, the Sub-Himalayan Bengal (Jalpaiguri, Alipurduar, and Cooch Behar districts of West Bengal) are no exception. Here, several indigenous communities have maintained their own therapeutic traditions. Among such communities, the Rajbanshis are considered to be the most significant one because of their wide knowledge of folk medicine, preventive food habits, folk culture, and health-seeking behaviour. In the pre-colonial and early colonial days, common and poor Rajbanshis of this region were solely dependent on their own community medicine.

R. K. Barman (✉)
Department of History, Jadavpur University, Kolkata, India
e-mail: rupkumarbarman@gmail.com

S. Ahmed and S. Sarkar (eds.), *Decolonizing Science and Modernity in South Asia*,
https://doi.org/10.1007/978-981-97-1829-0_6

However, with the introduction of modern medical practices, particularly with the development of indirect British rule in the Cooch Behar State (1773) and direct British rule in Jalpaiguri (in 1869), the traditional healing practices of this region began to be discouraged. In spite of it, the Rajbanshis of remote villages of Sub-Himalayan Bengal; have maintained their traditional knowledge of medicinal plants, herbs, and food practices. This article seeks to highlight the folk medicine of the Rajbanshis of colonial Sub-Himalayan Bengal and their continuity in the post-colonial period, along with the common diseases of the region and the attitude of the Rajbanshis towards them.

6.1 Geographical Location and Communities of the Study

The region of the present study is Sub-Himalayan Bengal, i.e., three districts of the northern part of West Bengal. These districts are Cooch Behar, Jalpaiguri, and Alipurduar. Socio-culturally, these districts were and still are closely attached to lower Assam and northern Bangladesh (particularly of the Rangpur division). So geographically this region is both inter-state and international in character.

This region possesses a rich plural culture because of the fusion of diverse racial and cultural communities as recorded in the pre-colonial Sanskrit, Persian, and other indigenous literature. The *Kalika Purana* (twelfth century CE) has categorically illustrated that the *Kiratas* or *Mlechchhas* were the original dwellers of Pragjyotisha-Kamarupa.[1] The *Yogini Tantra* (eighteenth century), on the other hand, has described them as *Kuvachaka* (evil speakers) or non-Aryan people.[2] The indigenous people as described in the classical texts are the Scheduled Caste and Scheduled Tribe communities of present nomenclature [such as- Koch, Mech, Kachari, Bhutia, Rabha, etc.].

The Rabhas and the Meches (Bodos) have their special concentration in Jalpaiguri, Cooch Behar, and Alipurduar districts. The Totos (of Totopara of Alipurduar), Garos, Tamangs, Limbus, etc. are other smaller indigenous communities of Sub-Himalayan Bengal. There are a few more immigrant tribal communities such as the Santhals, the Oraons, the Mundas, the Malpaharias, the Mahalis, the Lohars, etc., who had migrated to this region in the colonial period from Chhoto Nagpur region [mainly as tea plantation labourers].[3]

The Rajbanshis constitute the largest Scheduled Caste community of Sub-Himalayan Bengal. There are a few other Scheduled Caste communities in the region, immigrants (Namasudras, Malos, Kaibartyas, etc.) and indigenous (Damai, Sarkis, and Kami). The rest of the people of this region are OBCs (Muslims and Hindus), General (Hindus, Christians, and Jains, etc.), and tribal Buddhists.

[1] Panchanan Tarkaratna (ed), *Kalika Puranam,* Calcutta: Naba Bharat Publishers, 1384 B.S.

[2] Swami Sarveswarananda Saraswati (ed), *Yogini Tantra,* Calcutta: Naba Bharat Publishers, 1385, B.S.

[3] Barman (2007).

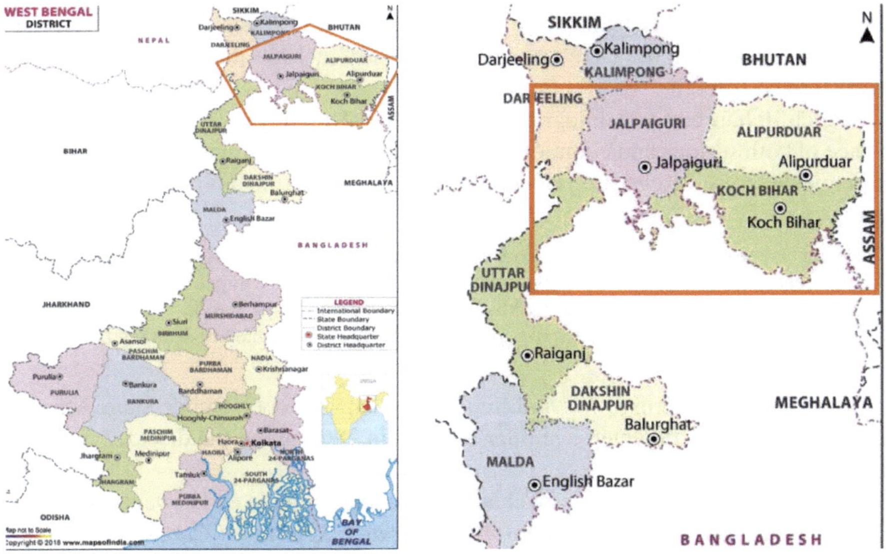

Map 6.1 The Sub-Himalayan Bengal and the region of concentration of the Rajbanshis

Among all these communities in this article, we are mainly concerned about the folk medicines of the Rajbanshis who are predominantly rural people (92.35%). In Cooch Behar, they share 37.67% of the total population [1,063,369 out of a total of 2,822,780 as per the 2011 census] of the district while in Jalpaiguri (including Alipurduar) they are 23.84% of the total population [922,603 out of 3,869,675 as per the census of 2011]. However, there are many Muslim Rajbanshis locally called Naishya Shaikhs. Culturally and linguistically, they are just like the Scheduled Caste Rajbanshis (Map 6.1).

6.2 Diseases of Sub-Himalayan Bengal

Sub-Himalayan Bengal especially the Duars region [Alipurduar and northern part of Jalpaiguri], is extremely popular among the tourists. Its hills and foothills are covered with jungles and sated with wild animals and trees. Its natural beauty has been a matter of attraction for outsiders since the early twentieth century. However, in the colonial days, certain common but fatal diseases of the Duars region were noted as unhealthy. Upendra Nath Barman [(1898–1988), (former MP of Cooch Behar (1952–62) and a well-known Rajbanshi political leader] in his autobiography (1985) had recorded that the educated people of colonial South Bengal used to think twice before joining any service in the Duars area because of its weather.[4] So the

[4] Barman (1895).

Government of Bengal had to introduce a 'special allowance' for the government servants deployed in the Duars region. It was only in 1969 that the Government of West Bengal lifted it.[5]

Cooch Behar, being an integral part of the Sub-Himalayan Bengal had/has experience of both serious and common diseases caused by the climate, geographical and biological features. Regarding Cooch Behar, W.W. Hunter had observed in the 1870s that:

> the chief diseases that prevail in Kuch Behar are dysentery, malarious fevers, spleen, goitre, and venereal diseases. The first two may be attributed to the dampness of the soil and humidity of the climate, together with the exclusive heat during the day in the summer months, followed by the sudden cold after sunset. Goitre is attributed to drinking the water of certain streams[6]

The Deputy Commissioner and Superintendent of Cooch Behar State, Mr. T.H. Lewin had also documented a similar account in 1876 about the diseases of Cooch Behar.[7]

The history of diseases did not change its character in the early twentieth century Cooch Behar too. From *the Annual Administrative Reports of Cooch Behar State*[8] and the monumental report of Chaudhuri (1903),[9] we do notice that cholera and smallpox were quite common in Cooch Behar[10] (which were locally called *Garmari* and *Thakurani* respectively).[11] Leprosy used to be found among the Rajbanshis on a large scale.[12] Among other common diseases, mention may be made of fever, hooping cough, dyspepsia, goitre or Derbyshire-neck (locally called *ghag*), blindness, deaf-mutism, and insanity. The rate of insanity was highest in Cooch Behar. It was common among the Rajbanshi Muslims.[13]

Cholera, smallpox, and malarial fever continued to be a matter of concern for the Cooch Behar State because of their fatal character and the inadequacy of medical facilities. Thus, in its *Annual Report of 1907–08*, the Cooch Behar State recorded that:

> the Sudder and the Mathabhanga Sub-Divisions were much affected by the outbreak of cholera during the year of 2167 occurred in the above two Sub-Divisions small pox was prevalent in the Sudder Sub-Division of 665 cases reported during the year as against 132 in the previous year, 619 occurred in Sadder. Cases of malarial fever numbered 10,668 against 9564 in 1906–07.[14]

---

[5] Mandal (2011).

[6] Hunter (1876).

[7] Lewin (1876).

[8] '*Annual Reports of the Cooch Behar State*' had been published regularly from the 1880s to 1946 which are available in North Bengal State Library (Cooch Behar).

[9] Chaudhuri (1903).

[10] W.W. Hunter, op.cit, *Vol. X.*, p. 441.

[11] H.N. Chaudhuri, *op.cit.*, p. 149.

[12] *Ibid.*, p. 150.

[13] *Ibid.*, p. 151.

[14] *The Annual Administration Report of the Department of General Administration & Criminal justice of the Cooch Behar State for the year 1907–1908* (1908).

Upendra Nath Barman's autobiography has also recorded cholera and smallpox as an epidemic in Mathabhanga Sub-Divisions in the first quarter of the twentieth century.[15]

Because of its close proximity with Cooch Behar Jalpaiguri (including present Alipurdur) had some similar features in regard to the common diseases. In the mid-nineteenth century, during his travel through Jalpaiguri (3rd–4th March 1854), Mr. J.D. Hooker found that Jalpaiguri was very much prone to malaria fever because of its climate and weather.[16]

The colonial administrative reports of Jalpaiguri starting from Hunter (1876–77) to Milligan (1919) have vividly illustrated the general health condition and common diseases of Jalpaiguri district. W.W. Hunter recorded in the 1870s that:

> the principal endemic disease of the district of Jalpaiguri is malarious fevers, both remittent and intermittent, with all their complications and squeal. These fevers are attributed to malaria generated by the decomposition of the rank vegetation which abounds throughout the District. They are most prevalent at the beginning and end of the rains, in the months of March and April, and in September and October….. Splenetic affections, enlargement of the liver, anemia, and anasarca, all of malarious origin, are common throughout Jalpaiguri.[17]

A first-hand report on the common diseases can be found in the report left by Sunder (1895). During his survey and settlement in Western Duars of Jalpaiguri district (1893–1895), Mr. Sunder categorically noted:

> Cholera occasionally occurs in the cold weather and in the months of April and May; spleen and goitre are common diseases, and many people die annually from fever and bowel complaints[18]

J.A. Milligan, the settlement officer of Jalpaiguri (1906–1916) had also illustrated that:

> till about ten years ago, the intensely malarious climate of the Western Duars was accepted with resignation by the inhabitants, European and Indian alike, as an inevitable part of their environment. The death rate must have been enormous, and few, if any, escaped that weakening of the constitution and the impaired vitality with prolonged saturation which malaria inevitably produces.[19]

The Malaria Commission of the Royal Society visited the Duars region of Jalpaiguri in 1901 to understand the severity of malaria in the Sub-Himalayan Bengal. The commission [consisting of Dr. Stephens, Dr. Christophers, and Captain S.P. Jones, I.M.S.] found that "the malaria endemicity of Duars was extraordinarily high and that black-water fever was common in that locality."[20]

However, malarious fever was not a serious problem for the indigenous communities of this region [like the Meches]. But the migrants from other parts of India

---

[15] Upendra Nath Barman, *op.cit.*, pp. 19–20.

[16] Kusari et al. (1981).

[17] W. W. Hunter, *op.cit*, p. 321.

[18] Sunder (1895).

[19] Miligan (1919).

[20] Gruning (1911).

and the Europeans were the worst sufferer of the malarious fever of Duars. It is to be noted that the population was increasing in the Duars region of Jalpaiguri with the growth of tea gardens and clearing of jungles since the 1870s. The Rajbanshis of Sub-Himalayan Bengal, however, were not the victims of the malarious fever. So J.F. Grunning had recorded in 1911 that—"the adult native population of Rajbanshis, Muhameddans, and Meches are relatively immune to fever."[21]

Diarrhea, dysentery, goiter (called *bawa* in Duars), scurvy, cholera, etc. were other common ailments in Jalpaiguri in the late nineteenth century. Cholera broke out as an epidemic in Jalpaiguri in 1869. Another attack of cholera took place in Jalpaiguri in 1870.[22] A severe attack of cholera causing the highest death rate in Jalpaiguri was recorded in 1924–25.[23] But the natives of the region could not develop proper preventive methods against cholera. However, the Rajbanshis have developed a specific folk medicinal practice in case of the other common ailments with their day-to-day experience. Their attitudes to the diseases along with their folk medicinal practices will be discussed in the subsequent parts of this chapter.

6.3 The Attitude of the Rajbanshis Toward Diseases

Indigenous communities have developed folk medicinal practices along with specific attitudes towards the diseases prevailing in their society/locality. Here, we do notice three specific attitudes—(a) healing through folk medicine, (b) ailment by magic and propitiation of spirits, and (c) discarding the incurable patients or keeping them isolated. Most of the Indian indigenous communities used to think that diseases and deaths are caused by certain spirits or supernatural power. Sometimes their faith got placed in their belief system.[24] These features had a considerable presence in the views of the Rajbanshis toward diseases.

From the pre-colonial sources, it has been found that discarding patients with incurable diseases and self–immersion in the sacred river were common among the Rajbanshis. The earliest reference to this attitude of the Rajbanshis was in the '*Darang Raj Vamsavali*' which described the death of Chilarai, the Koch military general of the sixteenth century. It describes that being affected by smallpox Chilarai died in 1571 CE on the bank of the Ganges.[25] He was discarded by his followers. So he dunked himself in the sacred river to get eternal peace.[26] According to *Darrang Raj Vamsavali:*

> And then the prince got a high fever because of smallpox. Pox spread all over his body. He gave up food and water and thought about the Ganges, the sacred river. He thought he would

[21] J.F. Gruning, *op.cit*, pp. 46–47.

[22] W.W. Hunter: *op.cit.*, p. 271.

[23] Abani Mohun Kushari et al., *op.cit*, p. 273.

[24] Pal and Jain (1998).

[25] Daivagna (1973).

[26] *Ibid. vs 588–90,* p. 120.

> die and it is the divine order since it has no remedy and it is incurable…With these words and by invoking the name of Narayan, the Lord of the Universe, he took a holy dip in the Ganges and died.[27]

So, it seems that self-immersion in the sacred river to get eternal peace had developed this tradition. Similarly, they had also developed the tradition of 'home isolation' for patients with infectious diseases. Thus H.N. Chowdhury recorded in 1903 that:

> when any member of a household is attacked with cholera, the house is shunned by the neighbours, and even the friends of the sufferer fear to approach him and nurse him. The victim is left in a hut with a pot of water to shift for himself, so long as strength lasts, and at last, dies a horrible death without sympathy and care. It is no wonder therefore that cholera often desolates entire households and towns or villages.[28]

On the other hand, the Rajbanshis had developed the tradition of propitiating certain spirits considered to be responsible for epidemics such as cholera and smallpox. Thus, Buchanan Hamilton's report (1808–1814) shows that "bites of serpents are cured in the name of *Bishahari*, and smallpox in the name of *Shitala*".[29] This trend has also been noticed in the twentieth century. The colonial records and memoirs show that the Rajbanshi villagers used to believe that the disease is the outcome of the bad gaze of the malevolent spirits/demi-gods/deities [called *deo, jokha, jokhi, thakurani*] such as—*Masan Deo, Padma, Kali, Chandi, Bishahari Kanduni Deo, Mina Thakurani, Bao Deo, Balaram-Hanuman-Ram, Chamok Deo, Gorakhnath, Basanta Thakurani*, etc.[30] And for their propitiation, the Rajbanshis followed magical means of healing and exorcism. At the same time, water and oil sanctified by village *ojha/gunin* (exorcist) were also used for healing. So Upendra Nath Barman's biography recorded that:

> Exorcism and propitiation of spirits [considered to be responsible for the outbreak of smallpox] were very much common in Cooch Behar State mainly because of traditions and non-availability of modern medicine.[31]

This trend actually was the outcome of the non-availability of modern medical facilities like hospitals, doctors, and dispensaries. After the formation of Jalpaiguri (in 1869), the district had only one hospital in Jalpaiguri town which could receive only 32.86 patients per day.[32] However, Jalpaiguri got a modern hospital in 1905 for which the average daily number of patients in indoors increased to 23.98 in 1907 from 9.12 in 1905.[33] The total number of indoor patients rose to 758 in 1905 while the number of outdoor patients in Jalpaiguri was 9215 in 1907. In the same year, the number of outdoor patients in Alipurduar and Falakata charitable dispensaries

---

[27] Translation by *self* from the *Darrang Raj Vamsavali vs*, pp. *588–590*.

[28] Martin (1976a).

[29] H. N. Chadhuri, *op.cit*, p. 150.

[30] Upendra Nath Barman, *op.cit.*

[31] *Ibid*, pp. 19–20.

[32] Abani Mohuni Kushari et al., *op.cit.*, p. 273.

[33] *Ibid*, p. 274.

was 7238 and 6839 respectively. The district had two other charitable dispensaries at Maynaguri and Kumargram which treated 5325 and 3033 outdoor patients in 1907 respectively.[34]

The condition of Cooch Behar State in the last quarter of the nineteenth century was comparatively better. *The Annual Administrative Report of 1893–94* has noted that Cooch Behar State had altogether eight charitable dispensaries. These were—(i) The Sudder Charitable Dispensary, (ii) Dinhata, (iii) Mathabhanga, (iv) Mekhliganj, (v) Haldibari, (vi) Phulbari, (vii) The Jail Hospital, Sudder, and (viii) The Police and Military Hospital, Sudder. Haldibari and Phulbari (Tufanganj) had no indoor facilities.[35] The number of patients, both indoor and outdoor, had increased considerably in the dispensaries of Cooch Behar State. Even the growing popularity of hospitals had encouraged the doctors of this small Princely State to perform 34 operations in 1907–08.[36]

But the government hospitals and a few charitable dispensaries in the Cooch Behar State and Jalpaiguri district were in no way sufficient for the rural people. The lack of modern medical facilities had compelled the Rajbanshis of rural Sub-Himalayan Bengal to be more dependent on magical means of healing and worshiping the spirits and demi-gods, and maintaining the folk medicine.[37]

6.4 Folk Medicine of the Rajbanshis

According to the World Health Organization (WHO):

> Traditional medicine is the sum total of the knowledge, skill, and practices based on the theories, beliefs, and experiences indigenous to different cultures, whether explicable or not, used in the maintenance of health as well as in the prevention, diagnoses, improvement or treatment of physical and mental illness.[38]

From this broader view, researchers on indigenous medicine have identified a few common features of folk medicine which are quite different from institutionalized classical medicine. The main features of folk medicine are—(a) it is oral and transmitted through oral traditions, (b) it is based on locally available fauna, flora, and minerals, (c) it encompasses important areas of healthcare like mother and childcare, treatment of common ailments, first aid, nutrition, home remedies. It also deals with broken bones, poisonous bites of snakes and other animals, veterinary care and treatment (d) 'Nature-specific community culture is related to healthcare, and (e) it also practices a magical-religious form of healing.

[34] *Ibid.*

[35] The Annual Administrative Report of the Cooch Behar State 1893–94 (1894).

[36] *The Annul Administration Report of the Department of General Administration & Criminal justice of the Cooch Behar State for the year 1907–1908* (1908).

[37] W.W. Hunter, *op.cit,* p. 323.

[38] World Health Organization (2013).

In such a context, we'll now analyze the folk medicinal practices of the Rajbanshis. We have found that geographical features have an immense influence on the pattern of their habitation, belief systems, folk medicine, and magical means for miraculous healing. So, we have classified the herbal treatment of the Rajbanshis into two groups—(i) the use of herbs for healing, and (ii) consumption of herbs and vegetables as preventive food habits.

6.5 Use of Herbs for Healing

It has been already exemplified that till the early twentieth century, the Rajbanshis of the isolated villages (except the educated sections) were solely dependent on their folk medicine. So, the Dewan of the Cooch Behar State wrote in 1903 that 'people very seldom use either Western or *Ayurvedic* medicines. They are ordinarily treated by the *Ojhas* or villages quaks, who in addition to offering *pujas* to offended *deo* or spirit, administer some native drugs.'[39] With the local drugs and herbs, the Rajbanshis used to deal with all kinds of ailments including smallpox, chicken pox, and measles.

In the pre-colonial and colonial days, pox (the Rajbanshis call them *Thakurani, Guti puskura*) was very much dangerous. They used to treat this syndrome by using certain plants and their sub-products. Consumption of a blend of juice of *neem* (*Azadirachta indica*) with honey in a specific dose in the early morning on an empty stomach was medicine for pox. To cure, ulcers generated from pox, they do use a blend of *neem* and mustard oil (extracted by traditional method). Concurrently, water with leaves of *chepti kalai* and *neem* is used for bathing the patient.

The most common disease in the Sub-Himalayan Bengal was/is the fever of different kinds. Thus, treatment of fever with available herbs has developed among the Rajbanshis. Adult Rajbanshis were immune to the malarious fever. But other forms of fever were quite common in Cooch Behar and Jalpaiguri. For cold fever, the Rajbanshis generally use a blend of *tulsi* (*Ocimum sanctum*) leaves and honey. Juice extracted from the leaves of *basak (Adhatoda Vasica)* is used to be considered as medicine for cold fever. The blend of *kalo jeera (Nigella sativa)*, garlic, and mustard oil was/is also used in cold fever. Leaves of *shiuli* (*Nyctanthes arbortristis*) are useful in case of black fever. During our fieldwork, we have noticed the healing of different kinds of fevers by the Rajbanshis including *Pala jwar*.

Diarrhoea and dysentery were/are quite common in the Sub-Himalayan Bengal because of the quality of water. In case of bowel complaints due to diarrhoea, the Rajbanshis use the juice of *patharkuchi (Kalanchoe pinnata Pers)* leaves and *gandhabhadal* leaves. In dysentery, they prescribe the use of the *thankuni (Centella Asiatica)* leaves, *boalida* leaves, a blend of *golmorich* (black pepper), ginger, and leaves of *sangibhar*, etc. For healing dysentery, they use two other methods.

---

[39] H.N. Chaudhuri, *op.cit.,* pp. 149–151.

(a) Blend or mixture of *Atia Kala or Beechi Kala* [a kind of banana full of seeds, Scientific name: *Musa Paradisiaca*] is prescribed to the patients.[40]
(b) Blend of green leaves of *bhant plant* (*Clerodendrum vicusum*) and green leaves of guava is prescribed (in tablet form) by the healers [see prescription].

For blood clotting from an injury, the Rajbanshis generally use the serum of *kalokachu* and leaves of marigold. The juice of *bisallakarani* is also very much helpful in case of bleeding from injury. The use of extracted juice from marigold leaves, *udarbhusi* plats and *Durbba* grass is also very much common among them. And it is very much effective.

Jaundice is another common complaint in the Sub-Himalayan region. To the Rajbanshis, jaundice is known in different names [like *Kumli, Kumali, Kamela, Haldiberam, Kaun*, etc.]. They have developed different methods for healing jaundice. The commonly used methods are as follows:

(a) They do generally use *jambural batabi lebu (Citrus maxima)* and the juice of bark of *sajinal sajna (Moringa oleifera)* tree, *swarnalata, neem, patal* leaves and *tauri kelai*, etc. as medicine.[41] They do also suggest the blend of green leaves of *kala khesari* and milk of goat in the early morning.
(b) Second method for healing jaundice is a combination of exorcism, medicine and therapy. In this method, they perform a ritual called *bharan*. This ritual is performed in the early morning and the patient is asked to keep a blend of leaves of *Bandar Nauka* plant on his head for two to three hours. In case of non-availability of *Bandar Nauka*, they prefer to use the blend of mustard oil, *bhogmana* (a kind of arum). Along with this blend they do prescribe incantation.

It has been already mentioned that hooping cough was a common disease in Cooch Behar. The Rajbanshis do generally use the blend of juice of *basak, labanga* (*Syzygium*), *elachi (cardamom)* and *kalojera (Nigella sativa)* in a prescribed dose as medicine for healing hooping cough. In case of ulcers and wounds, the mixture of *neem*, oil extracted from *endi* tree, *kelai khesary* and mustard oil blend are the traditional medicines for them.

In our fieldwork, we have noticed that the folk medicine of the Rajbanshis is largely based on the locally available plants and their sub-products including roots, fruits, flowers, bark, etc. While some of the plants are well-known to others, some of them are very much indigenous. On many occasions, we could not trace their botanical name. Among the known drugs *basak (Adhatoda zeylanica), neem (Azadirachata Indica), bhant (Clerodendronviscosum), ishanmul (AristolochiaIndica), satamuli (Asparagus Sarmentosus), pudina (Mentha Sativa), anantamul (HemidesmusIndica), gulancha (Tinospora Cordifolia) dhatura (Datura Stramonium), upang (Achyaranthesaspera), sialkanta (Argemone Mexicana), amla (Phyllanthus emblica), ajawan,*

[40] Interview with Jagannath Roy, Village & Post Office: Rasikbill, Police Station: Baxirhat, District: Cooch Bihar, West Bengal. Date of interview: 21.08.2017. Interview was taken by Dr. Debabrata Roy.

[41] Interview with Jatneshwar Barman, Village: Manta Para, Post Office: Bara Shalbari, Police Station: Baxirhat, District: Cooch Behar, West Bengal. Date of interview—19. 08. 2017. Interview was taken by Dr. Debabrata Roy.

*sajina (Moringa oleifera) ada (Zingaiber officinale), rasun (Allium sativum), tentul (Tarmarindus indica), kala Sarisaha (Brassica nigra), sada Sarisha (white mustard seed), Pipul (Ficus religiosa), amrul (Oxalis corniculata), haritaki (Terminalia chebula), erendi, anar, golmarich (Piper nigrum), long, jasthi madhu, bara elach, kabab chini, ban tulsi, ram tulsi (Ocimum americanum), kalo jira (*Nigella Sativa*), satamul (Aspaeagus racemosus), shiuli (*Nyctanthes arbor-tristis*), chhatim (Alstonia scholaris), halud (Curcuma longa), bel (Aegle marmelos), lajjwabati (*Mimosa Pudica*), shimul (*Bombax ceiba*), tejpata, (Cinnamomum tamala),* etc. are important in the folk medicinal practices of the Rajbanshis.

In our fieldwork we have also noticed the use of local plants in folk medicinal practices of the Rajbanshis. Among them mention may be made of *kankisa/dandakalas (Leucas cephalotes), shati/shuti, biskatali, kanta khuria, mongamari, mahabali/buch, kanai dinga [surimala/dingdinga], gultai, jharua narshing, dhola manamani, joka, pundi, darua haldi, bishmana, dharamana, charakmuni, bahutara, jatrasishi, havvina bish, gakghurkanta, jayatri, chai, kalamichri, damad, padma gulancha, pankira, bhuintinta, bakulkanta, hatubhnga, hanru gach, tokma, dhhulai, bhogmana, kur haldi, chanda, bhat, kechua bish, japsiri, bankapasi, hendar pata, rakhal nadu, kachuripana* (water hyacinth), etc.[42] The use of these plants in folk medicinal practices among the Rajbanshis is still prevalent.

6.6 Use of Mineral Drugs Among the Rajbanshis

Along with leaves, roots, bark, fruits, flower of different plants, the Rajbanshis of the Sub-Himalayan Bengal use different types of mineral drugs. Among them *phutkuri (Allum), rasanjan (Suphuret of Mercury), sankha (Arsenic), kharimati (Chalik), tuta (Sulphate of Copper), hira-khas (Sulphate of Iron), raskapu (Colomel), murdhan Shankar (Oxide of Lead), sorali (Nitrate of Potash), tabashir (Silicate of Potash), sajimati (Carbonate of Soda), sohaga (Bicobonet of Soda), Gandhak (Sulphur),* etc. are very much common [For details see Appendices].

6.7 Food Habits and Prevention of Disease

Food habit is an important part of folk medicine. The Rajbanshis have developed a specific use of herbs and vegetables and their sub-products in food habits as a preventive measure against certain diseases. Food habits of the Rajbanshis in certain cases prevent the disease or contribute to healing. Making *chheka* [or calcium carbonate] from the plantain tuber or bark of mustard seed was/is a common tradition among the Rajbanshis of North Bengal and Assam. *Siddle* [dried cake of dry fish with vegetable

---

[42] For details see, Barman (2019).

substance], *pelka* [light juice made with different leaves and vegetables], and *chheka* were closely associated with the food culture of the Rajbanshis.

6.8 The Rajbanshi Healers and Institutionalized Form of Folk Medicine

The healers or the practitioners of folk medicine among the Rajbanshis are basically ordinary people who are engaged in different occupations [as cultivators, agricultural labourers, domestic workers, teachers, housewives, and even teachers]. They use their knowledge of medicine in their day-to-day lives with a perspective of service or self-requirement. Both men and women carry the knowledge of medicine. It is transmitted from one generation to another through oral practices. In our fieldwork, we have found that the practitioners of folk medicine generally do not receive any fee for their services especially where sources of medicine [plants, leaves, roots, etc.] are easily available.

However, knowledge of the Rajbanshis on folk medicine is gradually getting an institutionalized form in Cooch Behar, Alipurduar, and Jalpaiguri districts. We have noticed that the knowledge of the Rajbanshis for healing of fractures and ailments of arthritis got an institutionalized form in certain places. Rajmohan Eshor, the Royal Physician of the Cooch Behar State [1540 Century to 1949 CE], had developed a special school of healing fractures and arthritis and a few diseases at Nishiganj of Cooch Behar. This knowledge has been transmitted to the present generation of that family. The branches of this family have opened up several clinics at Nishiganj as well as other places in Cooch Behar, Jalpaiguri, and Alipurduar. Patients frequently come from different places in North Bengal and Northeast India.

The knowledge of Rajmohan Eshore on the particular branch of medicine has also been transmitted to his disciples. It has been extended to almost the entire Sub-Himalayan Bengal. In our fieldwork, we have found several such clinics where patients are getting services by paying a suitable amount.

6.9 Magical Means of Healing and Exorcism

Truly speaking, the practice of magic for healing has been a common trend in India across the ages. In this context, North Bengal and North East India is no exception. Magical means of healing [including black magical practices and its tools, method and exorcists] of North East India have been specially referred to the pre-colonial literatures. Literary and historical texts as compiled in this region have left some impression about the practices of magical means as medical exorcism and as 'offensive charm'. Among these literatures the *Kalika Puranam*,[43] *Yoginitantra*,[44] *Buranjis and vamsavalis* (genealogies),[45] *Charita* literatures,[46] and chronicles have occasionally referred to the practice of black magic in this region. In this context, *Kalika Puranam* is noteworthy. It describes that by a magical means (*maya*), a human could change its shape.[47] A special charm called *valla* could behead the enemies.[48] *Chapter 38* of this text has also described in detail the magical arms for warfare. The *Gurucharita* (a biography of Sankaradev), on the other hand, repeatedly mentioned astrology and soothsayers.[49] A chronicle called *Rajopakhyan,* however, has given a detailed description of the practice of black magic by the Koches. It describes that Bisu, a tribal boy, could kill another Koch boy only with a wild leaf because of his magical power.[50] Bisu, later became the founder king of the Koch kingdom and started worshiping Devi, a special deity who is a synthesized form of tribal and non-tribal traditions.

The tradition of magical belief of the Rajbanshis continued to survive during the colonial period. Thus, the colonial records have left several references to this trend. Buchanan Hamilton during his visit (1808–1814) to Kamtapur fort had recorded that people of this region were extremely superstitious and were dependent on miraculous incidents. In the case of Dinajpur, Buchanan recorded that:

> No person teaches medicine in this district, and indeed proper physicians are very few in number. In all cases of sickness trust is chiefly placed in prayers and sacrifices, and in certain magical incantations, that are considered as effectual, and called *jharon Mantros*[51]

Buchanan has also described in detail different magical means for healing and charming and performers of magical rites in the early years of the nineteenth century.

---

[43] *Kalika Purana* is a Sanskrit text compiled in Kamarupa (Northeast India) in the thirteenth century CE. Along with the Sanskrit version, it is now available in different languages in printed form. I have used the text compiled and edited by Panchanan Tarkaratna. See *Kalika Puranam*, edited by Panchanan Tarkaratna, Calcutta: Nababharat Publishers, 1384, B.S.

[44] *Yoginitantra* is a Sanskrit text complied in Assam in the eighteenth century. This text has described the historical past of Sub-Himalayan Bengal and Assam. Here, I have used the text compiled and edited by Swami Swarbeswarananda Swaraswati. See, Swami Swarbeswarananda Swaraswati (ed), *Yoginitantra,* Calcutta: Nababharat Publishers, 1385 B.S.

[45] Writing of *Buranji* (or chronicles) was very much common in pre-colonial Assam from the thirteenth century to eighteenth century. A considerable number of *Buranjis* have been produced in the courts of Ahom rulers, Koch rulers and other king of Northeast India. *Vamsavali* means genealogy. Here, I have used *Darrang Raj Vamsavali* as a source. See Daivagna (1973).

[46] The biographies of Vaishnav saint of Assam especially Sankaradeva (1449–1568 CE) and his disciples are called *charitas*. Some of the *charitas* are available in printed form while many of them are still unpublished. Here, I have used—Thakur (2001).

[47] *Kalika Puranam, 38, vs. 38.*

[48] *Ibid, vs. 104,* p. 318.

[49] Thakur (2001).

[50] Munshi (1989).

[51] Martin (1976b).

The tradition of magical means of healing among the Rajbanshis was further recorded in the late nineteenth century by Hunter (1876).[52]

However, with the introduction of modern medical practices in different towns, people of this region, including the Rajbanshis gradually got acquainted with modern medicines, dispensaries, hospitals, and doctors. But the tradition of magical practice and propitiation of evil spirits for healing continued to survive among the Rajbanshis in the post-colonial period.[53] Thus Dr. Charu Chandra Sanyal (a notable physician of Jalpaiguri) recorded in 1965 that:

> Rajbanshis of North Bengal believe that diseases are due to evil spirits and exorcism is the method adopted to drive out the spirits for the cure of the diseases[54]

During our fieldwork, we heard about two types of exorcism—(i) magical means of healing, and (ii) Black Magic for charming. We have found that the Rajbanshi villagers held the view that certain evil spirits were/are responsible for the diseases of the Rajbanshis. Hence, their propitiation was/is very much common in Sub-Himalayan Bengal and Lower Assam. *Masan* (with large verities), *dhumbaba, poiry* (with large variation), *bhuts* (male spirits), *pettani* (female spirits), etc., were/are common spirits of the Sub-Himalayan Bengal [as associated with the lives of the Rajbanshi]. Naturally, their propitiation in different occasions has remained as a common tradition in the Rajbanshi-dominated villages. Even it is being adopted by the non-Rajbanshis in their day-to-day lives.

For the propitiation of the evil spirits, the Rajbanshis used to call an exorcist. The exorcists generally treat the affected person with different sanctified things, magic, and incantations as per tradition. It is also noticed that as protective means, the Rajbanshis do use different sanctified things. Among such things mention may be made of—*morcha kharu* (anklet), *hoidddar mala* (necklace with beads in the black string), iron (as a protective metal), *dhap*/talisman, etc. The Rajbanshis used to believe that these things had great value as protectives.

The practice of black magic, especially to interrupt the attention of the enemies, was/is another common trend among the Rajbanshis. They had developed the technique of throwing magical arrows (*ban* charmed arrows) at the enemies. Dr. Charu Chandra Sanyal had recorded the tradition of charming arrows practiced by the Rajbanshi charmers of the late twentieth century. According to him, the most common *ban* practiced by the Rajbanshis were/are—*Baro Gopalur Ban, Tepa Ban, Howatepa Ban, Meheri Ban, Masna Ban, Juguni Ban, Brahma ban*, etc.[55] At present, this magical charm has been extinct.

However, magical means of healing are still a common feature among the Rajbanshis [including the Rajbanshi Muslims] of Sub-Himalayan Bengal. Along with herbal medicine, magical charms for healing have been found extensively in our fieldwork. They do use different kinds of incantations called *mantras*. These *mantras*

[52] W.W. Hunter, *op.cit*, p. 326.

[53] Upendra Nath Barman: *op.cit.*, p. 65.

[54] Sanyal (2002).

[55] *Ibid*, pp. 168–179.

are still prevailing among them both in oral and written form. These are their self-compositions although, on many occasions, they do follow the trends of classical literature. We have collected the *mantras* from different places in Jalpaiguri, Cooch Behar, and Alipurduar districts. They use them in three ways—(a) use of incantations along with medicine, (b) Use of incantation as a magical means, (c) use of *mantras* for propitiating spirits believed to be responsible for diseases. The first category of mantras is basically used for improving the mental strength of the patient where medicine is used in a prescribed manner. The second category of *mantras* is basically a collection of *mantras* related to the process of sanctification of oil, water, etc. for healing the patients with a perspective of psychotherapy.[56]

6.10 Observations

As a geographical region, Sub-Himalayan Bengal is quite interesting because of the presence of different indigenous communities. They have been living here for generations with their own knowledge of flora, fauna, minerals, and other elements of nature. In our study, we have noticed that the geographical features have an immense influence on the settlement pattern, belief system, and folk medicine of the Rajbanshis. They have maintained mainly three trends. First: parts and sub-products of different varieties of plants and easily procurable minerals are used by the Rajbanshis for healing different diseases. However, due to deforestation and environmental changes many species of plants are no more available in this region. So, the expert Rajbanshi healers are now importing them from other states or collecting them from the wholesale market.

Because of traditions and professional ethics, the knowledge of the Rajbanshi healers is restricted within the community. It is being transmitted only through family traditions or through 'teacher–pupil traditions'. So, they generally don't disclose the method of preparation of medicine to outsiders.

The practices of folk medicine were basically non-profiting among the Rajbanshi healers. They maintained this trend of providing services to their community members for getting relief from different ailments. So, they generally don't receive a fee for their service. However, they do receive the charge for the medicine as per the capacity of patients. At present "Mushuru Baidya School of Healing of Nishigunj" is gaining an institutionalized form because of the proven quality of medicine.

Second: The Rajbanshi healers do use magical means for improving the mental strength of patients to fight against ailments. Here, sanctified oil, water, and *tabiz* (talisman) are prescribed to the patients along with the performance of the religious

[56] The *mantras* (incantations) have been collected from the Rajbanshi healers of different places. Here we may mention the name of Shri Anil Ray (67), [Village: Kongar Para, P.O. Shalbari, Baxirhat, Cooch Behar], Shri Jatneswar Ray (70) [Village: Manta Para, P.O. Bara Shalbari, Baxirhat, Cooch Behar], Shri Jagannath Ray (47) [Village: Rasik Bil, P.O. Rasik Bil, Baxirhat, Cooch Behar], Shri Majendra Nath Barman (64), [Village: Panchagarh, Mathabhanga, Cooch Behar], and Shri Upendra Nath Barman (76), [Village: Bhelapeta, Tufanganj, Cooch Behar].

rituals. However, the present generation of the Rajbanshis doesn't think that the spirits are responsible for the disease. They merely maintained the traditions of worshipping spirits as a part of their folk culture.

Third: we have found that the traditional food habits of the Rajbanshis and the 'diet chart' are quite healthy in nature. Preparation and consumption of certain dishes and foods are very much preventive and curative for ailments. Although the present generation prefers to consume the popular foods of the present day, the old generation of the Rajbanshis of Sub-Himalayan Bengal often talks about their own food!

References

Rup Kumar Barman, Contested *Regionalism,* Delhi: Abhijeet Publications, 2007, pp. 15–22.

Upendra Nath Barman, *Uttarbangar Sekal O Amar Jibon Smriti,* Jalpaiguri: Bijay Chandra Barman, 1895, p. 65.

Tanoy Mandal, *Rajbanshi Lok Chikisa*, Kolkata: Government of West Bengal, 2011, p.4.

W.W. Hunter, *Statistical Account of Bengal, Vol. X,* London: Trubner & Cong, 1876, p. 441 (*emphasis added*).

T.H. Lewin, *An Account of the Kuch Behar State*, Cooch Behar: The Cooch Behar State Press, 1876.

H.N. Chaudhuri, *The Cooch Behar State and its Land Revenue Settlement,* Cooch Behar: The Cooch Behar State Press, 1903.

The Annual Administration Report of the Department of General Administration & Criminal justice of the Cooch Behar State for the year 1907–1908, Cooch Behar: The Cooch Behar State Press, 1908, p. 24.

Abani Mohan Kusari (*et.al*), *West Bengal District Gazetteers: Jalpaiguri*, Calcutta: Government of West Bengal, 1981, p. 270.

D.H.E. Sunder: *Survey and Settlement of the Western Duars in the District of Jalpaiguri 1889-95*, Calcutta, Bengal Secretariat Press, 1895, p.7.

J.A. Miligan: *Final Report on the Survey and Settlement operations in Jalpaiguri 1906–1916*, Calcutta, Bengal Secretariat Book Depot, 1919, p. 24.

J.F.Gruning, *Jalpaiguri, Eastern Bengal and Assam District Gazetteers,* Allahabad: Pioneer Press, 1911, p. 47.

D.C. Pal and S.K. Jain, *Tribal Medicine*, Calcutta: Naya Prakash, 1998, p. 11.

Suryakhari Daivagna, *Darrang Raj Vamsavali*, edited by N.C. Sharma, Pathsala: Bani Prakash, 1973, *vs 588–90*, p. 20.

Montgomery Martin, *The History, Antiquities, Topography and Statistics of Eastern India, 5 Vols.*, London: W.H. Allen and Co, 1838; reprint edn., Delhi: Cosmo Publications, 1976a, *Vol. V*, p. 509.

The Annual Administrative Report of the Cooch Behar State 1893–94, Cooch Behar: The Cooch Behar State Press, 1894, p. 58.

World Health Organization, *WHO traditional medicine strategy: 2014 – 2023,* Hong Kong: WHO, 2013, p. 15.

Rup Kumar Barman, *Practice of Folk Medicines*, New Delhi: Abhijeet Publications, 2019.

Ramcharan Thakur: *Gurucharita*, edited by Harinarayan Dutta Barua, 18th edition Guwahati: Dutta Barua Publishing Company Private Limited, 2001, pp. 488, 577.

Joynath Munshi, *Rajopakhyan*, edited by Biswanath Das, Kolkata: Mala Prakashani, 1989, pp. 11–12.

Montgomery Martin: *The History, Antiquities, Topography and Statistics of Eastern India*, (in *5 vols.), Vol. 3, Dinajpur* (1838), reprint ed., Delhi: Cosmo Publications, 1976b, p. 720.

Charu Chandra Sanyal, *The Rajbanshis of North Bengal,* reprint edition, Kolkata: The Asiatic Society, 2002, p. 170.

Part III
Institutionalization and Professionalization of New Knowledge

Chapter 7
Evolution of Electrical Engineering in Colonial Calcutta: *Bhadralok* Articulations on Education and Industrialization Interface

Suvobrata Sarkar

> To generate electrical energy, electricity must be forced to move, that is, an electric current—a current of electricity—must be established. It has been noted that electricity may be thought of as a medium or agency for transmitting energy. When we generate energy we force some of these medium—electrons—to move. Electricity in motion constitutes an electric current.
>
> The establishment of a voltage is the first requirement for the generation of electrical energy. It should be understood that electrical energy cannot be generated directly. When it is desired to generate electrical energy, a voltage, difference of potential, electric stress or pressure must first be developed.
>
> Bhim Chandra Chatterjee, professor of electrical engineering, Benares Hindu University (1922)[1]

Our protagonist, Bhim Chnadra, thus wrote a series of notes to meet the requirements of the students of engineering of the Benares Hindu University. There the electrical engineering course had been directed to provide the students 'every opportunity to obtain the essential acquaintance with the practical and quantitative sides of the subject'. Both education and industry appear as the products of the same

[1] Chatterjee (1922).

Originally presented at the 26th International Congress of History of Science and Technology, Prague, 25–31 July, 2021 (Virtually via Zoom). The author is grateful to Dr. Prakash Kumar and Dr. B. Eswara Rao for insightful comments and suggestions. Parts of this chapter had been originally published in Suvobrata Sarkar, Let there be Light: Engineering, Entrepreneurship and Electricity in Colonial Bengal, 1880–1945 © Suvobrata Sarkar 2020—(ISBN: 9781108835985). Reproduced with the permission of the copyright holders and the publisher, Cambridge University Press.

S. Sarkar (✉)
Department of History, Rabindra Bharati University, Kolkata, India
e-mail: suvobrata.history@rbu.ac.in

S. Ahmed and S. Sarkar (eds.), *Decolonizing Science and Modernity in South Asia*,
https://doi.org/10.1007/978-981-97-1829-0_7

multi-faceted social, cultural, and economic background that accelerate the techno-scientific progress of any country.[2] Between the mid nineteenth century and the Second World War, the industrialized countries of Europe and America had made some provision, often of an elaborate kind, for the preparation of their engineers. What was new during this period was the emergence of institutions and courses that aimed specifically to prepare technical employees for work at manufacturing industries. Western techno-scientific discourse occupied an enormously significant place in the colonization of India. It is well-known that Western technology and ideas influenced various technological projects in the colony, including education.[3] This essay is basically on the education of electrical engineers as it had evolved and developed in the late years of the Raj, and the interface it had with entrepreneurship and industry in Calcutta.

Electricity was not a new subject in the nineteenth century. With the development of telegraph, it became an indispensable part of social life. Yet it was only at the end of the century, with the development of electric light and electric power, that its full practical capacities were beginning to be realized. The advent of a new technology necessitated the introduction of new institutional facilities. Referring to large-scale technologies such as the electric supply system in Europe and America, Thomas Hughes argues that when the new system gains momentum, a supportive context or culture forms rapidly. Human capital and institutions developed to keep pace with the advancement of the technology.[4] According to this hypothesis, such supportive systems are supposed to emerge in Calcutta as well. In the early twentieth century Calcutta, the electrification of the city, its industrial application, and transportation narrative all stand out as the biggest engagement for the bureaucrats, engineers, nationalists and ordinary Indians.[5] A supportive system such as community of experts—electrical engineers—and institutional facilities evolved along with the electricity system in the early twentieth century Calcutta. How did electrical engineering evolve as a new discipline with the advent of electricity and what role the electrical engineers played for the use of electricity in the industrial sector?

7.1 Institutional Training: Early Diffident Steps

Throughout the second half of the nineteenth century, Bengali elites were raising the demand for a full-fledged techno-scientific research system, a demand that was not accepted to by the imperial administration until, 1904, when the Indian Universities Act was passed that permitted post-graduate teaching and research in the humanities and the sciences.[6] What impact it had on the contemporary engineering education?

---

[2] Fox and Guagnini (1993).

[3] Macleod and Kumar (1995).

[4] Hughes (1983).

[5] Sarkar (2020).

[6] Raina (2011).

Taking stock of the much-diversified system of instruction, lengths of the courses and nomenclature adopted by the various engineering colleges,[7] the Act asked for uniformity in the system and stated that further provision was needed for instruction in mining and electrical engineering.[8]

When electricity system was started and was gaining popularity during the last decade of nineteenth century, a lack of expertise to organize the system was felt. As the network widened and as the power generation stations increased, it became inevitable for the colonial government to create an electrical engineering community within India as complete import of engineers from London was a costly option. In April 1893, the then principal of the college, J. S. Slater wrote to the Director of Public Instruction (DPI) about the 'liberal loan' from the Messrs Martin & Company, Calcutta based engineering firm, of certain electric machines with a view to begin electrical engineering there. The principal stated that this would enable him to demonstrate the students various electrical equipments which were hitherto unavailable to him due to high cost. As this branch of engineering was in an underdeveloped state due to unavailability of trained manpower, efforts were made to install small electric lighting plant at the college and train interested students in electrical engineering.[9] Hence the government started electrical engineering course, although very elementary level in the beginning. An official correspondence of 1896 reveals the following:

> The Lieutenant-Governor is endeavoring to establish at Sibpur a training class for electricians competent to find employment under the various firms who are now engaged in developing this class of business in Calcutta and elsewhere. It is largely with a view to the instruction of this class that the scheme was originally started by the College authorities.[10]

Thus, a full practical course in electrical engineering was already introduced at the Bengal Engineering College, Sibpur in the previous year (1895).[11] A proposal for a complete electric lighting installation at the college was initiated in the same year. Dr. P. Bruhl, professor of physics, prepared an estimate for the installation and to serve as a means of practical instruction to the students in electrical engineering. The modernization of the workshops through the introduction of electric power plant in place of the old steam power drives in 1904 again served the similar purpose.[12]

---

[7] (1) Thomason Engineering College Roorkee, (2) Bengal Engineering College, Sibpur, (3) Engineering College at Madras, and (4) College of Science, Poona. For more information see, Sarkar (2012).

[8] Home, Education, Deposit, Proceedings, October 1904, No. 20, National Archives of India, New Delhi (hereafter NAI).

[9] The following electric machines were received by the Bengal Engineering College, Sibpur-Ferranti (120 Incandescent 16 C.P. lights, one Dynamo Motor, two Brush Lamps, one Brush and eight light Dynamos, one Gramme Excitor for the Ferranti Dynamo. General, Education, File No. 4C/22—1 and 2, Nos. 4–5, May 1893, West Bengal State Archives, Kolkata (hereafter WBSA).

[10] PWD, Civil Works—Miscellaneous, April 1896, Nos. 1–7, NAI.

[11] *First Quinquennial Review of Education in Bengal, 1892–96* (1902).

To meet the expenses of electric installation students were levied one rupee each for ten months of the year.

[12] Sen (1991).

The history of electricity in Calcutta begins effectively in 1899: the commercial generation and distribution of electricity commenced under the Calcutta Electric Supply Corporation Ltd (CESC).[13] John Willoughby Meares, chief electrical engineer of the CESC and later electrical advisor to the Government of Bengal, closely associated with the Bengal Engineering College. Meares delivered several lectures at the college to inspire students in this special branch of engineering. One occasion, he thus motivated the students:

> I am very pleased to be able to congratulate the students who have taken up the special electrical classes on the success of those who have preceded them in obtaining appointments. This has been invariably the case, I believe, and the demand has actually been in excess of the supply up to the present time. And yet it may be safely be said that the electrical era in India has scarcely commenced; for so far as I know there are, excluding private plants, only three public electric-supply stations and one tramway system actually working, and about half-a-dozen more schemes in contemplation on under construction. This province is well to the fore in the matter, and Calcutta as is only right holds the first place with its extensive installations...[14]

A combined Electrical and Mechanical Engineering Department was started at the college in 1902. This department used to take both theoretical and practical classes in electrical engineering for the professional group of the Bachelor of Engineering (B.E.) degree course in civil engineering.[15] As a measure of developing the workshops, the principal sent up a proposal for the introduction of electric-power drive into the shops. The substitution of the old steam-power drive by modern electric drive was necessary from the point of view of training, as the major shops of those days, viz., the Calcutta Tramways Company, the Cauvery Falls in Madras, the Railway Workshops at Jamalpur, Lillooah, Khragpur, etc. had already decided to install electric drives. The proposal was accepted in 1902 and the new electric-power plant was inaugurated in March 1904, by Sir Andrew Fraser, the then Lieutenant-Governor of Bengal.[16]

In the year 1902 the students of the college started publishing a monthly journal in Bengali, *Shibpur College Patrika.* This journal published a wide variety of popular articles on science and technology and the contributors were mainly students of the College. In one such instance the author narrated the marvels of electricity:

> The main gossip of today's Calcutta is the electric tramways! This modern mode of transport is not horse-driven or human-driven, not even steam-driven—but electricity is the prime mover for tramcars. One become perplex while considering the vast range of activity of this new technology! In every aspect of our life electricity is now an obedient servant to mankind. Electric-stove for modern cooking (as substitute of fire), electric-fan during the hot summer season, and electric-lamp for illumination in the night—everywhere is electricity. To run the machines of a factory, electricity plays the pivotal role by producing hundreds and thousands of horsepower. If one wants to talk with a friend mile away from him, the

---

[13] Suvobrata Sarkar, *Let there be Light*, op.cit., p. 118.

[14] Meares (1902).

[15] Home, Education, A. Proceedings, October 1901, Nos. 80–84, NAI.

[16] *General Report on Public Instruction in Bengal for 1903–04* (1904).

> remedy is telephone—another offshoot of electricity. One cannot think beyond electricity in modern times.[17]

This venture however did not last long and with the departure of the editors (Atul Chandra Bagchi and Upendranath Kar) from the college, the publication discontinued in 1908.

In those days' electrical engineering mainly dealt with electrical power generation, transmission and utilization. With the multiplication of the number of electrical power houses, and with the increase of its industrial usage, the demand for men trained in the use of electrical machinery and acquainted with the fundamentals of this technology began to grow and the contemporary engineering colleges started upgrading their facilities for this particular branch of engineering. The Bengal Engineering College, Sibpur was also not an exception.[18]

The training in electrical engineering, in early twentieth century, was confined almost entirely to the apprentice (or subordinate) stage at the Sibpur college. Students in this branch first had to go through the ordinary four years' course for Public Works Department (PWD) overseer students, and then specialise in electrical work for some 15 months. In the first four years they spent much time in the various mechanical workshops and took such subjects as steam engine and applied mechanics. Thus, huge time was devoted to civil engineering and other subjects of little subsequent use to them in later life.[19]

In the meanwhile, the Government of India had deputed E. H. Atkinson and Tom S. Dawson to enquire about the state of technical education in the country with special reference to their industrial applicability. They visited various industrial concerns, both private and public owned, and prepared a huge report. The general assumption of the employers (i.e. various industrial firms) in Bengal, as mentioned in their report, was that the educated Bengalis were averse to manual labour as a regular employment even after their training in a technical institute. Their other major observations include:

> There is at present no demand with a university training in mechanical and electrical engineering…the university type of education unfits a Bengali for practical work, and although an engineering graduate would have a more complete scientific education he would be far behind the technically-trained student in practical ability and less valuable to an employer.[20]

Instead, they suggested for training as an apprenticeship on a living wage, work with their hands, and observe factory hours and rules. Such training would enable the Bengali jobseekers to convince the employers as there was a huge opportunity for the men with practical knowledge in mechanical and electrical engineering.

Despite the circumstances, B. Heaton, the then principal of the Sibpur college, proposed a three years overseer course in mechanical and electrical branch. The DPI,

---

[17] Shibpur College Patrika (1902).

[18] The Government of Bengal sanctioned a contingent grant for the electro-technical laboratory of the college in 1911 under the head 'Provision for additional teaching staff for Electrical Engineering Department'. General, Education, File No. 4C/37, Proceedings Nos. 221–22, June 1911, WBSA.

[19] *Report on Public Instruction in Bengal for 1910–11* (1911).

[20] Atkinson and Dawson (1912).

Bengal, was also in favour of creation of a professorship in mechanical engineering who would confine his attention to electrical engineering with the addition of the electrical portion of physics.[21] In January 1914, the Government of Bengal sanctioned the proposal.[22] The University of Calcutta granted affiliation to the B.E. degree in all branches of engineering, namely civil, mechanical, electrical and mining of the college in 1911. In the session 1912–13, the Electrical Engineering Department was separated from the Mechanical Engineering Department.[23]

7.2 Period of Consolidation

The Indian Industrial Commission visited the Sibpur college during 1916–17. B. Heaton, the then principal, in his testimony before the Commission, remarked- "The mechanical and electrical engineering course of its present lines is quite a recent departure…It is thought that if picked lads to be taken into the college after having completed four years of an apprentice course satisfactorily they will be of more immediate value to employees on passing out of Sibpur than those who have only been through a college course."[24] He further proposed to retain the Thomason College, Roorkee, for higher civil engineering course and convert the other three (Sibpur, Madras and Pune) into 'mechanical and electrical engineering colleges and into mining colleges where facilities could be offered for advanced courses of a university type.'[25]

Electrical Engineering as a subject was introduced but it took long time to initiate degree level course in the Sibpur college, a small community of Indian electrical engineers came up, but the colonial engineers retained upper hand. Finally, electrical technology was transferred and marketed but technology manufacturing was not encouraged. Given the disadvantageous position, very few Indian entrepreneurs tried to establish electrical industries on their own financial capacity or applied for state-aid. Indian Industrial Commission in its Report perhaps captured the lacuna best:

> In the past, the education of engineers has been too much influenced by the immediate requirements of the PWD, without regard to the future or to those other interests in India which can be handled only by engineers. The greater part of the work done in the college is the training of upper subordinates, lower subordinates, surveyors and draftsmen…In the four principal colleges, increasing attention has, in recent years, been paid to the provision of instruction in mechanical and electrical engineering; but the measures adopted are inadequate and are conceived on altogether too narrow lines to meet the needs, present and prospective, of a rapidly expanding industrial system.[26]

---

[21] General, Education, File No. 4C/59—1, Proceedings Nos. 50–53, March 1913, WBSA.

[22] General, Education, File No. 6E/19—1, Proceedings No. 61, January 1914, WBSA.

[23] Bengal Engineering College Centenary Souvenir (1956).

[24] *Indian Industrial Commission, Minutes of Evidence 1916–17* (1918).

[25] Ibid, p. 253.

[26] *Report of the Indian Industrial Commission, 1916–18* (1918).

Despite the limited opportunities, several Indians proved their credentials in the subject. During the deputation of W. H. Everett, professor of mechanical &electrical engineering, B. C. Gupta had overseen the Electrical Engineering Department and Power Installation of the Sibpur college. One Jitendra Nath Chakravarti was appointed at the department as demonstrator to the special class for training subordinate telegraphists of the Government of India. It was said that both were acclaimed for their teaching efficiency.[27]

The next important event in the history of the Sibpur college was the introduction of the degree level course. Due to financial stringency, the degree course in electrical engineering could not be introduced along with the mechanical engineering (1932). It was only in 1935 that the degree course in electrical engineering came into existence.[28] The College was so long generating its own electricity. To meet the increased load the electric supply system of the institution was re-modelled and supply of energy in bulk was arranged from the CESC from December 1946.[29] A 'Committee for the Development of Higher Engineering and Technical Education in Bengal' was formed under the chairmanship of the DPI (1945).[30] The Committee submitted its reports in two parts, one containing the 'Immediate Plan'[31] and the other 'Five Year Plan'. The object of the 'Five Year Plan' prepared by the committee was the ultimate attainment of converting the Bengal Engineering College into a most modern and well-equipped institution comparable to the best institutions of the world, with a greater range of degree courses and facilities for the post-graduate and research work. But the scheme saw the light of the day almost 10 years later, only after independence.[32] The collective adoption of the 'Immediate Plan' and the 'Five-Year Plan', by the committee, was with a view to fostering close co-operation and collaboration between industries and the Bengal Engineering College, Sibpur.[33]

---

[27] *Report on Public Instruction in Bengal for 1923–24* (1924).

[28] The first batch of students appeared at the degree examination of the University of Calcutta in 1936. The successful degree holders were 3 in number—Bandopadhyay et al. (1938).

[29] *Bengal Engineering College Centenary Souvenir*, op.cit, p. 44.

[30] The terms of reference of the Committee was 'to consider and make proposal for the development of higher engineering and technical education in Bengal, with special reference to the expansion of facilities now provided by the Bengal Engineering College, Sibpur, and the Asanulla School of Engineering, Dacca.' *Report of the Committee for the Development of Higher Engineering and Technical Education in Bengal*; quoted in *Report of the Director of Public Instruction, Bengal, 1945–46* (1946).

[31] While submitting its report under the 'Immediate Plan', the committee confined their recommendations to higher engineering training and the training of supervisory personnel, and aimed at increasing the number of qualified engineers and supervisors in the more important and immediately necessary branches of engineering, improving their quality and standard and also providing facilities for advanced instruction and research. Ibid, p. 10.

[32] The University of Calcutta approved new regulation for the degree of Master of Engineering (M.E.) in civil, mechanical, electrical and metallurgical engineering in 1953. The first batch of post-graduate students came upon in the year 1956. During the same time doctorate programme were also introduced to enhance the research activities of the college. *Bulletin Bengal Engineering College: One Hundred Twenty Fifth Anniversary Issue* (1981).

[33] *Report of the Director of Public Instruction, Bengal, 1945–46*, op.cit, p. 11.

It was not only the colonial government which stressed the need for better institutional facilities for effective maintenance of technological projects. The local erudite elites, who were also aware of the significance of investment in the human capital formation, recognized the necessity of the technological imperatives stimulated by higher technical education.[34] When their consistent demand failed to expand the scope of technological training, the Bengali intellectuals came up with their own model. The alternative crystallized as Jadavpur college in the first half of the twentieth century. The period was to prove significant for science and technology in India, in terms of its institutional manifestation. The prevalence of unemployment among the Bengali *bhadralok*, and the partition of Bengal (1905), which smashed the Bengali selfhood, intensified the already widespread nationalist movement.[35] The nationalist stand on science and technology in Bengal[36] can be understood by the study of the National Council of Education (NCE) and its progeny, the College of Engineering and Technology, Jadavpur.

The Jadavpur College offered a 4 years degree course in electrical engineering from 1910 onwards. The most notable incident of the period was a grant received from Radha Kumud Mukherjee and Benoy Kumar Sarkar (30,000 INR), which enabled the council to constitute seven technical scholarships for advance training in America.[37] The City and Guilds of London Institute recognized the college, and exempted students of electrical and mechanical engineering from appearing at the preliminary examination of the institute.[38] The Jadavpur students of the second decade of the twentieth century constituted practically the first batch of young men looking for careers in mechanical, electrical, chemical and allied industries after a systematic technological training.

In 1922, in view of the difficulty of securing competent teachers, especially in the mechanical and electrical engineering departments, it was decided to send four

---

[34] Pramatha Nath Bose (1855–1934), noted geologist and science enthusiast, was probably the first to talk about science-based industries and the need to remodel the university curriculum accordingly. In 1886 he published a pamphlet on techno-scientific education which is said to have inaugurated the movement for technical education in Bengal. Bose, thus said, "In regard to higher technical education, progress during the last two decades has, I regret to say, been practically nil in Bengal." Bose (1906).

[35] The ideal example is Sarkar (2010).

[36] In locating the history of engagement of science, technical education and industrialization, Dhruv Raina and S. Irfan Habib introduce 3 significant conceptual dimensions. First was based on an ethical critique of the Western sciences and industrialization. Second comprised a pedagogical programme, accompanying the systemization of traditional systems of knowledge. Third acknowledged the importance of modern science and technology in effecting social transformation but rejected the Western model of industrialization in favour of the Indian alternative. Raina and Irfan Habib (2004).

[37] Before leaving for America, these seven scholars entered into an agreement with NCE to serve it for at least of 7 years on their return to the country. *NCE Annual Report, 1909–10*, Calcutta: NCE, Bengal, not dated, p. 6.

[38] Availing the opportunity 6 students appeared in the grade II examination in which all of them excelled in the Direct Current (DC) and 2 in the Alternate Current (AC). *NCE Annual Report, 1912*, Calcutta: NCE, Bengal, not dated, p. 19.

students in Germany for higher training with a contract of service on their return.[39] Under the scheme, in 1923 three scholars were sent to Germany for higher training under the contract with the council.[40] The advantage of such scheme started coming soon. The Electrical Engineering Department was strengthened by the appointment of Suren Roy, Masters' from Harvard. He was the first scholar with an electrical engineering degree to be appointed by the NCE.[41] By 1925, the powerhouse, mechanical laboratory, chemical and physical laboratories, workshops, etc. were constructed which were of immense importance for smooth functioning of a technical institute. The council set up its own installation of electric lighting. The professors of the Electrical Engineering Department, with the help of their students, successfully installed the power utility system at the Jadavpur campus.[42]

7.3 Nationalizing Technology

The mechanical and electrical engineering departments of Jadavpur college were placed under the supervision of H. C. Das Gupta (graduate of Michigan) and Suren Roy (Masters' of Harvard)—both alumni of the college. However, the electrical engineering commanded the predominant choice of the scholars.[43] A. T. Weston, Director of Industries, Bengal, who paid a visit to the college in 1927, was satisfied with it's the overall progress towards advance training in technology. Weston remarked, "I can only say that in its equipment, machinery, power plant, varieties of actual engine and boiler plant, the institution has more resources than were available to me when I took the Honours Engineering Degree Course at the University of Manchester (1900–04). I am satisfied that both as regards equipment and the qualification of its teaching staffs the institution..."[44] Being an institute of nationalist character, such recognition from the bureaucracy is significant.

A college magazine, *Journal of the College of Engineering & Technology*, started in the year 1928.[45] It was a result of co-operative endeavour between the professors and students of the college. The journal was not a mere students' magazine or a mouthpiece of the institute for news and views. It represented a glance of the techno-scientific and industrial worldview of the Indian technologists and engineers. An article, by Nirmal K. Sengupta, a final-year student of electrical engineering (1937), is worth a mention. The following are the extracts:

---

[39] *NCE Annual Report, 1922*, Calcutta: NCE, Bengal, not dated, pp. 5–6.

[40] *NCE Annual Report, 1923*, Jadavpur: NCE, Bengal, not dated, p. 13.

[41] Sarkar (1946).

[42] *NCE Annual Report, 1926*, Jadavpur: NEC, Bengal, not dated, p. 9.

[43] By 1929, 215 students obtained the electrical engineering degree, compared to 117 in the mechanical engineering branch. Quoted in Benoy Kumar Sarkar, *Education for Industrialization*, op.cit, p. 147.

[44] Quoted in *NCE Annual Report, 1927*, Jadavpur: NCE, Bengal, not dated, Appendix H, p. 27.

[45] *NCE Annual Report, 1929*, Jadavpur: NCE, Bengal, not dated, p. 8.

> Now as to the different eras of thought about the nature of electricity; in the first of these eras, beginning probably with the earliest observations in the middle of the sixteenth century, electricity was regarded as a soul or spirit; the second era may be said to have been opened by Carden in 1551 and closed by Maxwell in 1865, during which electricity was regarded as a material fluid of one or two kinds; in the third era electricity with its various manifestations was regarded as some kind of another disturbance of a continuous nature; the fourth era is that which is still with us. It may be called the atomic or quantum period, in which it is noteworthy that but little attention, has been paid to the ultimate nature of electricity and a great deal to its structure. Of course knowledge of the structure is likely to give some useful hint in future as to the nature of electricity.[46]

The demand for the electrical, mechanical and chemical engineering among the Bengali elites was steadily on the increase. Electrical engineering remained the favoured subject: During 1930–39, 213 students turned out as electrical engineer, where the number of mechanical graduates and chemical graduates were 169 and 69 respectively.[47] The authority of the college was also in an expanding mood in the context of 'rapid development in ideas about the standard and scope of Technical Education'.[48] Accordingly the courses were reframed: the engineering department with a five years' degree course in (a) mechanical, (b) electrical and (c) chemical engineering; the junior technical department with a three years' course in mechanical and electrical engineering and survey and draftsmanship department with a two years' course.[49] The Electrical Engineering Department in the late 1930s consisted of the following main courses:

> Elements of Direct Current Engineering, Direct Current Machines, Direct Current Measuring Instruments, Principles of Alternating Current, Electrical Drawing, Alternating Current Machinery, Transmission & Distribution, Industrial Motors & their application, Telegraphy & Telephony, and Electrical Power Generation & Distribution.[50]

The techno-industrial boom of World War II was reflected at Jadavpur college as the authority decided to strengthen the mechanical, electrical, and chemical engineering departments with additional apparatus and appliances.[51] In the prevailing war-scenario, the government compelled to change their attitude towards the Jadavpur College—a nationalist institution. To meet the huge demand for technical manpower, at first the defence department recognized the chemical engineer graduates of the college and secondly the Director General of Posts and Telegraphs allowed the mechanical and electrical diploma-holders for recruitment to the grades

---

[46] Sengupta (1937).

[47] Benoy Kumar Sarkar, *Education for Industrialization*, op.cit, p. 179.

[48] *NCE Annual Report, 1932*, Jadavpur: NCE, Bengal, not dated, p. 11.

[49] *NCE Annual Report, 1934*, Jadavpur: NCE, Bengal, not dated, p. 7.

[50] *College of Engineering & Technology, Bengal, Session 1937–38*, Jadavpur: Superintendent, NCE, not dated, pp. 41–7.

[51] To meet the demands of the time, they also intended to introduce mining engineering, communication and automobile engineering, sugar and film technology in the existing curriculum of the college. *NCE Annual Report, 1940*, Jadavpur: NCE, not dated, p. 11.

of (a) engineering supervisors and (b) assistant foreman, examiners and instrument testers.[52]

The war-ecology thus enriched the college in the larger project of industrialization on national lines. The electrical engineering laboratory consisted of different types of modern electrical machines and instruments for demonstrating the application for the general rules and principles of electrical engineering. There was also a moderately equipped laboratory for communication engineering, where various experiments on radio, telegraphy and telephony were carried out. In the electrical workshop facilities were provided for repairing, designing and building up of electrical generators and motors, etc.[53]

At the end of our period of discussion, the course in electrical engineering of Jadavpur provided training in the fundamental subjects of the science of electrical engineering with particular application to the broad field of design, construction, operation and maintenance of electrical machinery, electric supply schemes, illumination, electric railways, communications and other branches involving the application of electricity. The theoretical work ran parallel with an elaborate course in the laboratories, which were provided with apparatus adapted to the needs of undergraduate study.[54] In the year 1945, the Massachusetts Institute of Technology (Boston), the University of Michigan, the Carnegie Institute of Technology (Pittsburgh) and Cornell University (Ithaca, NY) recognized the honours graduates of the Jadavpur college as qualified enough to join their post-graduate classes.[55]

The first few decades of twentieth century saw the massive electrification program, both thermal and hydro, in India. The construction of these power production sites was itself considered an industry. These were also centre of employment generation in two ways: direct and indirect. Directly it provided employment at the site itself and indirectly by providing electricity, as a motive power, to industries. Several electrical engineers of Jadavpur were involved in these projects in various capacities. They were associated with the industrial concerns such as Bengal Electric Lamp Works, General Electric Co, Indian Iron & Steel Co, National Iron & Steel Co, Siemens (India), etc. As the chief electrical engineer of Lahore Electric Supply Company, Banshidhar Pal had won good recognition. As foreman and shift engineer of the CESC at the Cossipore Power Station, Amulya Sen Gupta had made a name. An alumnus of 1924, Ugra Banerjee, electrical engineer, became an independent manufacturer. His Eugras Machinery Corporation won the estimation of many electric lamp factories.[56] Benoy Kumar Sarkar (1887–1949), rightly observed:

> The socio-economic structure of Bengal as of all India had been used to the profession of the lawyer, the medical man, the office clerk, and the school or college teacher of literary and

---

[52] *NCE Annual Report, 1941*, Jadavpur: NCE, not dated, p. 5.

[53] *College of Engineering & Technology, Bengal, Prospectus Session 1941–42*, Jadavpur: Secretary, NCE, not dated, p. 6.

[54] *College of Engineering & Technology, Bengal, Prospectus Session 1944–45*, Jadavpur: Secretary, NCE, not dated, p. 44.

[55] Benoy Kumar Sarkar, *Education for Industrialization*, op.cit, p. 155.

[56] Ibid, pp. 159–60.

philosophical subjects. The scholars that were turned out of the Government's Engineering College at Shibpur were chiefly civil engineers skilled in building of roads and houses for municipalities and elsewhere. It is the N.C.E engineers who commenced enriching the society with new professions nurtured by the tonic of machinery. The country began to get acculturated to technocracy, mechanics, chemical operations, electrical energies and so forth.[57]

7.4 Technological Expertise and Industrial Performance

Introduction of steam significantly affected industrial scenario of British India, but it was electricity which accelerated the production process and the development of large-scale industries. But economic historians of India have not given adequate attention to this technology so far.[58] The public of Calcutta experienced this new technology at least from the last decade of nineteenth century onwards and the credit definitely goes to Shibdas Sil, the proprietor of the 'Dey, Sil & Co-Electricians, Electrometallurgists and Brass Founders'. *The Statesman* reported in 1885: "In our Friday's issue we mentioned a novelty in marriage processions, which was the innovation of the electric light. The procession having taken place at night, the light, a *Serrin* 1500 cd-power, was most effective and successful. The electrician who arranged the apparatus is a rising young member of scientific Bengal, Mr. Sil, of the firm of Dey, Sil & Co., in Wellington Street…"[59] He was the pioneer Bengali entrepreneur in electrical industry. From an advertisement of the company published in 1895, one can get the following information:

> Electric Light Stationary- Electric Light Portable, Of 1500 candle-power, sufficient to illuminate brilliantly a quarter mile straight space, are supplied in Town and Moffusil for Illuminating Marriage Processions, court-yards, Halls, Gardens &c. These lights have been carried along with more than 200 Marriage Processions with unequalled success, since 1883, and have been honoured with numerous testimonials of Gentries, Nobilities, Rajahs, and Princes. Estimate for Permanent Electric Lighting Installations Arc and Incandescent System for Illuminating Palaces, Buildings, Halls, Parks etc. are furnished on application, Dey, Sil & Company.[60]

Apart from arranging for lights at meetings, parties, wedding processions etc., they manufactured a wide range of apparatus: carriage lamps lit by batteries of their own device, sewing machines and table fans worked by electric motors.[61] These were displayed at important gatherings, including the Second National Congress of 1886,[62] and the 1888 annual conversazione of the Mahomedan Literary Society, and won high praise from eminent scientists as well as the Viceroy himself.[63] This

[57] Ibid, p. 125.

[58] Morris (2008).

[59] 'A Novelty', *The Statesman*, 1 February, 1885.

[60] *The Amrita Bazar Patrika*. 1 December, 1895.

[61] Ray (2005).

[62] *The Statesman*, 1 January, 1887.

[63] *The Statesman*, 29 January, 1888.

pioneer Bengali electrical firm was in operation till 1912, although on a reduced scale.[64] Unfortunately we do not know anything about the educational background of Shibdas Sil. One should devote an entire treatise on character like Sil who, in the late nineteenth century itself, domesticated the modern technology—electricity!

Once the value of electricity was proved, thanks to the Industrial Commission's report, the colonial government recognized its potential in extracting resources from the colonies and intensified investigations on the large-scale hydro and thermal power sites throughout India. However, due to lack of experts in hydrology to identify hydroelectric power sites, unfamiliarity of electricity among the natives and economic difficulties, provincial governments did not show adequate interest in investigating power sites.[65] Nevertheless, the process started, and few reports were submitted. But experts called these reports unscientific and inaccurate.[66]

The constitution of the Hydro-Electric Survey of India (HESI) points to the colonial government's priority to exploring hydro-electric sites. But most of the industrial sites of the country, including Calcutta, were situated at great distances from those parts of the country where the natural features exhibited possibilities for the generation of waterpower.[67] In this context the Shillong Hydro-electric Ltd is worth a mention. Although situated in Assam, the protagonists of this hydro-electric scheme were mostly *bhadralok*. When the proposal first arose in the government circle in 1905, it was part of the Bengal Presidency. J. W. Meares, the then electrical adviser to the Government of Bengal, even identified the Beadon Falls and its adjacent areas as proposed hydro-electric site in 1908. But the proposal did not see light of the day then.[68]

Many years later Ramanath Roy, local entrepreneur, revived the plan and formed a joint-stock company, Shillong Hydro-electric Ltd. Subsequently Bidhan Chandra Roy, eminent doctor and celebrated chief minister of West Bengal, associated himself with the project and provided the necessary fund. His younger brothers Subodh Chandra Roy and Sadhan Chandra Roy (electrical engineer) were also closely involved. But the role of one Khitichandra Sanyal, an electrical engineer, was most crucial according to a contemporary source.[69] Work started with right earnest at the Beadon Falls in 1921 and two years later Shillong observed the first glimpse of electric lights in October 1923. In those days the work was an example of high standard of hydraulic and electrical engineering managed entirely by the Bengali entrepreneurs and technologists. However, like Shibdas Sil, we know little about the Shillong Hydro-electric Ltd.

Electrification was a science-based high technological innovation that demanded not only educated technicians, scientists, enormous amount of capital and industrial

---

[64] Ghosh (1991).

[65] Rao and Lourdusamy (2010).

[66] For example, J. W. Meares, an engineer of repute, complained about the indifferent attitude and scientific inaccuracy of the investigation reports. Meares (1935).

[67] *Report of the Indian Industrial Commission*, op. cit., p. 68.

[68] Suvobrata Sarkar, *Let there be Light*, op.cit., p. 169.

[69] Das (1927).

capability; but also, an enlighten public as consumers. It was true for Calcutta also. Although apprehensive in the beginning, the Bengalis not only welcomed electricity, but launched a monthly journal *Bijoli* on the subject by the third decade of twentieth century. In an editorial, the journal claimed the following:

> Realizing the need of a journal on electricity in vernacular, we started *Bijoli* few months back. We earnestly hope that our initiative would be recognized by the Bengali entrepreneurs who involve in the business of electricity and things electrical in and outside Bengal. We also hope that the journal would be helpful to them. We want to see our fellow Bengalis to excel in electricity industry and our journal would be the mouthpiece of such emerging entrepreneurial activities.[70]

The then electrical adviser to the Government of Bengal, S. W. Redclift, was all praise for *Bijoli* (1934): "The journal should have prove helpful in educating the public in the uses of electricity, especially in mofussil areas where the public electric supply undertakings are now working. When the uses of electricity become more widely known, these mofussil undertakings, in particular, should benefit by increased sales of current."[71] In an article the journal succinctly captured the victory of electricity over other forms of power.[72] Apart from the general topics on electricity, latest electrical innovations of the day were also featured in the journal. One Bhupen Ganguly, an expert in the science of electricity and managing director of Ms. Ganguly & Chatterjee Ltd, once wrote on 'Air Conditioning'.[73] Other significant articles were 'X-ray', 'Telephone', 'Street-lighting', 'Electricity in agriculture', and 'High unit rate of electricity in India' etc.[74] In *Bijoli* one can get a perfect blend of popular and technical writings on electricity.

While electric lamps were introduced in Calcutta by the foreign businessmen in the early twentieth century, the Bengali manufacturers invested in this field in large numbers in the 1930s. The most notable name in this sector was the Bengal Electric Lamps Works (BELW), popularly known as the Bengal Lamp. It was formed in 1932 as a private company at Kasba, in the southern fringe of the city and the proprietors were Suren Roy, Kiran Roy, Hemen Roy—three brothers. Both Suren and Kiran educated at Germany and later while Suren Roy joined as professor of electrical engineering, the younger brother Kiran became the secretary of the managing committee of the College of Engineering & Technology, Jadavpur.[75] At first they started on a small scale and initially turning out 200 lamps a day. The formative stage was naturally hard, and they had to struggle a lot to get a foothold in the market. In the latter half of 1932, it was converted into a public limited concern with an issued capital

---

[70] *Bijoli*, Vol. 1, No. 2, October 1933, p. 45.

[71] Quoted in *Bijoli*, Vol. I, No. 10, June 1934, p. 442.

[72] *Bijoli*, Vol. 1, No. 3, November 1933, p. 109.

[73] *Bijoli*, Vol. 1, No. 2, October 1933, p. 47.

[74] Basu (1933); again Manindra Basu in the same issue, 'Telephone', p. 182; Bankim Chandra Roy in the same issue, 'Bijoli Bati O Rajpathe Tahar Babohar', p. 190; Surekha Debi, 'Electricity in Agriculture', p. 151; Bhim Chandra Chattopadhyay, 'Bharate Bidyutik Uniter Uccahar O Tahar Bipod', *Bijoli*, Vol. 1, No. 7, March 1934, p. 272.

[75] Suvobrata Sarkar, *Let there be Light*, op.cit., p. 170.

of Rs. 3 lakhs.[76] Vacuum lamps of high-quality ranging from 25 to 60 watts were manufactured by the company, and soon lamps of low wattage, up to 5 watts could be made to meet the varied demands of the market.

Most of the directors and shareholders of this company were highly trained engineers and scientists. Several of them associated with the educational institutions and other industrial ventures of the city. From the declaration on the registration of the company, one can see the following names associated with the Bengal Lamp at various capacities:

> Some of the early directors and share-holders: S. K. Roy, Professor of Electrical Engineering, College of Engineering & Technology, Jadavpur (also the Managing Director of Bengal Lamp); J. N. Lahiri, Director, The Bengal Belting Works Ltd; A. K. Ganguli, Manager, The Scientific Instrument Co Ltd; U. Ghose, Proprietor, Bangia Diasalai Karjalay; H. C. Guha, Professor, College of Engineering & Technology, Jadavpur; H. L. Roy, Professor, College of Engineering & Technology, Jadavpur (later he became the Managing Director of Bengal Lamp).[77]

Gradually the company installed new machines and started turning out 1000 lamps every day. The production of vacuum lamps was followed and accompanied by that of gas-filled lamps by March 1934 which were initially of 60–100 watts, but later increased to 200 watts.[78] The steady progress of the Bengal Lamps is attested by the fact that within two years of its inception, the company could extend its market network from Bengal to such distant areas of Madras, Bombay, United Provinces and Punjab. Its lamps were of a high quality. These had to pass many outside tests such as those of the Government of India, lighting department of the Calcutta Corporation, the CESC Ltd and many mills. It was the intrinsic quality of their lamps that earned many medals and first-class certificates for BELW and enabled to develop despite keen competition from foreign and indigenous manufacturers.[79] One of its main customers was the Government. The number of lamps supplied to the government increased from 28,000 in 1935–36 to 80,000 in 1936–37, and the demand had been rising without a break. Prafulla Chandra Ray, great chemist and himself an entrepreneur, thus praised *Bengal Rosney*, the *Swadeshi* electric lamp manufactured by Bengal Lamp:

> I have been to the recent exhibition at the Sradhyananda Park and was delighted to find it illuminated throughout with "Bengal" electric lamps manufactured at Calcutta. The bulbs have stood the test of continued use for over a couple of months and are still going on all right. This is one of the few rare instances of a successful indigenous industry...Both in quality and in price the articles turned out compare favourably with imported goods. The organisers and workers are some of my past pupils and also of the Bengal Technical Institute who have got high scientific training to their credit.[80]

Bengal Lamps survived after independence (1947) also.

---

[76] Bhattacharya (1995).

[77] 'Memorandum of Association of the Bengal Electric Lamp Works, Ltd (1932).

[78] *All India Industrial and Commercial Directory, 1938-39, Bombay: no pub, 1938, p. 76.*

[79] Amit Bhattacharya, *Swadeshi Enterprise in Bengal*, op.cit., p. 131.

[80] *The Calcutta Municipal Gazette* (1933).

7.5 Concluding Remarks: Academia—Industry Interface

This essay treats the development of electrical engineering education and its relationship with industry in colonial Calcutta. The Bengal Engineering College, Sibpur, started providing basic instruction on the subject from the final decade of the nineteenth century. In a colonial situation, technological education was a contested terrain with contradictory positions of the bureaucrats, professional bodies, foreign business organizations, and, of course, Indian elites. The CESC, a London based power company, was instrumental in generation and supply of electricity in Calcutta—both urban and industry. From the beginning, they realized the need to create a pool of personnel (basically Indians) with elementary knowledge on electrical engineering. This was basically to economize the cost of power generation and supply in India. For supervision, needless to mention, they had European engineers with hefty salary and benefits. Thus, J. W. Meares, chief electrical engineer of the CESC, occasionally delivered lectures on the subject at the Sibpur college. The evolution of electrical engineering was gradual and half-hearted there—the colonial British government was reluctant to provide higher level of instruction.

The Indian elites appreciated the utility of this 'new' knowledge in emerging sectors of manufacturing and identified, on several occasions, the imperfect training of Indian engineers, including electrical. The Jadavpur college, from the inception, used to provide advanced training in electrical engineering. The development of their provision for higher technical education occurred through the adaptation of models that had been pioneered in the advanced countries—Germany, America, Japan, among others. Investments in education, especially techno-scientific, were now seen to be of great importance. The electrical power was to be the key to a modern industrial economy excited Bengali *bhadralok*.[81] Another significant dimension of the period was close coordination between academia and industry. For a quick example, the Bengal Lamp, specializing in the production of electric bulbs, was the brainchild of three Bengali brothers. As the electric blubs were slowly becoming indispensable among the growing middle class, taking the advantage they were instrumental in ushering a new era in the electric bulb manufacturing industry of the country. At the same time, they were closely associated with the Jadavpur engineering college, which is one of the preferred destinations of the post-secondary Indian students even today (now Jadavpur University). Jatindranath Basu, professor at the Jadavpur college during early twentieth century, rightly commented:

> Many consider electrification is luxury even today. But this is not true. We know electrification leads to rapid industrialization. The recent industrial growth of our country creates huge demand for smooth power supply. Use of diesel engine is economical compared to steam engine on several reasons. To fulfil the huge industrial demand, we need plenty of such engines for power supply without interruption. Here arises an important question: how

[81] In an article, read at the annual conference of the Chaitanya Library, the author emphasized, "...Machine helps to boost the productivity of any industry. The mystery of modern machine can be solved, although we are sceptical, through foreign visit. We will only advise our countrymen to learn the art of electricity—because future rests here. The current industrial trend indicates the acceptance of electricity as main motive power over the others." Mukhopadhyay et al. (1906).

long we will depend on the foreign counties for supply of these engines? Or shall we try to manufacture such engines in our country itself?[82]

References

Arthik Unnoti, Vol. 3, No. 3, *Aswar* 1335 B. S. (1928), p. 221.

E. H. Atkinson & Tom S. Dawson, *Report on the Enquiry to bring Technical Institutes into closer touch and more practical relations with the Employers of Labour in India*, Calcutta: Superintendent Government Printing, 1912, pp. 89–90.

Ramendra Nath Bandopadhyay, Sibapada Chattopadhyay and Md. Mahibul Majid. *Report on Public Instruction in Bengal for 1937–38*, Calcutta: Bengal Secretariat Press, 1938, p. 34.

Manindra Chandra Basu, 'X-Ray', Bijoli, Vol. 1, No. 4, December 1933, p. 169

P. N. Bose, *Essays and Lectures on the Industrial Development of India*, Calcutta: Newman & Co, 1906, p. 60.

Bengal Engineering College Centenary Souvenir, Calcutta: no pub, 1956, p. 56.

Amit Bhattacharya, *Swadeshi Enterprise in Bengal 1921–47*, Kolkata: Setu Prakashan, 1995, p. 130.

Bhim Chandra Chatterjee, *Notes on the Elements of Applied Electricity*, Benares: Satyanarayan Chatterjee, 1922, 131.

Bulletin Bengal Engineering College: One Hundred Twenty Fifth Anniversary Issue, 1981, p. 14.

Ananda Kishore Das, 'Shillong Hydro-electric Limited', *Arthik Unnoti*, Vol. 2, No. 2, *Jaistha* 1334 B. S. (1927), pp. 145–148.

First Quinquennial Review of Education in Bengal, 1892–96, Calcutta: Government Printing, 1902, p. 106.

Robert Fox and Anna Guagnini (eds.), 'Introduction', *Education, technology and industrial performance in* Europe, Cambridge: Cambridge University Press, 1993, p. 5.

General Report on Public Instruction in Bengal for 1903–04, Calcutta: Bengal Secretariat Press, 1904, p. 30.

Sidhartha Ghosh, *Kaler Sahar Kolkata*, Kolkata: Ananda, 1991, p. 118.

Thomas P. Hughes, *Networks of Power: Electrification in Western Society, 1880–1930*, Baltimore and London: John Hopkins University Press, 1983, p. 140.

Indian Industrial Commission, Minutes of Evidence 1916–17, Vol. II, Calcutta: Government Printing, 1918, p. 250.

Roy Macleod and Deepak Kumar (eds.), 'Introduction', *Technology and the Raj: Western Technology and Technical Transfers to India, 1700–1947*, New Delhi: Sage, 1995, pp. 11–21.

J. W. Meares, *Electric Traction: Six lectures delivered in March 1902 at the Civil Engineering College, Sibpur*, Calcutta: Bengal Secretariat Press, 1902, p. 1.

Meares, *Water Power: Its Estimates and Development with Particular Reference to India. Practical Treaties for Civil and Electrical Engineers*, Calcutta: Thacker, Spink & Co, 1935, pp. 7–8.

'Memorandum of Association of the Bengal Electric Lamp Works, Ltd', *Declaration on Registration of Company*, 24 March, 1932; Office of the Registrar of Companies, Kolkata.

Morris D. Morris, 'The Growth of Large-Scale Industry to 1947', in Dharma Kumar (ed.), *The Cambridge Economic History of India*, Vol. II, New Delhi: Orient Longman in association with Cambridge University Press (Reprinted Edition), 2008, pp. 599–600.

Dhruv Raina and S. Irfan Habib, *Domesticating Modern Science: A Social History of Science and Culture in Colonial India*, New Delhi: Tulika, 2004, pp. 83–112.

[82] *Arthik Unnoti* (1928).

Arthik Unnoti means 'Economic Progress'. This was a journal on economics, and the applications of industrial research and scientific inventions for social welfare and material progress. Benoy Kumar Sarkar, the great educationist, was the founder-editor.

Dhruv Raina, 'Institutions and Knowledge: Framing the Translation of Science in Colonial South Asia', *Asiatische Studien 'Etudes Asiatiques*, Vol. 65, No. 4, 2011, pp. 945–68.
Srinivasa Rao and John Lourdusamy, 'Colonialism and the Development of Electricity: The Case of Madras Presidency, 1900-47', *Science, Technology and Society*, Vol. 15, No. 1, 2010, p. 34.
Jitendranath Ray, *Banglar Kalkarkhana O Karigari Bidyar Ihihas*, Kolkata: Dey's Publishing, 2005, p. 155.
Report of the Indian Industrial Commission, 1916–18, Calcutta: Superintendent Government Printing, 1918, p. 106.
Report on Public Instruction in Bengal for 1923–24, Calcutta: Bengal Secretariat Press, 1924, p. 27.
Report of the Director of Public Instruction, Bengal, 1945–46, Calcutta: Bengal Secretariat Press, 1946, p. 9.
Report on Public Instruction in Bengal for 1910–11, Calcutta: The Bengal Secretariat Book Depot, 1911, pp. 35–7.
Suvobrata Sarkar, *Let there be Light: Engineering, Entrepreneurship and Electricity in Colonial Bengal, 1880–1945*, Cambridge: Cambridge University Press, 2020, pp. 115–89.
Suvobrata Sarkar, *The Quest for Technical Knowledge: Bengal in the Nineteenth Century*, New Delhi: Manohar, 2012.
S. N. Sen, *Scientific and Technical Education in India*, New Delhi: Indian National Science Academy, 1991, p. 394.
Shibpur College Patrika, Vol. 1, Issue 8, August 1902, pp. 231–2.
Sumit Sarkar, *The Swadeshi Movement in Bengal*, New Edition, Ranikhet: Permanent Black, 2010 (first published 1973).
Benoy Kumar Sarkar, *Education for Industrialization: An Analysis of Forty Years of Jadavpur College of Engineering and Technology, 1905–45*, Calcutta: Chuckervertty, Chatterjee & Co, 1946, p. 248.
Nirmal K. Sengupta, 'What is Electricity?', *Journal of the College of Engineering & Technology*, Vol. IX, No. 2, December 1937, p. 121.
Sri Tarak Nath Mukhopadhyay, 'Bharatiya Shilpa, Krishi O Banijjer Sanrakshan O Unnoti', *Janmabhumi*, Vol. 14, No. 10, *Baishak* 1313 B.S. (1906), p. 365.
The Calcutta Municipal Gazette, Ninth Anniversary Number, 25 November, 1933.

Chapter 8
Dental Public Health and the Institutionalisation of a Science: R. Ahmed Dental College and Hospital, Calcutta, C.1920s–1950s

Sahara Ahmed

> A nation's morals are like its teeth: the worse they are decayed the more it hurts to touch them.
>
> G. Bernard Shaw (1856–1950)
>
> Although medical education was started about the year 1858 no effort was made by the Government to inaugurate dental education, although dentistry formed a branch of medicine. It was not till the year 1920, that the first dental college was established in India at Calcutta. What an uphill task it was, with an apathetic public, a step-motherly government and an ignorant intelligentsia, the work was very difficult.
>
> Dr. Rafiuddin Ahmed (1890–1965)[1]

The Calcutta Dental College (later renamed as R. Ahmed Dental Medical College and Hospital) was an institution which was pivotal in crafting the modern discipline of medicine in the arena of oral care. Catering to both the educational and scientific clinical aspects, it was the locus of a new form of pedagogy, as well as the apparatus for ushering in a new knowledge of the body and innovative practices of medical intercessions. A peek into its early history brings out the multiple layers of the forces at work in the interactions of modern medicine and the other forms of medical expertise in vogue at that particular juncture. One distinguishing feature of this institute was that the founder father was a staunch believer in the principles of western medical practices, more so in the American rather than the British. His encounter was not with the indigenous medical practitioners for oral care but was

[1] Mukherjee et al. (1993).

This article is an outcome of a University Research Project (URP) funded by Rabindra Bharati University to which I owe my deepest gratitude.

S. Ahmed (✉)
Department of History, Rabindra Bharati University, Kolkata, India
e-mail: sahara.ahmed@rbu.ac.in

S. Ahmed and S. Sarkar (eds.), *Decolonizing Science and Modernity in South Asia*,
https://doi.org/10.1007/978-981-97-1829-0_8

a lone journey of an indigenous practitioner vouching for and transposing western medical practices which was till then the reserve of the colonial government.

Research on colonial medicine have engendered studies on the role of the colonial state, and bequeathed works on academic, legal, and other public institutions. Changes in economy, polity and society had its subsequent impact on medicinal practices and trajectories of such histories have been perused considerably.[2] A plethora of empirical studies on the encounters of modern medicine with other medical systems have generated interest amongst the academia.[3] Economic and political aspects have been attributed utmost importance in these studies. Extensive studies on epidemiologies of various diseases have received significant attention and particularly in the recent past these studies have been revisited and revamped.[4] The centre-periphery dichotomies have been examined by several studies and the way they shaped the delineations of medical practice in the colony.[5] The literature produced from such a vast repertoire is by all means numerous but despite this explorations of institutional histories have been few and limited in scope, and we possess a few descriptive accounts but an extensive macro and micro-level research on medical institutions provide room for further investigation.

'Scientific dentistry' was introduced in India in the nineteenth century by the British and modern dental care and the study of this discipline was initiated a hundred years ago. Biomedica or modern medicine was initially started in India for East India Company's employees and later on for officials of various British administrations. Medical officers were trained and mostly British doctors were invited from United Kingdom. The medical doctors expanded their private practice to public and in the middle of the nineteenth century the British Government established the medical colleges in India to train South Asian doctors, who could supplement the practice of British physicians. Hospitals and dispensaries (clinics) were built in the districts and in towns. Dental care was restricted to the extraction of teeth at such hospitals by medical men/doctor and assistant. In spite of this for long dentistry predominantly remained in the hands of the unqualified. The year 1920 is a milestone in the history of dental care with the establishment of the first dental college in Kolkata. Significantly, this perspective found expression in the endeavours of Dr Rafiuddin Ahmed and the institute he established in India in Kolkata in 1920 which came to be known as the R. Ahmed Dental Medical College and Hospital later on. Prior to this dentistry was in a 'primitive' state.[6] The establishment of an institute was instrumental in the development of the science of dentistry. Formerly, the foreign dentists catered to

[2] Commemorative Volume on the Occasion of the TerJubilee Year of Foundation of Medical College, Bengal, Medical College Ex-Students Association Calcutta, p. 9.

[3] There are several such studies: Bala (1991), Gupta (1976), Science and the Raj (2006), Leslie (ed.) (1992), Panikkar (1992), Pati and Harrison (2001), and Kumar and Basu (2013).

[4] Buer (1926).

[5] Curtin (1968) and Klein (1973). See also Harrison (1994) and Bala (2012), Watts (1999), Arnold (2000), Samanta (2002), Sinha (1998), Samanta (2017), and Dasgupta (2011).

[6] Kochchar (2000).

the recuperation of the British army.[7] Dr. Ahmed advocated the American model of dentistry in lieu of his training rather than the British model and which laid emphasis on the study of basic medical subjects.[8] The duration of the medical course was two years but through the years it witnessed many transitions which would come under the purview of this essay.

In 1920 when modern dental education was entirely unknown, a few men under the leadership of an idealist started the dental college in Calcutta overcoming all odds. It is not generally known that Dr R. Ahmed Dental College, now a part of our national heritage, was his personal gift to the nation.[9] The trials and tribulations of the young man who left India at the age of nineteen for higher studies in the United States with only rupees sixty in his pocket is a fascinating episode.[10] It was his commitment to the cause of his country that gave his romantic adventure and suffering the true human dimension. He ignored the allurement of a very lucrative practice and started the first dental college in Calcutta in 1920.[11] His sole ambition was to bring to India that education in a specialized subject which they had received abroad and make that knowledge available to a large number of our students so that thereby the mass of the people may benefit. Dr. Ahmed earned both fame and money, but he did not remain content with the reputation he had thus acquired. It was his ardent desire to do something for dental education in India. He realised that though there were some dentists in big cities like Calcutta and Bombay, the number of dentists for the country as a whole, was negligible. There was an influx of Chinese dentists in the country who he believed did more harm than good to the innocent masses. Besides, there were no dentists, in district places of most of the provinces. Thousands suffered from dental diseases in silence and many had to travel a distance of two hundred miles just for an extraction of a painful tooth.

For the systematic training of young men in the dental science, there were no institutions in the country and the Indian Government was apathetic to this problem. Apprenticeship with qualified or unqualified dentists on payment of fancy fees was the only course left for the few young men who wished to study this science. This method of education was far from satisfactory. Dr. Ahmed founded the Calcutta Dental College and Hospital in 1920 on systematic lines as in other countries of the world and thus served our country by this generous and magnificent act. He donated Rs. 25,000 to start with and devoted much of his valuable time in establishment of this much needed institution. By this noble act a new avenue was opened for young men and in 1928 he published a 'Student Handbook on Operative Dentistry' for ready guidance.

Practice of scientific dentistry was something unheard of in those days. With the exception of a few qualified European dental surgeons whose activities were limited to the upper strata of the society. Dentistry was more in the nature of trade or a craft

---

[7] P.C. Kochchar, The Army Dental Corps and Military Dentistry, op.cit., p. 25.

[8] *Ibid.*, p. 26.

[9] *Ibid.*, p. 27.

[10] Home Medical Proceedings, WBSA, DEC 1920 B38-50, p. 18.

[11] *Ibid.*, p. 19.

than a profession in those days. Getting people interested in a profession which did not exist, creating facilities for training in a career whose future was uncertain was an uphill task.

Despite these and many more hurdles Dr R. Ahmed was determined to put the name of dentistry on the educational map of India.[12] Perfectly aware that there would be lack of suitable candidates for admission in the school due to lack of proper consciousness regarding dental health amongst the masses he designed his programme of teaching dentistry in a manner suitable for gradual expansion in future to cater to the needs of the public and that of the profession. From a handsome gift of rupees ten thousand donated by Dr Ahmed himself, necessary furniture and equipment were purchased for a well-designed dental college of the time.[13] Some unwavering and enthusiastic companions, men of medicine, supported him in his noble venture. Dr R. Ahmed was, however, willing to bear full responsibility, both at theoretical and clinical levels.

8.1 Humble Beginnings and Pedagogy

His dental chamber was situated at 12/1, Esplanade East and within a year he attained fame for himself. Many young boys went to him either for a job or to work as apprentices.[14] Profulla Kumar Chatterjee of Barisal was first to work as apprentice under him and within one year three or four boys joined him. This gave Dr Ahmed an idea of opening a dental institution. He soon realised the situation and tried to open a new line for these Indian boys, but found that the educational authorities did not think it worthwhile to make any provisions for the training of dentists in India. Annexed to his chamber he hired a room for training these boys in technique as well as in basic theory. It was an experiment from 1920 to 1923 and during this time 11 students came from different provinces in India.

The names of the students who took their training during that period were: (1) Prafulla Kumar Chatterjee from Sylhet, Assam, (2) D.N. Bose, Indore, (3) M.G. Chanda from Jammu and Kashmir, (4) P. Sharma from Mandalay, (5) Edwin Marks, I.M.D. from Lucknow, (6) A.M. De Greuther, I.M.D. from U.P. and (7) J.E. D'Souza, M.B.B.S, Jhansi from Central Provinces. Three more students besides these proceeded to U.S.A for degree courses.[15] In 1923 he found that unless he obtained a separate building and got other professional men to help him in teaching, he would not be able to accommodate the increasing number of students anxious to learn

[12] Home Medical Proceedings, July 1920 A 38–50.

[13] Home Medical Proceedings, Aug. 1925 B 45–50.

[14] Milton B. Asbell, *Dentistry: A Historical Perspective*, History of Medicine, Shelf mark W/ASB.

[15] *Annual Report of the Calcutta Medical Institute of Hygiene and the Carmichael Hospital for Tropical Diseases for the year, 1924*, Calcutta, Bengal Government Press, 1925.

dentistry.[16] He, therefore, hired the premises at 261, Bowbazar street and donated about Rs. 10,000 from which 5413 were spent for furniture, instruments and other material. Since then, he gathered round him other dentists and physicians to teach various subjects based on the American, British and continental colleges of dentistry was formulated incorporating the best from all. The management was vested in a body consisting of six members with Dr R. Ahmed as president, Dr S.C. Das, Dr S.C. Kaviraj, Dr T. Ahmed, Dr P. Das as members and Mr Amin Ahmed as Honorary Secretary and treasurer.[17] The constitution and bylaws were drafted eventually the object of that organisation was pithily summed up thus,

(a) To further the cause of dental education in India.
(b) To establish and maintain the dental training institution called the Calcutta Dental College and Hospital and similar other institutions throughout the country.
(c) To seek the help and co-operation of individuals, corporate bodies and the government for the furtherance of the cause of the Calcutta Dental College and Hospital.
(d) The organisation shall be non-political and non-sectarian.

The founding members in consultation with the teaching staff framed the rules and regulations of the college.[18] They maintained, that,

> We have not failed to learn a lesson by study of history of dental education in other parts of the world. Even in some of the very advanced countries of the world, as for example the United States of America, the curriculum of a dental student extended up to only 2 terms of nine months each till as late as 1900. Of course, dentistry since then has become a branch of the university studies and the course lengthened. We in India, who are just starting our dental education, cannot afford to jump to the top of the tree all at once, we must also evolve slowly till we come up to the standard of the more advanced countries. Therefore, we have started two years intensive course in dentistry to begin with. So far, we have been content to get our dentists from the clerks or mechanics, who served for a time under European and American dentists and combine dentistry with a trade such as spectacle fitting or watch making. That has got to be changed. If we had started at once with a four years course, we believe we would have been doomed to failure. We have started with a 2 years course which will keep on increasing year by year till our public get used to the fact that dental training is as arduous as a training for any other profession.[19]

Hence it was decided that the general course leading to the diploma of Licentiate in Dental Science (L.DSc) will be given to those candidates who will pass through the full course of two years.[20] However, the idea generally was to become a dentist after a very short training. The short course of one year as it was given 1920–24 was

---

[16] *Confidential Reports in Native Papers for the week ending the 29 June, 1895,* p. 25, National Archives of India (hereafter NAI).

[17] *Ibid.*, p. 23.

[18] *Ibid.*, p. 24.

[19] *Annual Report of the Calcutta Medical Institute of Hygiene and the Carmichael Hospital for Tropical Diseases for the year, 1924,* Calcutta, Bengal Government Press, 1925.

[20] *Ibid*,

also continued. No theoretical lectures were given to them.[21] The complete course of study was offered from 1924. The minimum entrance requirement was matriculation examination of an Indian university or its equivalent.[22] One year course was also given to licentiates and graduates in medicine of State Medical Faculty or university. Twenty-two students joined the college amongst them 3 were girls and 19 boys. Out of these 22 students only 4 students joined the short course. These boys and girls came from every province in India. One Ms Amina Ida Sakir came from Mauritius. The names of those 4 students who joined the regular course and passed out in 1925 were,

1. Ms. Surga Mohmed Ali
2. Sydney James Bellgard
3. Arthur Duncan Campbell
4. Vashudev Mukund Pathak

The tuition fee was rupees five hundred per year payable in four instalments and Rs. 250 was charged for the short course. The students had to buy a set of working instruments which was a sort of investment as they could utilise them in their private practice. It was reiterated that dental education is one of the most expensive educations in the world. Unlike other professions dentistry requires from its votaries a certain amount of skill, nimbleness of fingers.[23] Besides, it was decided that each student must be given a work room bench, a dental chair, a cabinet and many other dental accessories in order that the student may learn the practical part thoroughly.[24] Therefore, it costs more to train per head in dentistry than other professions where one need hear only lectures.[25] Moreover, the number of students was less, so the fee of Rs. 500/- was presumed to be reasonable in comparison with the fees charged in England which was 100 pounds and, in the U.S., it was 250 dollars a month.[26]

8.2 Faculty Structure and Equipments

Regarding the teaching staff, it was not very difficult to get medical men to teach medical subjects. However, the difficulty came in getting qualified dental surgeons to teach the various dental subjects. Dr P. Das who was trained in Edinburgh was the only dentist to help Dr Ahmed in dental subjects. He used to teach Prosthetic Dentistry including crown and bridge work while Dr Ahmed used to take up the rest

---

[21] *Confidential Reports in Native Papers for the week ending the 19 June, 1925.*, p. 34. NAI.

[22] *Ibid.*, p. 24.

[23] *Bengal Public Health Report for the year 1926*, Calcutta, Bengal Secretariat Press Book 1927, pp. 22–23.

[24] *Annual Report of the Calcutta School of Tropical Medicine, 1924*, Calcutta, Bengal Government Press, 1927.

[25] *Annual Report of the Calcutta Medical Institute of Hygiene and the Carmichael Hospital for Tropical Diseases for the year, 1924*, Calcutta, Bengal Government Press, 1925, pp. 23.

[26] Ibid., p. 14.

of the dental subjects starting from Dental Anatomy to Orthodontia.[27] In 1925, Mr. Sultan Mahmood came from Philadelphia, U.S.A. after acquiring training in Mechanical Dentistry and was appointed as Prosthetic Tutor of the Mechanical Laboratory. During the later part of 1925, Dr M. H. Sajjad, D.D.S. (Penn., U.S.A), Dr H. L. Dhingra, D.D.S. (Indiana, U.S.A.) came to Calcutta and joined the College. That was a great relief to Dr Ahmed. But to get a qualified full time dental surgeon on the staff was a very expensive proposition, as any dental surgeon, who was qualified abroad was likely to earn much more in private practice than what the College could offer him.

The faculty comprised of 6 part time professors and 10 demonstrators who devoted their unflinching service to the college for a paltry honorarium. Students had to work from 8 a.m. to 12 p.m. and 1 p.m. to 5 p.m. The hospital had to remain open from 8 a.m. to 12 p.m. and from 2 p.m. to 5 p.m. every day except Sunday. The hospital was equipped with 6 crank chairs.[28] Each dental chair had a wooden locker adjacent to it that the students utilised as a dental bracket for instruments.[29] An ordinary aluminium spittoon was placed on the left side of the chair on a wooden stool, which was cleaned by a sweeper after the work was finished.[30] A nominal fee was charged from the patient as the cost of material used.[31]

8.3 Structure of Curriculum

The subjects taught during the two years were,

> 1st Year: (1) Anatomy, (2) Physiology, (3) Materia Medica, (4) Dental Metallurgy, (5) Bacteriology and Pathology, (6) Chemistry, (7) Operative Dentistry (Technique work and lecture).
>
> 2nd Year: (1) Prosthetic Dentistry (lectures and hospital work), (2) Operative Dentistry (Technique work and lectures and hospital work), (3) Orthodontia, (4) Local Anaesthesia and Exodontia, (5) Radiology, (6) Dental Jurisprudence, (7) Professional Ethics.

From the session beginning July 5th, 1926, the course was raised to three years. The course was raised to three years.[32] The curriculum was developed as follows:

> 1st Year: (1) Physics, (2) Chemistry, (3) Anatomy, (4) Physiology, (5) Materia Medica, (6) Dental Metallurgy, (7) Operative Dentistry (Technique and lectures).

[27] *Ibid.*, p. 15.

[28] The Indian Dental Journal, The Indian Dental Association, Calcutta, 1925–46.

[29] Gelber (2005). The 1879 register is referred to as the dental register.

[30] *Ibid.*, p. 470.

[31] Merrel (1992).

[32] Mukhopadhyay (2014).

2nd Year: (1) Anatomy, (2) Physiology, (3) Dental Therapeutics, (4) Pathology, (5) Bacteriology (6) General surgery, (7) Local and General Anaesthesia and Exodontia, (8) Operative Dentistry (technique, lectures and clinics, (9) Prosthetic Dentistry (Technique, Lectures and clinics).
3rd Year: (1) Oral Surgery, (2) Pathology, (3) Radiology, (4) Dental Jurisprudence, (5) Ceramics, (6) Operative Dentistry (clinical work and lectures), (8) Professional Ethics, (9) General and Dental Hygiene.

The buildings at 261, Bowbazar Street soon proved inadequate to accommodate this burgeoning branch of medicine and in October, 1926, the College was shifted to 33, Bowbazar Street. Due to the amplifying number of students in 1926, they enquired about the course of the study but when they found that the course duration was three years they were disappointed.[33] Although the number of students was increasing every year, it was too small to meet the demand of the country.[34] The income accruing from the students and the patients was too insignificant for maintaining college expenses.[35] Hence, the executive committee decided to continue the short course and to open a mechanical course for training Dental Mechanics.[36] The teaching was imparted by practical demonstration and technique work on models. The duration of course was six months. The fee for the mechanical course was Rs. 20 per month. The number of boys in the short course and the mechanical course was double the number in the regular course. The student strength of these two courses attributed to the college its intransience. In 1928, Dr. S.K. Majumdar, D.M.D. (Harvard, U.S.A), joined the college. During the same year Dr Sajjad left Calcutta and the subjects, Orthodontia, Anaesthesia and Exodontia were the responsibility of Dr Majumder. One year later, i.e., in 1929, Dr M. Ahmed, B.D.D.S. (Penn., U.S.A.), came to Calcutta and joined the College. He was taken as Professor of Prosthetic Dentistry and Superintendent of the Hospital. The College and Hospital improved under their tutelage.[37]

Besides the theoretical lectures and the technique work on models and mannequins each student had to complete the following amount of work in the hospital under the guidance of the Superintendent and House Dental Surgeons and demonstrators[38]:

Gold Filling: 2
Amalgam or Porcelain filling: 100
Prophylaxis: 100
Gold or Porcelain crown: 3
Bridge Work: 20
Root Canal Filling: 4
Upper and Lower full dentures: 50

---

33 Berry (1965).
34 Jayade (1988).
35 Kochar (2000).
36 Ring (1992).
37 Jayade (1988).
38 *Ibid.*, p. 2.

Partial dentures: 50
Extractions: 400
Gold or other metal inlays: 4

Students were not allowed to work in the hospital unless they finished their technical training work in the hospital in the second year and satisfied the superintendent regarding their proficiency. One student was allotted for the day to examine outdoor patients, under the supervision of the House Dental Surgeon.[39] The patient was then assigned to some student who was free. At the beginning the number of patients was not high but as soon as the college gained popularity the number was sufficient to keep the students engaged.[40] Sometimes patients were even refused for want of accommodation.[41]

In 1931 Dr W.F. Hendrickson, M.D., D.D.S. (Penn., U.S.A.) while working as Dental Surgeon at Messrs Smith Bros. Ltd., agreed to join the College as a permanent Superintendent on a moderate remuneration.[42] Up to the year 1931 the College could save a nominal sum from the tuition fee and from the Hospital. Hence, the Corporation of Calcutta was approached for some pecuniary help but they did not attach any importance to it, till Dr R. Ahmed was elected a Councillor of the Corporation in 1932.[43] He convinced the mayor and other members of the Corporation and succeeded in getting a small grant of Rs. 1000 per year from 1933 on the condition that the children of the Free Primary Schools would have to be treated free in the dental hospital.[44] The University of Calcutta was also approached but they did nothing although they were sending the students of the schools and colleges through the "Students Welfare Committee" to get their attendance in Calcutta Dental Hospital.[45]

It was during the time of Dr Hendrickson that great improvements were made in this institution. He made a scheme to improve the hospital and laboratory.[46] Accordingly dental brackets and spittoons fitted on a stand were ordered for the hospital, furniture and other necessary equipment were purchased to equip the existing three laboratories, viz., (a) Physics and Chemistry; (b) Histology and Pathology; (c) Prosthetic Laboratory.[47] Three more demonstrators were appointed. The students had to

---

[39] Berry (1965).

[40] *The Indian Dental Journal*, The Indian Dental Association, Calcutta, 1925–46, p. 25.

[41] *Indian Dental Journal*, The Indian Dental Association, Calcutta, 1936, p. 15.

[42] *Bengal Public Health Report for the year 1926*, Calcutta, Bengal Secretariat Press Book 1927, p. 10.

[43] *Bengal Public Health Report for the year 1926*, Calcutta, Bengal Secretariat Press Book 1932, p. 25.

[44] *Annual Report of the Calcutta Medical Institute of Hygiene and the Carmichael Hospital for Tropical Diseases for the year, 1932*, Calcutta, Bengal Government Press, 1935, p. 15.

[45] *Ibid*., p. 20.

[46] *The Indian Dental Journal*, The Indian Dental Association, Calcutta, 1925–46, p. 22.

[47] *Ibid*., 23.

attend the dissection class for practical Anatomy at the Calcutta Medical School.[48] The students were working more sympathetically under strict discipline. In fact, he took so much interest that within one year he changed the whole atmosphere of the College and Hospital.[49]

The total number of students in the regular course was about 75 in 1932. Thus, the Executive Committee decided to abolish the special short course and the mechanical course. In the same year another move was made to more commodious quarters at 249B, Bowbazar Street premises was kept as the habitat of the College due to the fact that this street was located in a densely populated neighbourhood, because there had never been any dearth of a daily supply of patients and instructive cases.[50] Unfortunately, by the end of 1933, Dr Hendrickson had to go back to the U.S.[51]

The College did not have sufficient funds to train graduates abroad. During the same year Dr R.N. Ghosh and Dr R. Chanda, graduates of the College returned to Calcutta after taking their training in L'Ecole Dentaire de Paris and joined the College as Deputy Superintendents of the Hospital.[52] Later on some energetic boys went to France and the United States of America, started their practices in Calcutta and served the Calcutta Dental College and Hospital as professors, Dr N.C. Berry, D.D.S., graduating from the University of Pennsylvania, U.S.A., joined the College in 1935 as Professor of Dental Ceramics and later on as Professor of Prosthetic Dentistry, including crown and bridge work. Around this time, Dr P.K. Dasgupta, D.D.S., graduating from University of Pennsylvania, U.S.A., worked as Professor of Radiology and Electrotherapeutics. Two years later Dr S.P. Mewar, D.D.S. graduating from University of Pennsylvania, U.S.A., joined the College as Superintendent of the Hospital. During the same year, Dr T.C. Das, D.D.S., graduating from North-Western University of Chicago, U.S.A., joined the College as Professor of Orthodontia.[53]

8.4 Government Initiative

Tremendous amount of persuasion and concerted effort by the authorities of the institution culminated in the Government of Bengal conceding to the appointment of a Dental Education and Registration Enquiry Committee in November, 1934 comprising of,

1. Col. R.P. Goil, I.M.S., Principal, Medical College.
2. Sir Kedar Nath Das, Principal, Carmichael Medical College.

---

[48] *Bengal Public Health Report for the year 1926*, Calcutta, Bengal Secretariat Press Book 1927, p. 10.

[49] *Ibid.*, p. 24.

[50] *Government of India, Education, Health and Land, Sanitary Branch, June 1923, Proceedings No. 21, F2/33,* NAI.

[51] Berry (1965).

[52] *Ibid.*, p. 110.

[53] Jayade (1988).

3. Dr R. Ahmed, D.D.S., Principal, Calcutta Dental College.
4. Dr H.A. Taylor, L.D.S., R.C.S., Dental Surgeon, P.G. Hospital.
5. Dr B. Randel Jennings, L.D.S., R.C.S., Professor of Dental Surgery, Medical College, Hospital.
6. Dr S.K. Majumdar, D.M.D., Professor of Dental Surgery, Carmichael Medical College.
7. Dr G.E. Constantinides, D.D.S., Superintendent, Calcutta Dental College and Hospital.

The Committee appointed a sub-committee, consisting of four Dental Surgeons, Dr Taylor, Dr Jennings, Dr Majumdar and Dr Ahmed. This sub-committee submitted a report to the Dental Registration and Education Enquiry Committee and ultimately to the Government of Bengal recommending a standard of dental education and a curriculum of study in the Dental College.[54] The minimum entrance qualification was decided as Intermediate Science Examination of any Indian University with Chemistry, Physics and Biology. At the suggestion of Sir Kedar Nath Das it was also decided that the examinations would be held for the Diploma under the auspices of the Bengal State Medical Faculty to which body some dental members would be appointed. The Committee suggested that a Dental Registration Act be passed in the province.[55] A separate Board of Dentistry also should be formed, which would be responsible for only registration of graduates, collection of fees and other office work. The affiliation of the Calcutta Dental College and Hospital to the proposed Board of Dentistry also should be formed, which would be responsible for only registration of graduates, collection of fees and other office work.[56] The affiliation of the Calcutta Dental College and Hospital to the proposed board was recommended, subject to its curriculum conforming to that required by the board, viz:- Pre-educational qualification and four years study according to the syllabus outlined by the sub-committee.[57] The Committee suggested that the medical subjects would be taught in the Calcutta Medical College and the dental subjects in the Calcutta Dental College.[58]

8.5 Commodious Quarters

The Calcutta Dental College and Hospital authorities were willing to conform to the curriculum required by the Enquiry Committee and the proposed Dental Board. Accordingly, the college was reconstituted. In 1935, the College was removed from 249/B, Bowbazar Street to 114, Lower Circular Road. The building with an area of 2 bighas (acres) of land was taken on a 10 years lease, the hospital building was

[54] *Annual Report of the Director of Public Health for Bengal*, Bengal Secretariat Press, 1921, p. 10.

[55] *Ibid*. p. 11.

[56] *Government of India, Education, Health and Land, Sanitary Branch, June 1923, Proceedings No. 21, F2/33*, p. 15, NAI.

[57] *Ibid*., p. 15.

[58] *Ibid*., p. 16.

built at College's expense. Lecture halls were well furnished with chairs and writing desk and Ediadiascope for projection. The hospital was equipped with 25 pump chairs and brackets, X-ray apparatus and other accessories. Many important books were purchased and journals subscribed for the library. Some rare specimens, casts of deformities and irregularities of the teeth and other instructive models were collected to enlarge the College Museum. The entrance requirement was raised from matriculation to the intermediate science from an Indian university.[59] The course of study was made four years according to the syllabus outlined by the Dental Registration and Education Enquiry Committee.[60]

About Rs. 20,000 was spent for such improvements with the expectation that the government would give all possible assistance for the improvement of the College as they found that they could style themselves as dentists even without studying in the College for four years and spending so much money. As there was no future prospect, excepting one's own private practice, the guardians hesitated to send the boys into the branch of science.[61] Government seemed to be indifferent in the matter and did not encourage the intending students by giving scholarships or any employment. No protection was given to the graduates of the College. Naturally the number of new students was very small. After a hard struggle for 2 years the College authorities decided to lower the entrance requirement from Intermediate Science to matriculation examination.[62] The Corporation of Calcutta was again approached for increment and their grant was increased from Rs. 1000 to Rs. 2950 annually.[63]

Ahmed's pioneering effort in dental education was very soon rewarded by short increase in the number of admissions and the training programme was taking shape gradually. In less than six years from the beginning of the college most of the basic subjects were introduced in the syllabus alongside the existing dental subjects and the total training period was raised from 2 to 3 years.[64] By this time, through his own initiative and interest of several other qualified dental surgeons teaching of clinical and practical dentistry at Calcutta Dental College were well coordinated with didactic courses. As the number of students and patients increased, he had to shift the college to more commodious quarters.[65] By 1928 under his able leadership, a well-planned dental school in spite of financial hardships was imparting scientific training to the best of its ability as a pioneer centre of education in dentistry in this country.[66]

Now that he had turned an idea into an efficient, well organised, pioneer institution through his own singular efforts, he tried to get the educational authorities in Bengal

---

[59] *Annual Report of the Director of Public Health for Bengal,* Journals/Bengal Secretariat, Bengal Secretariat Press, 1921, p. 21.

[60] *Ibid*., p. 21.

[61] Berry (1965).

[62] *Ibid*., p, 22.

[63] *Ibid*., p. 24.

[64] Milton B. Asbell, *Dentistry: A Historical Perspective, History of Medicine*, Shelf mark W/ASB, pp. 25–30.

[65] *Ibid*., p. 25.

[66] *Ibid*., p. 26.

interested in helping him to establish dentistry in its due place along with teaching of medicine. His idea was to introduce teaching of dentistry in the University or at best at the State Medical Faculty level.[67] His plea was turned down as early as it was made. This attitude very often pained him but did not discourage him and he used to say that, "Education is the responsibility of the state, an individual cannot discharge such responsibility, but if no one is willing to carry the cross, I will as long as I can."[68]A mission he accomplished successfully. The years from 1928 to 1938 saw continued progress of Calcutta Dental College marked by its fruitful activities and advancement into the field of dental education.

The then government of Bengal showed little interest in this institution and never encouraged it in any way except occasional words of appreciation from the visiting officers of the government. Unhampered by such obstructions he went on recruiting young dental teachers willing to join him in his missionary work. These noble young men joined his work willingly and without any considerable financial benefit for themselves, just out of sheer pleasure of working for the profession.[69] These additions in his staff lightened his burden and gave him an opportunity to devote more of his time to muster public and official opinion for recognition of dental education at university level.[70] After several unsuccessful appeals and representations to the government for recognition of dental education at par with medical education, Dr R. Ahmed's singular efforts were partially successful.[71] A commission was formed by the government of Bengal to examine the state of affairs in dental education and for necessary recommendation.[72]

This commission reported that the entrance qualification to the dental college should be intermediate science with biology followed by four-year course in dentistry leading to an examination by the State Medical Faculty of Bengal.[73] The prescribed syllabus in dental subjects was more or less on the pattern of L.D.S Courses offered in England in those times. It may be noted with satisfaction that the syllabus, framed more than 75 years ago compares very favourably with any B.D.S. syllabus of Indian universities of today.[74] However, the introduction of higher secondary as minimum qualification proved to be a serious setback as the annual admission fell to 4 students only from 25 students admitted in the previous year. At his request the State Medical

[67] *The Indian Dental Journal*, The Indian Dental Association, Calcutta, 1926, p. 28.

[68] *Ibid.*, p. 23.

[69] Resolution passed by the sixth Conference of Medical Research Workers held at Calcutta between the 18th and 20th December 1928, *The Indian Dental Journal*, The Indian Dental Association, Calcutta, 1928, p. 124.

[70] *Ibid.*, p. 125.

[71] *Ibid.*, p. 30.

[72] *Ibid.*, p. 31.

[73] *Reports of the Medical Officer of the Privy Council and local govt. Bengal*, Bengal Secretariat Press, 1938, p. 15.

[74] *Ibid.*, p. 16.

Faculty lowered the entrance requirement to matriculation keeping at par with its own medical schools.[75]

The school was shifted to its present address in 1936. Increase in number of teaching staff, increased floor space in it, new quarters and reorganization of the different departments made it possible for the school to meet the requirements of the State Medical Faculty.[76] During the second world war the institution suffered a serious setback due to economic hardship, high cost of living, low student enrolment and various other difficulties.

8.6 State Affiliation

Since the birth of this institution the full financial burden was borne by Dr R. Ahmed himself with the exception of an early grant of Rs. 3000 from the Calcutta Corporation in token of services rendered to the school children of Calcutta Corporation Schools from the late 1930s.[77] In the post independent period an increased awareness of dental health problems and special interest of former chief minister and late Dr B.C. Roy in dental education led to the transfer of responsibility for the institution to the government of West Bengal.[78] Dr R. Ahmed made a gift of the college and all its properties to the state government in 1950, leading to the introduction of Bachelor of Dental Surgery (B.D.S. course) by Calcutta University in 1952.

The Calcutta Dental College and Hospital was soon a responsibility of the Government of Bengal. When Dr. Ahmed was asked about this he said "What more appropriate gift can I leave to my Province than a Dental College? A College which I believe will be the finest in India in the days to come. It is yours and mine—it is ours."[79] Dr. B. C. Roy, himself a renowned doctor by profession, and the then chief minister of West Bengal, in his speech succinctly summed up the need for concerns for public health with the aid of the hospital and others of its kind,

> I thank you for giving me this opportunity of speaking to so many of our dentist friends who have met today, through my Chief Secretary who has agreed to deputise for me on this occasion. You are responsible for the health of a very important part of our body mechanism. Food presented to the gastrointestinal tract could only be digested properly and assimilated satisfactorily if it reaches the stomach after being effectively masticated and mixed with saliva during such mastication. Only those who have reached or are approaching the "second childhood" realise what value to put to a properly qualified dentist who can, by making a proper denture, not-only preserved the facial features but also provide a regular set of teeth

[75] *Bengal Public Health Report for the year 1926, Calcutta*, Bengal Secretariat Press, 1938, p. 18.

[76] *Ibid.*, p. 19.

[77] Resolution passed by the sixth Conference of Medical Research Workers held at Calcutta between the 18th and 20th December 1928, *Indian Dental Journal*, Indian Dental Association, Calcutta, 1938, p. 20.

[78] *Government of India, Education, Health and Land, Health Branch, May 1925*, Proceedings No's 20–21, pp. 18–19.

[79] Mukherjee (1993).

> to bite and chew so that digestion of food and preservation of health for such people can be ensured.....The West Bengal Government have felt the need for this approach to Dentistry. It has therefore been decided that Government should take up the Calcutta Dental College and Hospital which represents the life longs sacrifice and service to the cause of Dentistry of Dr. R. Ahmed and his associates. (Loud applause) Government desire to convert his in to a fully equipped institution for training and research in Dentistry. The Dental Profession and the people owe a deep debt of gratitude to Dr. Ahmed for his *making over the Institution free of charge to Government*. Such example have not been unknown in this Province and in India but they represent a spirit of sacrifice and service which, I trust, the Dentists will try to emulate for all time to come.[80]

The Bengal Dental Board held its first meeting, in the hall of the State Medical Faculty of Bengal in July, 1910. The meeting was attended by Dr. B. R. Jennings, LD.S., R.C.S. (Chairman) Dr. R. Ahmed, D.D.S., F.I.C.D, Dr. M. Ahmed, B.A., L.M.P., D.D.S., Dr. Bankim Mukherjee, L.D.S, R.C.S., Dr. S. P. Neogi, L. D.Sc., D.E.D.P., D.D.S. Dr. M. Haque, L.D.Sc., Dr. Q. M. Musa, M.B., Ch.B. Dr. S. K. Ghosh, MBIBS MALCS. TR. C. S. This was the first meeting of the Board, after its constitution under Bengal Government notification No. 1026 Medl., dated the 13th. May 1940.

The Chairman in welcoming the members observed that this was a noteworthy event in the history of Bengal in the matter of development of Dentistry according to Western Medical science. He felt confident that the Board would do all in their power to strengthen the foundation laid by their leader and ratified the Bengal Dentists Act, and gradually hailed a sound and worthy structure on it.[81] He then gave a resume of the efforts made in the past by several eminent persons, which matured after a lapse of over 20 years. This was the Dentists Act of 1939.[82]

The work of the Ministry of Health was recorded in the official Annual Report as well as the Medical Report.[83] A brief indication of the organisation, functions and manner of working in relation to health services of the principal departments which did not form an integral part of the ministry of Health Act would facilitate a clearer visualisation of their relationship to the Ministry.[84] These departments issued separate reports.[85] The medical work of the Board of Education was then done under the authority of the Minister of Health, in accordance with the provisions of Sect. 3 (1) (d) of the Ministry of Health Act 1919, and there was a constant interchange of opinion between the Ministry and the Board; proposed decisions that might affect the policy

---

80 *Confidential Reports in Native Papers for the week ending the 29 June, 1952*, Bengal Secretariat Press, Calcutta, 1952.

81 *Indian Dental Journal*, Indian Dental Association, Calcutta, 1938, p. 10.

82 Government of India, Education, Health and Land, Sanitary Branch, June 1939, Proceedings No. 21, p. 31.

83 Annual Reports of the All-India Institute of Hygiene and Public Health, Calcutta, Bengal Secretariat Press, 1947.

84 *Report of the Committee on the organisation of Medical Research under the Government of India*, Calcutta: Department of Education, Health and Lands, 1939.

85 *Annual Report on the Working of Hospitals and Dispensaries under the Government of Bengal for the year 1933*, Calcutta: Bengal Government Press, 1939–41.

of the Ministry, or vice versa were reviewed by the officers of both departments.[86] Since its inception under the Education (Administrative Provisions) Act of 1907, the School Medical Service was expanded gradually and has been viewed with growing appreciation by the local authorities and by the public. At that juncture it corresponded to the medical and dental treatment of those requiring it.[87]

The staff of the Local Education Authorities responsible for the carrying out of this programme included 318 school medical officers (Whole time or part-time), 686 assistant school medical officers (whole time or part-time) and 626 specialists (part-time), making a total of 1,630 medical men concerned under the undertaking. There were also 1,575 whole time and 452 part-time school nurses. The number of school clinics throughout the country was (approx.) 700. These figures indicated a considerable increase on those for the "war years" and there was little doubt that the School Medical Service had recovered from the temporary setback it received during the war. It was true that a number of authorities had not found it possible to resume fully all the arrangements that were in force before the war and that temporary expedients for carrying on some of the work were still functional, both most authorities had now re-established their medical staff completely.[88] This was significant since in accordance with Section I of the Education Act of 1918, it was the duty of every County and County Borough Local Education Authority,

> So far as their powers extend, to provide for the progressive development and comprehensive organisation of education in respect of their area, by submitting to the Board, schemes showing the mode in which their duties and powers under the Education Acts are to be performed and exercised. These schemes will fail in their purpose no matter, how perfect they may be in other respects, unless they have full regard to the healthy, responsive and alert physical and mental condition of the children and young persons for whose benefit they are formulated. It is now accepted that the body of the child must be sound and unimpaired by defect or deformity if the full effect of the education provided is to be secured and that its training and development are an integral part of all true education. Local Education Authorities throughout England and Wales are now in possession of an adequate and comprehensive instrument for achieving these great national ends and the inestimable good which they have accomplished in ten years of exceptional difficulty affords good promise of their wise and far-seeing use of the new and enlarged opportunity which now lies before them.[89]

Akin to this a Central Dental Health Publicity Bureau was instituted and a special fund for the Bureau was arranged for. The members were able to collect about Rs. 2000 on the occasion of the Annual General Meeting. Dr. B.C. Roy subscribed to the view that the problem of prevention of dental diseases should be dealt with by opening dental clinics for children in schools. That required complete collaboration between the dentists and the authorities-local or governmental. The West Bengal Government

[86] *Confidential Reports in Native Papers* for the week ending the 29 June, 1935.

[87] *Annual Reports of the* All-India *Institute of Hygiene and Public Health*, Calcutta, Bengal Secretariat Press, 1947.

[88] *Report of the Committee on the organisation of Medical Research under the Government of India*, Calcutta: Department of Education, Health and Lands, 1919.

[89] *Proceedings of the govt. of India, Education.* dept. Meld. June–Aug, F-3–2/47, A I, July 1957, S I, NAI.

had felt the need for this approach to dentistry. It had, therefore, been decided that Government should like to take up the Calcutta Dental College and Hospital.[90] Dr. B.C. Roy lauded Dr. Ahmed as an example that others should emulate since he represented a spirit of sacrifice and service.[91] Replying to felicitations, Dr. Ahmed said that he prided himself in being a Bengali and the honour they were showing him really went to Bengal which was the pioneer province was the study of dentistry in this country.[92]

Dr. Ahmed recalled under what great difficulties the work of the Calcutta Dental College and Hospital was carried on during the past thirty years.[93] They all knew how difficult it was to make the previous governments take interest in such works, either the British or the Muslim League Government. There had been a change in the atmosphere with the inauguration of the National Government and he felt that the progress of dental education in India would be greatly accentuated.[94] In his address he reiterated that the universities of Lucknow, Agra and Madras were thinking of opening dental colleges attached to their respective universities. He was reassured by the chief minister of West Bengal, Dr. Roy and he believed that in course of a year or so they would have a Dental Faculty attached to the Calcutta University which would take over holding of the examination in Dentistry, now being conducted by the State Faculty of Medicine.[95] Dr. Ahmed hoped that there would be a bachelor's degree in dentistry in the Calcutta University.[96] He received the reassurance of the Vice Chancellor of the University in this connection who was very sympathetic and was thinking of setting up a Faculty of Dentistry in the University.[97]

The dental profession and the people, Dr. Roy remarked owed a deep debt of gratitude to Dr. R. Ahmed for his making over the institution *free of charge* to Government. Calcutta Dental College and Hospital represented the life-long sacrifice and service to the cause of dentistry of Dr. Ahmed and his associates.[98] It was obvious that this was a massive proposition for the state government too. Even if finances had been arranged, they had not the requisite personnel for a very large number of colleges since the establishment of a dental college was an expensive project.[99]

Dr. R. Ahmed, Chairman of the Reception Committee, of the fourth Annual General Conference said,

---

[90] *Proceedings of the govt. of India, Sanitary. Land Dept. Meld. June – Augh, 1952, F152-154, Part* A, March 1960, NAI.

[91] *The Indian Dental Journal, 1948–49*, vol 2–5, no. 3–4, The Indian Dental Association, Calcutta, 1948, p. 35.

[92] *The Indian Dental Journal, 1949–50*, The Indian Dental Association, Calcutta, 1949, p. 39.

[93] *Ibid*., p. 40.

[94] *The Indian Dental Journal year 1948–49, vol 2–5, no. 3* Calcutta, 1949, p. 40.

[95] *The Indian Dental Journal year 1948–49, vol 2–5* Calcutta, 1950, p. 42.

[96] *Ibid*., p. 42.

[97] *The Indian Dental Journal year 1948–49, vol 2–5* Calcutta, 1950, p. 50.

[98] *Ibid*., p. 50.

[99] *Ibid*., p. 51.

In the fitness of things that the first dental college in India was established in Calcutta in 1920. Calcutta, therefore, might be called the birth place of modern dental education in India. In dental development Bengal could rightly claim to be the pioneer. In 1939 they had the first provincial Dental Act among all the provinces of India. Dental education was regularised by having a standard curriculum of studies and having the Calcutta Dental College and Hospital affiliated to the State Medical Faculty of West Bengal...Under Dr, B.C. Roy's aegis we hope to have dental education in West Bengal provincialized and brought upto international standards. We have had a very uphill task during the last 29 years, but the future will we are sure, be much easier than the past, due to interest of the National Government in all phases of nation building work. I hope I will live to see the day when every medical college in India will have a dental college attached to it, there will be a provincial dental service and a full-time dental staff in every hospital in the land. I may not be alive to see the inauguration of a National Health Act, but I am sure our successors will see that too, because we are on the right track now.[100]

The list of students who passed out from Calcutta Dental College and served in many provinces in India, for the year 1954.[101]

Sl. no	Name	Designation	Working
1	Dr. A. Ganguly	Professor	R.G. Kar Medical College
2	Dr. R.N. Ghosh	Associate professor	Do
3	Dr. M.K. Jon	Lecture	B.S Medical College
4	Dr. Arup Kumar Dar	Principal	Dr R. Ahmed Dental College & Hospital
5	Dr. S.P Mewar	Professor	Do
7	Dr. P.K. Basu	Associate professor	Do
8	Dr. M.K. Basi	Do	Do
9	Dr. P.N. Kundu	Do	Do
10	Wr. P.B. Guhn	Assistant professor	Do
11	Dr. A. Roy Choudhury	Do	Do
12	Dr. G.C. Des	Do	Do
13	Dr. M.P. Dutta	Do	Do
14	Dr. K.P. Roy	Do	Do
15	Dr. A.K. Mitra	Do	Do
16	Ur. A.L. Son	Do	Do
17	Dr. K.L. Boneeje	Do	Do
19	Dr. A.k. Dis Cup to De	Do	Do
20	Dr K.S. Roy Choudhury	Do	Do

(continued)

[100] *Govt. of Bengal legislative council*, West Bengal State Archives, Proceedings No. 54–59, p. 10–15.

[101] *Ibid*., p. 10.

(continued)

Sl. no	Name	Designation	Working
21	Dr. B. B. Dutt	Do	Do
22	Dr S.N. 'Talukolar	Do	Do
23	Dr. 'T.K. Dutto Clinical Titor	Do	Do
24	Dr. M.N. Sikuar	Do	Do
25	Dr. T. Sur Roy	Do	Do
26	Dr. (Mis) Gita Dutta	Registrar	Do
27	Dr. B.C. Bera	Do	Do
28	Dr.M.M Dutta Gupta	Do	Do
29	Dr M.A.G. Moyeen	Do	Do
30	Dr V. P Jaly	Do	Do

The list of students who passed out from Calcutta Dental College and served in many provinces in India, for the year 1967.[102]

Year 1967

1	Dr. S. Bhatterjee	Lecturer	Do
2	Dr. S. Bhattnagar	Do	Do
3	Dr. S. S. Achriya	Do	Do
4	Dr. G. T. Simon	Do	Do

The popularity of the college can be gauged from this that students hailed not only from India, Burma and Ceylon but from distant countries like China, Australia, South Africa and South America. With Dr. Ahmed as Principal and his able staff, the college flourished each year. A college report of 1936 says that the college had then 25 pump chairs, an X-ray unit and a very well-equipped laboratory. Dr. Ahmed now has the satisfaction of having properly trained several hundred young men who are practising in different parts of this country. Thus, he is the pioneer of dental education in India. In appreciation of his work, the Association of the United States of America honoured him with a gold medal and the title of F.I.C.D. The spread and popularity of dental education in India in the last 65 years has shown that the undaunted spirit of one man, his implicit faith in the future of dental education and his tireless efforts to keep alive his dream, has materialized in tremendous progress in the field of dental education in India today through the collective efforts of all the leading dentists, states, universities and the Dental Council of India.[103] According to an account which stated that, "We have today thirteen dental colleges affiliated to various universities where about 500 students seek admission in the dental course every year. In the foundation of this grand edifice of dental education is the sweat,

[102] *Ibid.*, p. 10.

[103] *Ibid.*, p. 495.

toil and labour and not to say the money and efforts of Dr Ahmed." it was attested that his and his teams efforts did not prove ephemeral.[104] The dental profession, education and dental health service leaves it to posterity to remain indebted to Dr R. Ahmed, the father of dentistry in India, for not only institutionalising the science of dentistry but also for making the study and practice of dental health an integral part of the study of medicine and public health

References

David Arnold, Disease Resistance and India's Ecological Frontier, 1770–1947, in Biswamoy Pati ed., Issues in Modern Indian History for Sumit Sarkar. New Delhi: Popular Prakasan, 2000

Poonam Bala, "Nationalizing Medicine: The Changing Paradigm of Ayurveda in India," in Poonam Bala ed., Contesting Colonial Authority, Medicine and Indigenous Responses in Nineteenth and Twentieth Century India. New Delhi: Primus Books, reprint, 2016, 1–12, originally published by Lanham MD: Lexington Books, 2012

P. Bala, Imperialism and Medicine in Bengal, Sage Publications, 1991

N.N Berry, A tribute to Dr Rafiuddin Ahmed. *All India Dental Association 1965* April (37); 109–10

M.C. Buer, Health, Wealth and Population in the Early Days of Industrial Revolution, Routledge, 1926

Philip D.Curtin (1968) "Epidemiology and the slave trade," Political Science Quarterly 83 (2): 190–216

Gelber, Stanley. 125 Years of Developments in Dentistry, *British Dental Journal* (2005); 199, 470–73

B. Gupta, 'Indigenous Medicine in 19th and 20th Century Bengal in Asian Medical Systems, University of California Press, 1976

Mark Harrison, Public Health in British India: Anglo-Indian Preventive Medicine, 1859–1914, Cambridge: Cambridge University Press, 1994, 164

V.P. Jayade, A goal for orthodontists. *Journal of Indian Orthodontics Society 1988*; 18(2):1–5

Ira Klein, "Death in India: 1871–1921," Journal of Asian Studies, no. 32 (1973):639–59

P.C. Kochchar, *History of army dental corps and military dentistry.* Lancer Publishers and Distributers, 2000, pp. 35–40

P.C. Kochchar, *History of the Army Dental Corps and Military Dentistry*, New Delhi, 2000, p. 25

Deepak Kumar and Raj Sekhar Basu ed., "Medical Encounters in British India, Oxford University Press, 2013.

Leslie (ed.), Asian Medical Systems: A Comparative Study; C. Leslie and A. Young (eds), Paths to Asian Medical Knowledge, University of California Press, 1992

H.B Merrel (1992) "Noise Pollution and hearing loss in the dental office", *Dental Assisting Journal*, 61(3):6–9

Subir Mukherjee, *The Man Who Dared, A Commemorative Volume on the Life and Work of Dr. Rafiuddin Ahmed*, Dr R. Ahmed, Memorial Committee/I.D.A. West Bengal State Branch, April, 1993, p. 52

Aparajita Mukhopadhyay, Colonised Gaze? Guidebooks and Journeying in Colonial India, *South Asia Journal of South Asian Studies*, vol. 37: Number 4, Routledge, 2014, pp. 656–669

K.N. Panikkar, 'Indigenous Medicine and Cultural Hegemony: A Study of the Revitalization Movement in Keralam', Stidies in History, Volume 8 (2), 26 -Aug 1, 1992

[104] *Annual Report on the Working of Hospitals and Dispensaries under the Government of Bengal for the year 1933,* Calcutta: Bengal Government Press, 1939–41, pp. 35–40.

B. Pati and M. Harrison, Health, Medicine and Empire: Perspectives on Colonial India, Orient Longman, 2001
M.E. Ring. Dentistry: *An illustrated history. Harry N Abrams*, New York. The CV Mosby Company, St Louis, Toronto, Princeton 1992, p. 14.
Arabinda Samanta, Malarial Fever in Colonial Bengal: Social History of an Epidemic, 1820–1939. Kolkata: Firma KLM, 2002
Arabinda Samanta, Living with Epidemics in Colonial Bengal, 1818–1945, Routledge, 2017
Deepak Kumar, Science and the Raj, 1857–1905, Delhi, Oxford India Paperbacks 2006, 2nd edition
Sandeep Sinha, Public Health Policy and the Indian Public: Bengal 1850–1920. Kolkata: Vision Publications, 1998
Uma Dasgupta, ed., Science and Modern India: An Institutional History, c. 1784–1974, Vol. XV, Part 4 in D.P. Chattopadhyay (General Editor) History of Science, Philosophy and Culture in Indian Civilization (Series), Pearson Longman, Delhi, 2011.
Sheldon Watts, "British Development Policies and Malaria in India 1897–c.1929," Past and Present 165 (1999): 141–81

Chapter 9
Electronics and Computing: Experiments in Government Entrepreneurship

Dinesh C. Sharma

The organization of science in the post-independent period in India was marked by large investments in the establishment of national laboratories, R&D centres, higher education institutions for engineering, management and medicine, as well as production facilities in key areas such as pharmaceuticals and defense. This manifested in the rapid establishment of national and regional laboratories under the Council of Scientific and Industrial Research (CSIR) and other agencies like the Department of Defense Research and Development (DRDO). In addition, scientific Departments were formed in the central government and scientists made secretaries with executive powers similar to those of civil servants. Top scientists thus became a part of the policymaking apparatus of the government. The Indo-China war in 1962 was a wake-up call for the government as it exposed deficiencies in the technological preparedness of the armed forces, among other things. The armed forces experienced an acute shortage of certain electronic components used in imported equipment during the war, and they had to be developed by scientists at the Tata Institute of Fundamental Research (TIFR) at short notice.[1] Procuring strategic electronics in open markets was both costly and difficult. The Atomic Energy Establishment and TIFR had already experienced this and therefore, had developed in-house groups for electronics fabrication and production.

Given the shortage of equipment faced during the war, the Ministry of Defense prepared a list of electronic instruments required by the armed forces. In DAE, Homi Jehangir Bhabha asked A S Rao, Director of Electronics Group at AEE, to prepare a similar list. Rao already had a proposal ready for setting up an electronics

[1] Srikantan (2006).

D. C. Sharma (✉)
Independent researcher, New Delhi, India
e-mail: dineshcsharma@gmail.com

S. Ahmed and S. Sarkar (eds.), *Decolonizing Science and Modernity in South Asia*,
https://doi.org/10.1007/978-981-97-1829-0_9

production unit for commercial exploitation of the expertise developed at Trombay.[2] Cabinet secretary S S Khera, who was a member of AEC, suggested that DAE should form a committee to take a comprehensive look at India's electronic requirements for defense as well as civilian sectors. This resulted in the formation of the Electronics Committee under the chairmanship of Bhabha in August 1963 with Vikram Ambalal Sarabhai (Chairman, Indian National Committee of Space Research) A S Rao of AEE and S Bhagvantam (Scientific Advisor to Minister of Defense) as members. The panel was asked to suggest measures for "planned development of electronics so that the country as a whole may become self-sufficient in this field in the shortest possible time."[3]

The Electronics Committee was the first attempt at "planned development" of a new and emerging high technology industry in India. The resolution setting up the committee observed that "electronics is the nervous system of modern technology and has assumed an important role in monitoring and control of the production process in engineering, chemical and metallurgical industries. It is vital for atomic energy, communication and defense."[4] The committee's report formally submitted shortly after Bhabha's death in 1966 laid out a ten-year blueprint for the development of an indigenous electronics industry based on research and development, design, training and select foreign inputs. A significant part of the report dealt with achieving self-sufficiency in the production of medium and small-sized computers within a decade. This was in the context of the monopoly enjoyed by IBM and ICL in the supply of data processing equipment and mainframe computers in India, and their growing clout with government and academic institutions. Science bureaucracy was keen to restrict the two companies particularly because they were dumping in India used equipment and depriving Indian customers of the latest technology.[5]

The earliest attempts at government entrepreneurship in electronics and computing were made at ISI by its founder-director P C Mahalanobis. It was his dream to not only develop and fabricate computers but also manufacture them on a large scale. One of his close associates C R Rao attributed Mahalanobis' interests in business activities to his lineage of a family with business acumen". Mahalanobis had started fabrication of desk calculators at ISI early on when he found that there was some demand in the country for desk calculators after the Second World War and importing them was difficult.[6] When ISI acquired a large Soviet computer, Ural, between 1955 and 1958, it came with a set of machine tools (lathe, drilling machine, cutter etc.). Right from the start of negotiations with the Soviets, Mahalanobis had nurtured an idea of manufacturing calculating machines using equipment provided

---

[2] Unpublished interview with A S Rao by S P K Gupta, December 1972–June 1973 in New Delhi and Hyderabad.

[3] *Electronics in India: Report of the Electronics Committee*, Government of India. New Delhi: 1966, p. 4.

[4] Ibid.

[5] Sharma (2009).

[6] Radhakrishna Rao (1973).

by the Soviets. Two Soviet experts who evaluated the production plan of ISI recommended that the target should be to manufacture 5000 manually-operated desk calculators per year in the first phase, while motor-driven and other calculating machines should be manufactured in the second phase.[7] In October 1960, ISI applied for the grant of a license to set up a manufacturing unit and the license was granted in December 1960.

As the next step, Mahalanobis wanted to convert the production unit at ISI into a commercial venture. A new company, Sankhya Yantra (Private) Limited, was registered in December 1962 by C D Deshmukh (who was President of the ISI Council) and Mahalanobis as two original promoters in their personal capacity.[8] The ISI Council nominated members to the Board of Directors. The company, however, became a bone of contention between ISI and the government. The government raised objections to the ownership of Soviet equipment being transferred from ISI to the company. The First ISI Review Committee—chaired by Humayun Kabir in 1966—made adverse comments on the manufacturing project. It observed that designing and manufacturing computers and components could not be a legitimate activity of an academic institute.[9] Such activities, in its view, were started without proper planning and detailed consideration of the implications and felt that large-scale manufacturing activity would retard ISI's work in more vital areas. Sankhya Yantra was folded up, with its assets transferred to the Garden Reach Workshop, a public sector unit under the Ministry of Defense.

9.1 Failed Experiments in Hardware and Software

While the idea of Sankhya Yantra was nipped in the bud, it did not deter DAE and TIFR scientists from planning their foray into hardware production and software development. TIFR's attempt to enter the manufacturing of core memories and integrated circuits was a bold move, in this context. Instead of opting for an IBM computer, the institute chose to buy a computer developed by a new entrant in the market—Control Data Corporation (CDC). For the American firm, the TIFR deal opened the doors to the large Asian market that was dominated by IBM. It was so important that CDC featured a picture of its system in Bombay on the cover of its annual report 1964.[10] The machine arrived in Bombay at a time when Bhabha was working on the future policy framework for the electronics industry in the country and was envisioning a leading role for the atomic energy establishment in

[7] Letter from P C Mahalanobis to V Sergeev, dated July 29, 1962. ISI Archives.

[8] *Indian Statistical Institute: Thirty-First Annual Report: April 1962-March 1963*, ISI Kolkata, 1963, p. 388.

[9] *Report of the Indian Statistical Institute Review Committee*, Government of India. New Delhi: 1966, pp. 23–24.

[10] Letter from James G Miles, Vice President, CDC to H J Bhabha, dated August 18, 1964, D-2004-01345, TIFR Archives.

this sector. In 1965, he visited CDC plants in America to get firsthand knowledge of new technologies like integrated circuit-based systems and production processes that involved procuring a large number of parts and sub-assemblies from other firms.[11] For example, Bhabha found that the entire drive unit with its base was obtained from a neighbouring firm, which, in turn, obtained the motor from General Electric.

CDC Vice President James G Miles indicated the company's interest in taking up the manufacture of computers in India. This encouraged Bhabha to think of a plan to "develop a national computer industry under the aegis of the Atomic Energy Commission"—a sort of 'government entrepreneurship' with foreign collaboration. He wanted CDC to take an early decision as "other computer interests in the US are pressing very hard to establish computer production facilities in this country, and if CDC wants to enter the field, which in my opinion has vast future possibilities, it should not let others get too far ahead of it."[12] CDC accordingly informed the Indian government about its intention to manufacture magnetic core memory planes for use in high speed electronic digital computers, pending a final decision on joint venture partner and location.[13]

The CDC decided to start with a small plant for the stringing of ferrite core computer memory stacks—a labour intensive operation—in India, as a first step in 'an ambitious plan to develop a computer manufacturing industry in India.'[14] All the raw material was to be imported and finished products exported back to CDC. However, CDC was concerned about a "considerable amount of red tape in India" and wanted to be assured of uninterrupted import of raw material and export of memory planes once the manufacturing started.[15] Specifically, it pressed for 'guaranteed expeditious handling by export control authorities and ready access to air transport'. The proposed Indian operation was designed to feed the production of CDC computers in the US worth $20 million a month. As per CDC's projections, the Indian unit could earn up to $750,000 a year through exports. The foreign exchange was to be used for importing into India computer systems which could be sold in Rupees. The Indian unit, according to CDC, could serve as a nucleus of large-scale computer manufacturing operations providing employment to many and making available the latest computers in India.[16] Given the scarcity of foreign exchange, Bhabha wished a portion of the dollar earnings from the CDC project be utilized for procurement of

[11] Note from H J Bhabha to M G K Menon and others on his visit to CDC plants, dated March 12, 1965, D-2004-01352, TIFR Archives.

[12] Letter from H J Bhabha to W C Norris, Chairman, CDC, dated March 13, 1965, D-2004-01052, TIFR Archives.

[13] Letter from R M Price, director of international operations, CDC, to T Swaminathan, Secretary, Ministry of Industry and Supply, dated July 2, 1965, D-2004-01052, TIFR Archives.

[14] Letter from J Schmey, Manager, Far East Operations, CDC, to H J Bhabha, dated November 16, 1965, 4(1,2/15), D-2004-01351, TIFR Archives.

[15] Note from MGK Menon to H J Bhabha, dated November 12, 1965, D-2004-01351, TIFR Archives.

[16] Letter from J Schmey, manager of Far East operations of CDC, to H J Bhabha dated November 16, 1965, 4(1,2/15), D-2004-01351, TIFR Archives.

spares for already installed CDC computers at TIFR and also to fund non-computer imports.[17]

Armed with the commitment from CDC Bhabha approached Tata Sons to join as a partner in this venture, seeking an investment of Rs 500,000 through a trust fund to be created for this purpose. J R D Tata agreed, in principle, and deputed TIFR physicist M G K Menon (on behalf of the Tatas) to the U.S. to explore possibilities of working with CDC. Menon saw CDC's peripheral equipment division, industrial data processing division, plant for the development of new computer prototypes. On the financial side, Menon found that CDC had huge capital locked up in computers on lease and had to resort to extensive borrowings for further expansion, and because of shortage of funds, it may have problems in undertaking a new investment in India.[18] CDC felt that the proposed operation 'will not be very successful if carried out by government agencies in the public sector since it was a highly competitive and technologically rapidly moving industry. Miles, therefore, suggested that while TIFR should be used as a catalytic agent and for high-level consultation, the actual work would have to be carried out by a competent private sector industry such as the Tatas.[19]

After Menon's visit, CDC began talking to the Tatas directly and invited JRD to visit CDC plants and hoped that together they would establish a computer industry in India.[20] Over the next three years, Tata Sons and TIFR continued to work out details of the manufacturing plan with CDC but it did not materialize. CDC was apprehensive about bureaucratic red tape and restrictions on foreign investment in India, which it had expressed earlier. The proposal, as Menon recalled later, did not fructify because the government did not approve it.[21]

Another hardware project TIFR scientists proposed to the Tatas related to the production of integrated circuits. The physics group brought this up before the Electronics Board at its first meeting in December 1969, but the idea did not find favour with Sarabhai and other members of the Board. Sarabhai suggested that as a matter of policy, design and development groups must prepare necessary documentation including detailed specifications, mechanical and electrical drawings, manufacturing procedures, test procedures etc. as the final stage of their work. This was necessary for technology transfer. Manufacturing could then be done at another unit.[22] Homi N Sethna, a member of the Board, said an engineer or a scientist who developed laboratory scale know-how was not always the person who could commercialize it. Menon, however, opined that "the manufacture of even specialized electronic

[17] Note for H J Bhabha on Core stringing operation in Collaboration with CDC, dated January 17, 1966, from M G K Menon, 31(1,2/10), D-2004-01352, TIFR Archives.

[18] Report on Setting up of Production Facilities in India in Some Areas of Computer Industry, by M G K Menon, dated August 18, 1966, D-2004-01346, TIFR Archives.

[19] Ibid.

[20] James G Miles, Vice President, CDC letter to JRD Tata, dated July 27, 1966, D-2004-01046, TIFR Archives.

[21] M G K Menon, Interview with the author, New Delhi, April 4, 2007.

[22] Minutes of 7th meeting of the Electronics Board in DAE on November 21, 1969, TIFR Archives.

equipment or components on a less than fully commercial scale need not always be uneconomical."[23] Sethna pointed out that expansion of production from pilot plant to commercial scale had often been accompanied by a large-scale increase in unit costs.

The atomic energy establishment had taken a lead in developing and installing modern computers in the mid-1960s and had taken up software writing activities as well. A fairly large number of engineers and scientists had been exposed to skills in application software writing at TIFR as well as the AEE. TIFR proposed a separate corporation for taking up work relating to software writing. It was supposed to be called 'Computer Corporation' or 'Software Corporation'.[24] It was proposed to employ 150–200 engineers and mathematicians from TIFR and other academic institutions in the proposed corporation. Sarabhai also saw the potential for software exports from the corporation. "Given adequate organizational support and financial backing, production of software is bound to pick up and also prove to be a good foreign exchange earner, although market surveys and contacts would have to be properly organized to enable our software to enter foreign markets", Sarabhai informed a meeting of economic secretaries in July 1970.[25] This was the first time 'software exports' appeared at the government policymaking system.

Sarabhai did not want agencies of DAE to monopolize the proposed corporation. In reply to a query at the secretaries' meeting, he said 'the large computers under the control of DAE would be available to all software specialists and users, so long as spare time was available and booked by them. In other words, there is to be no reservation or monopoly of such users in favour of the proposed software corporation staff. Ample opportunity would be given to competitive Indian specialist groups too, were they to come up as desired.'[26]

Sensing that its plan to make a strong entry into electronics was facing resistance from other quarters in the government (notably departments of defense supplies and statistics), the atomic energy establishment organized a national conference on electronics at the TIFR in 1970 as a follow up of the Bhabha Committee recommendations. Here scientists forcefully articulated their idea of 'government entrepreneurship,' while arguing that since the process of transferring technology from research laboratory to industry is very complex, scientists themselves should be allowed to commercialize their ideas:

> ...the best way may be for the originator of the idea himself to follow his brainchild into development, testing and final production. The research organizations must be closely associated with the industry right from the start, so that these laboratories benefit from licensed know-how, and will be in a better position to modify or improve the existing production line later on. This close association is possible only if the research organizations are allowed

[23] Minutes of the meeting of the Electronics Board held on Feb 24, 1969, D-2004-01050, TIFR Archives.

[24] "Government to set up computer corporation", *Times of India*, November 9, 1967.

[25] Minutes of the meeting of the Committee of Economic Secretaries held on July 7, 1970, D-2004-01356, TIFR Archives.

[26] Ibid.

to establish field units in industries. These field units may consist of a few scientists of the research organization working in the factory, making use of the services and facilities there.[27]

9.2 ECIL and CMC—Mixed Bag of Success

One of the best examples of government entrepreneurship, as defined by atomic energy scientists in the Electronics Conference of 1970, is the Electronics Corporation of India Limited (ECIL). It's an example of the 'innovation chain' which starts with fundamental research and is carried forward through applied research, prototype fabrication and finally large-scale production.[28] The seeds of the corporation were sown when Bhabha convened a small electronics group at the TIFR to develop and produce radiation detectors in the early years of the nuclear programme. Over the years, this group grew into a fairly large activity so it was transferred to AEE at Trombay. Here it expanded to become a large Electronics Division with about 600 people, producing control and electronic instrumentation for atomic energy as well as other applications. In 1967, the Electronics Division was hived off into a separate public sector enterprise—ECIL—located at Hyderabad.

Around 1965, the Electronics Division started developing a digital computer for real-time applications like control of nuclear reactors. Real-time computers, as distinguished from general-purpose machines, are connected with manufacturing or operational plants such as in petroleum, steel plants or nuclear power reactors or air traffic control, for real-time data capturing and analysis. The group took up the development of logic circuits with a typical signal propagation time of 50 ns and an expandable memory system of 2 microseconds cycle time. A central processing unit was developed to provide hardware capability to handle a set of 45 instructions.[29] The work on this system was completed towards the end of 1968, and the know-how was transferred to ECIL. The system was named Trombay Digital Computer or TDC-12. Technologically it was a much superior machine to those developed earlier in India, as it was based on solid-state technology. TDC-12, modelled after commercially available—and hugely successful—PDP-8 computer of Digital Equipment Corporation (DEC), was the first India-developed computer to have deployed semiconductor devices and made use of standard components.[30]

Within the atomic energy groups, differences cropped up between Sarabhai and A S Rao over the path the electronics production unit should take. Rao had led the electronics development activity of AEE since 1955. Sarabhai proposed foreign collaboration for the production of electronic equipment at ECIL to fill the technology gap and to ensure that it became a commercial success, but Rao wanted complete

[27] *Perspective Report on Electronics in India*, Electronics Commission, 1975, pp. 28, 29.

[28] Menon, M G K. First Vikram Sarabhai Lecture, December 31, 1972, Ahmedabad.

[29] Appendix-2 of Technology proposal for a real-time computer for space science and technology, ECIL February 1969, D-2004-01,050, TIFR Archives.

[30] ECIL Case Study, Indian Institute of Management, Ahmedabad (notes made available by S Manikutty, IIMA).

self-reliance.[31] The Electronics Committee too had suggested foreign collaboration as the second option, next to imports, for the production of high-speed mainframes computers.[32] Both Bhabha and Sarabhai were aware that it was not possible to grow in high technology areas in isolation. In both atomic energy and space programmes, they had opted to get technical assistance and training from their Western counterparts. In Sarabhai's vision, India should have opted for export-oriented electronics development, rather than mere import substitution, so that it could act as a motivating force to close the technology gap between India and the advanced world. He felt that "export thrust could put product quality and production economies in a healthy perspective, and in the process bring advantages to Indian buyers who would then buy more; thus, further boosting the electronics industry."[33] At that time, Rao's will prevailed but eventually, ECIL had to go in for several technical collaborations with foreign companies.

Having grown in the research environment in AEE, Rao promoted research for new product development at ECIL while placing little emphasis on functions like marketing, costing and finance. Marketing especially was given low status. Rao believed that if a good product was made, it would sell, and the role of the marketing personnel was to sell what was produced.[34] Products were developed not based on their market potential, but based on the technological challenge they posed. Production plans were tailored to suit production requirements rather than based on assessment for prospective sales by the marketing department.

The result was that by 1975, a 32-bit mainframe computer hoped to have been delivered by then, was nowhere in sight. TDC-312 and TDC-316 were scientific computers with hardly any application software was available for them. A case study on ECIL by IIM professor, S Manikutty summarized the company's problems this way: "these machines could not be used in most electronic data processing applications because such applications needed more software and input/output devices. The business language compiler COBOL was not available. ECIL wrote its version called E-COBOL, but this did not become popular among users as it was a non-standard language. As a result, the computers of ECIL had very few applications packages making them unsuitable for business and even scientific applications."[35] The company tried to change the course after the departure of Rao in 1978, but it was too late for it to catch up. Personal Computers had come to the market by 1980, and new Indian firms like DCM, ORG, Wipro and HCL had aggressively marketing their products, educating their customers and coming out with application packages that suited their needs. Between 1978 and 1980, these firms had installed 725 new

[31] Sundaram et al. (1998).

[32] *Electronics in India: Report of the Electronics Committee*, February 1966, Government of India.

[33] Rao (1972).

[34] ECIL Case Study, Indian Institute of Management, Ahmedabad (notes made available by S Manikutty, IIMA).

[35] Manikutty, S "ECIL: A case study of environmental change and organizational response in the public sector", a dissertation submitted in partial fulfillment of the requirements for the Fellow Programme in management, IIMA 1987, pp. 160–5—quoted in Subramanian, C R. 1992. *India and the computer: a study of planned development*. Oxford University Press, pp. 189.

computers, compared to 180 sold by ECIL right from its inception in 1967. Its market share had shrunk to 15 per cent from being the sole Indian computer company.

Perhaps the only successful example of 'government entrepreneurship' is the Computer Maintenance Corporation [renamed as CMC Limited in 1984]. The idea of setting up this corporation was conceived much before IBM wound up its operations in the country. The three-member committee, appointed by Sarabhai, to look into the functioning of IBM and ICL in 1968, held extensive talks with the two companies. "It became clear quite early on that there was no possibility that they (IBM) would agree to Indianize their equity. And they were unwilling to stay on merely to sell their equipment if they could not also provide services for supporting them," as the committee's chairman Narasimhan would recall later.[36] The Computer Division at DoE came up with the idea of setting up a public sector company to support and maintain all imported computers. This, as Narasimhan would point out later, was "an indispensable prerequisite if mainframe computer availability in India were to be delinked from dependence on IBM and ICL."[37]

The government approved the formation of CMC in 1975 and the corporation started functioning in 1977 as a public limited company. But its actual business got off only in June 1978, when IBM finally ceased its manufacturing and maintenance operations in India. It took over maintenance of nearly 800 IBM installations all over the country. The formation of CMC was necessary to reassure the vast number of IBM and ICL customers that there won't be any disruption of maintenance services after the multinationals cease their operations in India. Still, these customers were apprehensive about getting commercial maintenance services through a state-owned setup. These doubts were justified because IBM gave maintenance a very high priority in its business working in India. Its central stores in Bombay, strategically located near the Santa Cruz airport so that spares could be shipped quickly, used to store nearly one million spare parts of 30,000 types at any given time and had a modern communication system. When IBM left India in 1978, the country had 154 mainframe computers—the bulk of them (119) were of 1400 series—and 357-unit record machines. Another 250-unit record machines were from other foreign manufacturers.[38] The maintenance of about 20 mainframes installed by ICL was also passed on to the new corporation. So, CMC had a good set of customers from day one of its formation though what it had to maintain was largely junk. As part of its exit deal with the government, IBM agreed to pass on maintenance of its machines as well to sell all its spares and parts in its inventory in India to CMC, and also 200 of its trained maintenance engineers. Prem Prakash Gupta, who was heading marketing operations of ICL in India, was picked up as managing director of the new company. The choice of an ex-ICL executive as Managing Director and the availability of a large number of ex-IBM engineers helped CMC make a head-start in the maintenance market. This way it could acquire a service-oriented culture right from the beginning.

---

[36] Narasimhan R "My association with CMC during its first 10 years" (made available by R Narasimhan).

[37] Ibid.

[38] Narasimhan (1978).

Since CMC was formed to provide support to obsolete mainframe computers of any age, any type, anywhere in India, it was saddled with a sunset business. Moreover, data processing machines were being used mostly for routine functions like accounts and payroll automation. CMC wished to change this by trying to convince users to deploy computers for mainline functions such as passenger and freight traffic movement in railways and management of generation, transmission and distribution in electricity boards.[39] A project funded by the UN, Interact [International Education and Research for Applications of Computer Technology] presented CMC with an opportunity to work on new ideas. It catalyzed its transition from hardware maintenance into software and services as well as also turnkey projects for domestic and international markets. Capabilities in systems analysis, design, development, implementation and support of large online systems that CMC acquired through this project helped it develop, deliver and support large projects such as railways freight operations, passenger reservation, banking computerization, stock exchange online systems, enterprise resource planning for large companies and bulk management of containers at ports.[40]

The government also proposed that CMC could take up marketing of scientific, business and educational computers of ECIL, and also diversify into consultancy, systems engineering, software development and turn-key supply of real-time computer-based systems.[41] Armed with success at home in large online projects and an encouraging policy environment, CMC entered international markets. Like the railway project at home, a highly successful project executed by CMC was timetable scheduling and signal data generation system for the London Underground. With these early successes in domestic and international markets, the company emerged as a top software and services firm. It remained among the top ten software exporters in the 1980s and early 1990s. But it could not retain its position for long as the market opened up and expanded. CMC slipped to 23rd rank in 2001 from 16th in 2000 in the listings of the *Dataquest* magazine.[42] Gradually, the government divested its equity in CMC and Tata Consultancy Services (TCS) acquired a controlling stake of 51 per cent in 2001.

9.3 Discussion and Conclusions

Several internal and external factors were responsible for research laboratories becoming centres of production and trying to take up entrepreneurial ventures based on the knowledge they generated and the expertise they developed. The early electronics and computer fabrication work at ISI and TIFR began in response to the needs of their core research activities – large statistical projects at ISI and building of

[39] Ramesh P Jhunjhunwala, Interview with the author, Hyderabad, January 2007.

[40] Ibid.

[41] Sondhi Committee and Menon Committee Reports: Government's Decisions (1981).

[42] Narayan (2001).

nuclear reactors at TIFR. They had to undertake the fabrication of their instruments and computers mainly due to the foreign exchange crunch and the strategic nature of their research. In the process, the two research centres began helping other centres and meeting the needs of research institutes, universities, private industry and armed forces, besides developing a pool of technical workforce. As the next logical step, they thought of setting up full-fledged production units. Some of them succeeded, others failed.

Such ventures were in line with industrial and economic policies focused on import substitution and self-reliance. On the other hand, the private sector worked under policies that restricted production capacities and discouraged foreign collaboration. The private industry also had no incentive to spend on R&D nor did it possess adequate capacity to absorb technology. All these factors emboldened national laboratories and scientific institutions to become entrepreneurs to complete the innovation chain that began on the laboratory bench. P M S Blackett, British Nobel laureate and a scientific advisor to Prime Minister Nehru, had observed that "the problem of continuity of the innovation chain is a complex one in India where little R&D is done by firms either public or private, and nearly all in the government research establishments. This being so, some means must be found to take care of the later stages of the innovation chain."[43] It was to complete the vital missing end of the innovation chain that laboratories became entrepreneurs. This happened not just with the entities associated with the atomic energy establishment but with CSIR labs as well. The National Physical Laboratories (NPL) established the Electronics Component Unit and the Radio Components Unit to commercialize its knowhow, while the Regional Research Laboratory (RRL Hyderabad) took up commercial production and marketing of active carbon based on the process it developed.[44]

R Narasimhan, who developed the TIFRAC computer at TIFR and later became the first Chairman of CMC, summed up the need for government entrepreneurship this way:

> Reforms and changes can be left to free-market forces provided the market, in fact, is free.... In countries like ours, where the initial conditions of the market are loaded against the vast majority, allowing free rein to market forces would in practice amount to exploitation of the disadvantaged many by the advantaged few. In these circumstances, what is needed on the part of the government is not mere regulation, but positive, entrepreneurial action....... I would consider the roles of agencies like Atomic Energy, Space, Electronics, University Grants Commission and Planning Commission, and their associated executive departments, where they exist, and the Department of Science and Technology, as fundamentally entrepreneurial in nature.[45]

When DoE was formed in 1971, scientists and technocrats running the show took upon themselves the task of developing an electronics industry in India. They started looking at what all can be done in R&D, design, production, export and import; and in various sectors such as defense and space. A large number of projects were

---

[43] Blackett (1973).

[44] Annual Report 1960–1961, Regional Research Laboratory, Hyderabad, pp. 53–55.

[45] Narasimhan (1978).

funded for research, development and design, production units were set up, corporations were set up for electronics exports and imports—all government-controlled. DoE formed a commercial firm or a development organization in each sector of the industry—hardware (ECIL), software and maintenance (CMC), international trade (Electronic Trade and Technology Development Corporation or ET&T), microelectronics (Semiconductor Complex Limited, SCL), education (National Centre for Software Development and Computing Techniques), market research and analysis (Information, Planning and Analysis Group, IPAG), government computerization (National Informatics Centre, NIC), electronic data processing (Regional Computer Centres, RCCs) and production of components and products like television sets (State Electronics Development Corporations).

The government entrepreneurship, however, had a mixed bag of success. The ventures conceived at TIFR before the formation of DoE did not take off because of the lack of adequate capital, planning, managerial skills and restrictive government policies. The institute was looking to the house of Tatas for funding and industrial support for its plans into microelectronics and computer memory production, but the Tatas as well as CDC were apprehensive of long-term business prospects. These projects had longer gestation periods and were fraught with technology risks. The only venture of this period that became a full-fledged business concern was ECIL—based on its ability to serve users outside the DAE family, despite floundering in initial years.

One of the main reasons for the poor performance or failure of government entrepreneurial ventures in the 1970s and 1980s was the lack of business and market orientation. This, in turn, had something to do with the legacy and lineage of each of the organizations. Most of them emerged from the same parent as DoE—TIFR and AEE (renamed Bhabha Atomic Research Centre or BARC). All key decision-makers and later administrators in electronics in the 1960s and 1970s came from these two bodies—Bhabha, Sarabhai, M G K Menon, A S Rao, R Narasimhan, Ashok Parthasarthi, N Seshagiri, S Ramani and so on. Only Sarabhai constantly emphasized market orientation to electronics development but there were few takers for his line of thinking in his lifetime. The focus in the production unit was also on R&D. For instance, all the key division heads of ECIL came from BARC or TIFR in the formative years. They had no business orientation and continued to treat ECIL as a technology development organization that had to fulfil specific needs of national projects.

ECIL saw itself as a development organization, not as a commercial entity. "Why did we not go into electronic data processing earlier? Why did we insist on self-reliance? It was our basic orientation. We saw ourselves as doing development. Commercial considerations were secondary", this is how S Srikantan, head of ECIL's computer division till 1981, summarized the path company took in its first decade of operation.[46] Such orientation became a liability for the company. In the PC era, marketing, systems design, applications software and customer support were

[46] ECIL Case Study, Indian Institute of Management, Ahmedabad (notes made available by S Manikutty, IIMA).

becoming crucial factors for success. Sridhar Mitta, who returned to India from the US after a PhD from Oklahoma State University, joined ECIL in December 1973 as a technical manager. The fact that ECIL was the only company in India which was engaged in designing, manufacturing and selling computers attracted Mitta to it. He was made in charge of designing computers for defense and space applications such as satellite tracking, a satellite launch, message switching and missile check out. Though the work was exciting, the attitude of division heads—all earlier scientists at BARC—towards new technology as well as wider applications of computers was discouraging.[47] He left ECIL soon and joined Wipro.

On the other hand, CMC was different because it did not have the same lineage as ECIL It was the only organization created by DoE that did not have an ex-atomic energy person as to its head. Though Narasimhan was appointed Chairman of CMC, an experienced industry person Prem Prakash Gupta was picked up as Managing Director. The choice of Gupta was deliberate and strategic. CMC was set up with a specific objective—to continue maintenance to a large base of ageing IBM and other foreign computing machines. If CMC had failed in this primary objective, it would have brought a bad name for DoE. That's why perhaps Menon did not want to take a chance by appointing a TIFR scientist as the Managing Director. Gupta with his experience with ICL, CDC and Ferranti, proved to be the right choice. He attracted the type of talent necessary for running CMC. The corporations headed by scientists lacked clearly defined business goals, strategic plans and never had profit as their motive. However, CMC too suffered from some of the same ills as other ventures of 'government entrepreneurship'—lack of market orientation. As Gupta put it: "We never did any marketing. That was the problem with all public sectors. We thought the business will come on its own. Marketing was not just on...... one of the greatest weaknesses of CMC was marketing. We also wondered. After all, we did the (railway) reservation system, it made a huge impact. But we could not move forward."[48]

Other countries in Asia—Japan, Korea, Taiwan—also initiated and supported R&D programmes in electronics and computing but the support was mainly to the private sector. In certain core technology areas, the development effort was co-funded by governments and industry. Fully owned government agencies were engaged only in areas like trade development, testing and export promotion. The governments in these countries also created domestic demand for products made by local industries through tailor-made procurement policies. This is in contrast to the situation in India where procurement policies were tilted in favour of public sector companies. Korea set up the Korea Institute of Electronics Technology in 1976 in the area of semiconductor technology, much after India thought of SCL. But the objective of the Korean outfit was not the production of chips, but to fabricate wafers for industry, supply tools for domestic industry, provide testing services and do contract

---

[47] Dr Sridhar Mitta, Interview with the author, Bangalore, March 2006.

[48] Dr P P Gupta, Interview with the author, New Delhi, June 2004.

research for the industry.[49] Private companies like Samsung benefited greatly from these public-funded services.

The policy planners in India were aware of such trends through personal contacts with industry leaders from America as well as through the study of literature and personal visits. Even though the advocates of state entrepreneurship knew of their shortcomings, they went ahead and entered the commercial arena. After a visit to the plants of International Computers Limited, a state-owned computer enterprise in the UK, R Narasimhan wrote to M G K Menon in 1972: "...... weak points of the British industry are not the lack of intellectual ideas and the absence of professional competence. The weak points are the lack of organizational and management enterprise to productionize ideas and the absence of aggressive marketing skills to sell ideas and products even when these are demonstrably technically superior. The Atlas computer when developed by Ferranti was far ahead of its time and because of various organizational problems ran into severe difficulties. Want of management foresight and governmental support resulted in the emigration of many of the people associated with Atlas design to the USA. Both CDC and DEC made use of them in developing the software and operating systems for their new computers and were thus able to capitalize on the professional know-how they could purchase cheaply. There is a lesson in this for us, I think."[50]

Though government entrepreneurship ventures were commercial failures, nearly all of the DoE-sponsored, funded and supported organizations helped in developing capabilities in advanced areas of research and design. It helped create nuclei to foster the seeding of human capital and nodes around which networks grow. In this context, the contributions of ECIL and CMC are noteworthy. When private sector companies appeared on the scene in the 1970s, they benefited greatly from the experienced talent pool of the public sector. The public sector companies and research organizations also built and nurtured networks with academic institutions such as the Indian Institutes of Technology, the Indian Institute of Science and leading universities. These institutions, in turn, created sustainable student networks for the future.

The presence of large public sector undertakings catalyzed the development of clusters of ancillary industries around them. The development of the electronics components industry in Hyderabad—mainly to feed ECIL—is a good example of this. Similarly, the presence of SCL at Mohali helped Chandigarh develop into a destination of high-tech industries and research institutions in the decades to follow. The state electronics development corporations, another initiative of DoE, boosted the spread of electronics production and developed skills. The Uttar Pradesh Electronics Corporation provided much-needed capital to the team led by Shiv Nadar to start Hindustan Computers Limited (HCL) which became a leading hardware and software firm in the 1980s and a major exporter of software services in the later decades. The Karnataka State Electronics Development Corporation founded the Electronic City which became the hub of software exports eventually. The ventures of government

[49] Evans (1995).

[50] Note from R Narasimhan to M G K Menon. 1972. D-2004-01356, TIFR Archives.

entrepreneurship thus proved to be a critical part of the long chain of events ending in India emerging as an important player in the global technology and services business in the new millennium.

References

P. M. S. Blackett, *Science and technology in an unequal world* in Jawaharlal Nehru Memorial Lectures, Bhartiya Vidya Bhawan, Bombay, 1973, p 15.

Peter Evans, *Embedded Autonomy: States and Industrial Transformation*, Princeton University Press, 1995, pp 174–175.

Narasimhan, R "Technological Change and Government Entrepreneurship", Fourth Dr Vikram Sarabhai Memorial Lecture, August 22, 1978.

Narayan, Rajeev. "Tottering to Disinvestment", September 28, 2001, accessed at www.dqindia.com.

Calyampudi Radhakrishna Rao, "Parasantha Chandra Mahalanobis, (1893–1972)." Royal Society Publishing, 1973, p. 482.

Rao, A S "Sarabhai and Electronics", *Electronics Today* Vol 5(2) 1972 p 49.

Dinesh C Sharma, *The Long Revolution; The Birth and Growth of India's IT Industry,* New Delhi: HarperCollins Publishers India, 2009, pp 92–93.

Sondhi Committee and Menon Committee Reports: Government's Decisions", *Electronics-Information and Planning*, Vol 8, No 8, May 1981.

B V Srikantan, "Sixty years of the Tata Institute of Fundamental Research 1945–2005: The role of young men in creation and development of this institute", *Current Science*, Vol 90, no 5, 10 march 2006, p. 1022.

Sundaram, CV et al; *Atomic Energy in India: 50 Years*, DAE, 1998, pp 46.

Part IV
Metropolitan Methods, Colonial Practices

Chapter 10
The Emergence of Indian Science Community: 1870s to 1940s

Venni V. Krishna

Social history of science in India, from the last quarter of nineteenth century to around 1940s, present us an episode in history that has emerged between leading Indian scientists, social thinkers and the nascent national movement against British colonialism. Among various science institutions established by Indian scientists, The Indian Association for the Cultivation of Science (IACS), founded by Mahendra Lal Sircar on 15 January 1876, stands out with a special significance. For Sircar, just as with other leading scientists, pursuit of science was a moral force something akin to Baconian and Comtean ideas of scientific morality. In fact, science was a 'metaphor of nationalism'. For several leading scientists such as P. C. Ray, J. C. Bose, C. V. Raman, M. N. Saha, among others, cultivation of science in India and its professionalization was an expression and an integral part of the national movement. As Chakrabarti (2004: 151) points out, 'political nationalism had no meaning without science as its guiding spirit'. Relentless efforts by scientists during last two decades of nineteenth century witnessed a series of activities and institutional developments that have laid a strong foundation on which the nascent Indian science community emerged by 1930s.

This chapter is a revised version of the article, 'Portrait of the Scientific Community in India: Historical Growth and Contemporary Problems' in Jacques Gaillard, V V Krishna and Roland Waast (eds), Emergence *of Science Communities in the Developing World*, Sage Publications, 1997. I have gained immensely from discussions held with Deepak Kumar, Irfan Habib, Dhruv Raina and Shiv Visvanathan at various points. I also wish to acknowledge the support of SUD, ORSTOM, Paris for providing the opportunity to interact with a number of researchers working on scientific communities in developing countries. Thanks are due to J. Gaillard, R. Waast, R. Arvantis, Y. Goudineau and Y. Chatelin. I also thank Dr. Ashok Jain and Prof. M. N. Panini for their suggestions in the finalization of this paper.

V. V. Krishna (✉)
School of Humanities and Languages, University of New South Wales Sydney, Sydney, NSW, Australia
e-mail: v.krishna@unsw.edu.au

S. Ahmed and S. Sarkar (eds.), *Decolonizing Science and Modernity in South Asia*,
https://doi.org/10.1007/978-981-97-1829-0_10

With the rise and peaking of *Swadeshi* and national movements in the first decade of twentieth century and the arrival of Gandhi from South Africa around 1915, all leading Indian scientists got involved in this struggle in varying forms. Creating a space for Indian science within the international sphere or world of science with advancements of knowledge from Indian institutions was one of the major objectives during 1920s and 1930s. So far, social scientists including historians in India have mapped the growth of modern science but have not paid adequate attention to the whole issue of emergence of an Indian scientific community.[1] Did the advent of modern science in India entail the advent of professionalization and the formation of specialist groups? What was the social character of science under colonialism? What were the social identities of scientists at large? And how did the Indian scientific community emerge? Who are the lead actors or scientists? And what types of research labs and institutions they have created to advance knowledge and assign national identity within the international sphere of science? These are some of the issues that will be explored in this chapter through systematic evidence both of quantitative and qualitative nature. Given the limited space, this chapter highlights some important contours of early social history of science in the emergence of Indian science community. But before we get down to address the main theme, it is pertinent to understand the concept of science community and what means professionalization of science in India within the colonial context.

10.1 Professionalization and Scientific Communities

The growth of science organized in terms of specialist groups or small communities sharing a set of 'social' and 'cognitive' values to explore and advance systematic knowledge is a recent historical development.[2] The centrality of science in the transformation of modern societies was, to a large extent, the result of what Ben-David

[1] Visvanathan (1985) is, however, an exception. Adhikari (1987) deals at length with growth patterns of science in terms of 'extractive and servicing mode', 'intelligensia-generated mode' and 'state-organized mode' which I share with her in so far as I am talking in terms of colonial science and non-colonial independent tradition in science (see also Krishna 1992). However, given Adhikari's objective to specify its (science) different organizational modes of existence and their social dynamics of growth and change, the interaction perspective was not specifically directed to establish the emergence of an Indian scientific community. Jairath (1984), on the other hand, relies too much on Basalla's model and falls short of informing us about the emergence problem within the colonial context. There is another aspect of Jairath on which I differ which is taken up in the concluding section. The 'colonial' phase of Indian science in the late nineteenth and early twentieth centuries, covering industrial, agricultural and university-based education and research is discussed at some length in Ramasubban (1977: Chap. 3). Ramasubban and Singh (1987: 166) rightly imply the development of the Indian scientific community in its embryonic stage with the formation of the Indian Science Congress in 1914. In this paper I go on to establish its origins much earlier, in the late nineteenth century, sharing their idea of the scientific community as 'young' in its historical growth by the 1920s.

[2] See Whitley (1974) for 'cognitive' and 'social' institutionalization. For a critical review of these concepts see Blume (1974) and Bourdieu (1975).

(1971) refers to as the growth of professionalization associated with the emergence of effective scientific communities. Even though science appeared in its institutionalized form from as early as the late sixteenth and seventeenth centuries, the transformative role of science did not come about until the emergence of professionalized communities in nineteenth century Europe and twentieth century North America (Lucier 2009). For instance, the Royal Society in U.K. was formed as an 'invisible college' of natural philosophers and physicians in 1660, but it was only in the second half of eighteenth century that the British government begun to support the Royal Society and it begun to institutionalize and professionalize on firmer grounds. The British Association for the Advancement of Sciences came into existence in 1831, the National Academy of Science of USA in 1863 and the Kaiser Wilhelm Institute in Germany in 1911.

When we speak of a scientific community, the size of professional grouping becomes more meaningful in terms of what Whitley (1976) defines as disciplines, specialities and research areas, which may hold together between three or four to a few dozen scientists. Whereas Kuhn (1962) drew attention to paradigm-based specialist communities, Robert K Merton (1942) underlined the importance of certain functional normative 'ethos of science' which bind and drive scientists towards communities. The relevance of scientific communities makes sense both from relatively speaking Universalist, discipline-oriented scientific research areas as well as from a national-local milieu. As Thomas Schott (1991: 442) draws attention, national science community can be understood as:

> The scientists within a country form a national scientific community, a community within the world scientific community. They enter the national scientific community through the relatively similar scientific education which they have undergone and their acquisition of the shared culture of science—at an elementary level. They perform their research in the framework of national institutional arrangements for research such as universities with similar pattern, the same national associations and journals, supported by the same national foundations and the same bodies which set the national science policies; thus they perform their research within a common institutional and intellectual setting.

One can differentiate national science community at a meso-organizational level as well as micro-institutional levels represented through higher educational and scientific research institutions. The concept of national scientific communities also signifies the formation of national identities in the practice, production and advancement of scientific knowledge. The formation of national identities, however, is a historical process (Gaillard et al. 1997). As will be argued in this essay, the formation of Indian science community within a colonial science context and its hegemonic structure came about as a result of a relentless intellectual struggle waged by a number of elite group of mainly Indian scientists over three quarters of a century beginning from 1870s. As argued elsewhere, (Gaillard et al. 1997), the twin processes of institutionalization and professionalization are important factors in the formation of national science communities. The former refers to funding, infrastructure, training of scientists, buildings, higher educational institutions and support from different patrons for a specialized field of science. Professionalization, on the other hand refers to higher institutionalization of advancement of knowledge in particular areas, formation of research groups with a network of formal and informal relationships and communication, formation of journals, professional bodies, scientific societies, academies and more importantly a sense of identity and in some instances national pride in the world of science.

The drive towards professionalization and the emergence of scientific communities in nineteenth century Europe shows that these developments have come about in a somewhat 'organic' form catalyzed by the prevailing political structures. Even though the Euro-centered investigation of Ben-David and others offer little insight into the understanding of non-Western cultures, such as India, the social context mapped by them is not without relevance in so far as the development of institutional factors are concerned. Despite the implantation of modern science from about the eighteenth century onwards, colonial structures separated institutionalization from professionalization. In other words, even though the colonial administration is credited with the institutionalization of modern science in India, it created several stumbling blocks and even aborted attempts at professionalization of science in India.

The period chosen for study is instructive for two main reasons: the emergent nationalism, among other economic issues, opened a discourse for the first time on the role of modern science and made it an integral part of the struggle for independence; and, the period witnessed a significant break from the phase of colonial scientific enterprises as a result of the indigenous participation of resources in science and educational infrastructure.

10.2 Goals and Social Character of Colonial Scientific Enterprise

In 1967, Basalla's published a key article on 'The spread of Western Science'. This created a form of intellectual 'storm' in the social history of science. By early 1990s, a flourishing network of scholars around the globe came under the banner of 'Science and Empires'. Several meetings and conferences were organized to systematically explore Basalla's three stage model of science diffusion, particularly in Africa, Asia and Latin America.[3] In this Eurocentric model, Basalla put forward a thesis that non-scientific society provided an important source for the spread of European science. The term non-science societies refer to the absence of western science (Kumar 1986, 1995) and included India along with other eastern countries such as Japan, South America, Africa, China etc.

From a different theoretical perspective, while Basalla's spread of Western Science in three stages represented a non-moving metropolis, MacLeod (1987) came out with five stage metropolitan science focusing on British imperial science. His scheme begins with (i) Metropolitan science characterized by Banksian (Jospeh Banks

---

[3] See Patrick Petitjean et al. (1992) along with Michel Paty, Catherine Jami and others coordinated Science and Empires group.

voyage to Australia) maritime trade, discovery of raw materials etc.; (ii) Colonial science wherein metropolitan dominance persists with control from above; (iii) Federative science which characterized by cooperative, imperial research, higher education and professional legitimation with greater local participation in world markets; (iv) Efficient imperial science involving codified science, specialization of disciplines and dominated by experts with management of resources; and (v) Empire or Commonwealth science signified by metropolitan trusteeship and coordination of fundamental research etc. Following Basalla, MacLeod and other perspectives on colonial science, several scholars contributed a large number of studies.[4]

Some historians have defined the practice of research after the mid-nineteenth century as colonial science (MacLeod 1975, 1987; Kumar 1986, 1995). Colonial science by definition meant a 'derivative' science identified with 'fact-gathering' activity. When viewed from the metropolis it was a 'low science'. MacLeod refers to psychological connotation to science in colonies as the work 'done by lesser minds working on problems set by savants in Europe'.[5] The way science was organized in the colonies was indeed a planned activity from the metropolis, the periphery being assigned the subordinate task of 'data gathering', while the actual theoretical synthesis (pure or fundamental research) took place in the metropolis. Devoid of its intellectual essence, the goal of scientific practice in the colony was not the advancement of science but the exploration of natural resources, flora and fauna (Mukherjee 1989) to feed the intellectual and industrial 'revolutions' in the metropolis. As argued elsewhere (Krishna 1992), the definition of colonial science fits well with the activity undertaken by scientific enterprises such as the geology, education and survey departments. The Asiatic Society of Bengal cooperated with the British Geological Society to promote Indian resources development. The data gathered and sifted from the colonies not only enhanced the horizons of British geology but served as an important basis for colonial policies on minerals, coal mining, agriculture, transport surveys and communications (Stafford 1990). Kumar (1995: 229, 233, 238), in his influential work on *Science and the Raj*, clearly spells out the controlled conditions under which science was organized and pursued in the colonies like India.

Colonizers were fully aware of the importance of science as a very effective instrument of colonization and control. …The logic of the metropolitan-colony relationships was thus not in favor of the latter getting anything like a higher form of techno-scientific education. Unlike Australia and Canada, what Victorian India

---

[4] While I do not accept this connotation given by MacLeod, I am in general agreement with scholars on colonial science in so far as it is considered one among other phases of science in India for the period up to the late nineteenth century. The colonial phase could be considered a dominant phase of science in India but all other scientific activities including local, indigenous traditions cannot be subsumed under the umbrella of colonial science. For examples see Science and Empires Group, Petitjean et al. (1992) and references to colonial science in Kumar (2001).

[5] While I do not accept this connotation given by MacLeod, I am in general agreement with scholars on colonial science in so far as it is considered one among other phases of science in India for the period up to the late nineteenth century. The colonial phase could be considered a dominant phase of science in India but all other scientific activities including local, indigenous traditions cannot be subsumed under the umbrella of colonial science.

Table 10.1 Scientific staff in colonial scientific enterprises in 1920 and their status

Name of the service	Imperial grade		Average pay in INR	
	European	Indian	European	Indian
Botanical Survey	2	0	1000	0
Geological Survey	16	0	1010	0
Zoological Survey	3	1	970	700
Agricultural Service	38	5	1000	460
Forest Service	9	1	1040	660
Medical and Bacteriological Service (on civil employment)	24	5	1220	520
Indian Munitions Board	11	1	780	300
Meteorological Department	10	2	970	770
Veterinary Department (civil)	2	0	1100	0
Educational Service	34	3	910	470
Indian Trigonometrical Surveys*	46	0	–	–
Total	195	18		

**Note* All of these officers except one were Royal Engineers and held military ranks. The provincial service, also highly paid, consisted of 112 officers, of whom 80% were Europeans and Anglo-Indians without any academic qualifications. *Source* P. C. Ray (1920)

received was a low form of scientific and technical education administered under "controlled conditions". Fundamental research work was simply out of question.... the requirements of the colonial government in all its ramifications made science dependent and greatly limited in scope. The seed of science does not grow unless the social ground is well prepared for it through economy activity.

One notable feature of the colonial scientific enterprise was that it was entirely government controlled. Scientific personnel were employed by the East India Company (before 1857) or by the British government from both military and civil services. Undue preference was shown to scientific personnel of European origin both in the recruitment and promotion to higher positions (Kumar 1983). The organizational basis for the emergence of a scientific community therefore, greatly depended on the flexibility of the scientific organizations. As late as 1920, P. C. Ray, the doyen of chemistry in India, presented figures on the scientific personnel employed in colonial scientific enterprises. Out of eleven scientific services, including the educational service, P. C. Ray could count only eighteen Indians out of 213 scientific personnel (Table 10.1).

Highly qualified and deserving Indian scientists were discriminated against and relegated to positions below their entitlement. As Patrick Geddes, the biographer of J. C. Bose, observed, 'the Higher Education Service is accessible only by nomination; and these posts, with extraordinarily rare exceptions, had not been given to Indians, even of the highest European qualifications' (Geddes 1920: 34).[6] Whereas

[6] As quoted in Choudhuri and Singh (2018).

the Europeans employed in the education department were placed in the elite Indian Educational Service (IES), Indian scientists were placed in the Provincial Educational Service (PES) and given half the salary of their counterparts in the IES. The first and perhaps only Indian who found a berth in the IES was J. C. Bose, but his monthly emoluments too were half those of a European's salary in the IES.

P.C. Ray on his return from England in 1888 with a Ph.D. in Chemistry had to wait for one year to be employed in the PES, whereas British chemists with similar qualifications and experience were immediately placed in the IES by the Secretary of State. Ray's complaint against the unequal treatment evoked the retort, 'there are other walks of life open to you. Nobody compels you to accept this appointment' (Ray 1958: 65). H. B. Medlicott, Head of the Geological Survey of India (GSI) held that 'Indians are incapable of any original work in natural science'. He wanted to wait till the scientific chord among the 'natives' was touched and added that, 'if indeed it exists as yet in this variety of human race so let us exercise a little discretion with our weaker brethren and not expect them to run before they can walk'.[7] The most glaring example of applying discriminatory policies in the GSI was on P. N. Bose. In 1903 Thomas Holland superseded P. N. Bose for the position of Director of GSI even though Holland was ten years junior in the service. In protest against the injustice meted out to him, Bose retired from the service the same year.[8] Unfortunately, the tone continues to prevail even during 1930s when C.V. Raman was nominated for the most coveted Nobel Prize. As Choudhuri (2018: 226) points out, it is significant to note that the nomination statement of Raman begins with the phrase 'although trained entirely in India has made remarkable additions to our knowledge of sound and light'.

A small section of Indian scientists associated with colonial scientific organizations resorted to struggle from within against this blatant racial discrimination. By the last quarter of the nineteenth century, due to these inherent tensions the social organization of colonial science showed definite cracks. Towards the turn of the century, pressures towards professionalization of science and scientific autonomy were struggling to find expression as part of the emerging Indian national consciousness. A small section of the scientific and political intelligentsia set an agenda to fight colonial science, on the one hand, and to create alternative structures to professionalize and integrate modern science within the framework of nationalism, on the other. This development led to divisions within the scientific establishment as focus with a clear-cut agenda.

[7] See Review of Agriculture survey No. 25, September 1830 as quoted in Kumar (1986).

[8] Bagal (1955) in an illuminating biography of P. N. Bose also records that Bose played an important part in the creation and expansion of Tata Iron and Steel Works through his discovery of iron ores in Gurumahishini, Durg district.

10.3 Social Divisions Among Scientists and Their Orientations

From a sociological point of view, one can identify three broad but distinct categories of scientific and technical personnel and associated institutions from about the third quarter of the eighteenth century. The *first* category relates to the transplanted or settler scientists employed by the British government. The scientific and technical personnel belonging to this category identified themselves with the colonial administration. They were basically the 'gate keepers' of colonial science who controlled and regulated research strategies to serve colonial ends. They promoted discrimination against native Indian scientists as exemplified in the instances cited earlier. They operated on several fronts such as in education, industry, finance and science departments.

There were several 'Gate Keepers' like H. B. Medlicott and T. Holland who restricted the exploration of science in colonial India to survey and applied research and found every scheme and opportunity to curb and abort attempts to promote professional science linked to advancement of knowledge. For instance, we may take the name of William Roxburgh, Superintendent of the Calcutta Botanic Gardens; E. C. Buck, Secretary, Department of Revenue and Agriculture; Nathaniel Wallich, Superintendent of the Museum of the Asiatic Society; John Eliot, Director General of Observatories; and several administrators of the British Raj in India during 1880 and 1900 acted as 'Gate Keepers' in one form or other. One major contribution of colonial science and the survey research, which mostly is dominated by applied aspects by dozen or so colonial science enterprises, is that modern science was institutionalized in India and an organizational structure for modern science cultivation was established. However, as will be argued later, the professionalization of science and advancement of knowledge from India had to wait till the contribution of third category of scientific personnel, that is, 'native scientists', such as M. L. Sircar and others, and their associates. One can count David Hare, Father Laffont, Patrik Geddes, Sister Nivedeta, among several others in this third category.

In the *second* category were scientific and technical personnel who were called upon by the colonial administration to execute specific tasks. They had no commitment to the promotion of scientific disciplines or scientific societies, and their goal was limited to the accomplishment of their assigned tasks. When these British scientists completed their assignments or attained the age limit, they returned to their country taking with them a vast treasure of skill, tacit knowledge and experience gained in the colonies. In the words of P. C. Ray, 'the present system arrests Indian intellectual growth' (Ray 1920: ix). They can be referred to as 'scientific soldiers'.[9] In the whole Empire, as MacLeod (1975: 348) observes, 'the adventures of Indian civil servants and "scientific soldiers" gave them experience second to none in the

[9] Using scientific soldier in a metaphorical sense. As a soldier some tasks are given to be accomplished and when the task is over, the soldier is sent or assigned other tasks. There is no permanency of a place and task for a soldier.

lessons of administrative organization and coordination'.[10] There are several examples of these soldiers. Dr. William Brook O' Shaughnessy (1809–1889); William Pole (1814–1900); and Colonel Alexander Strange (1818–76) are typical examples of scientific soldiers.

As a product of continuing British colonial policies, particularly discrimination in science, the *third* category of scientific personnel became prominent after the 1870s. They were mainly native Indian scientists supported by a small group of Western settler scientists, missionaries and Jesuits, who relentlessly worked towards the professionalization of science in India. Their numbers run into hundreds, but mention may be made of few important personalities. David Hare, Watch Maker made a big fortune and comes to India. He founded the Hindu College and established a School Book Society in Calcutta and stayed back. He interacted with Raja Ram Mohan Roy and several Bengal based nationalists and supported them in a number of ways. He died in Calcutta and his grave can be found in the College Square, Calcutta. Father Eugene Lafont, the founder of St Xavier College and donor of IACS, Calcutta; Sister Nivedita close associate of Swami Vivekananda; William Carey and Marshman of Serampore missionaries, P. C. Ray, J. C. Bose, C. V. Raman, M. N. Saha, Ashutosh Mukerjee, M. L. Sircar and Visvesvaraya.[11] Basalla (1967) and many other historians club together Medlicott, O'Shaughessy and J. C. Bose as colonial scientists. Sociologically speaking, these three categories of scientists had their constituencies, their goals, network of relationships, social identities and scientific programmes. While scientists in the first two categories (Gate Keepers and Scientific Soldiers) were part and parcel of the colonial scientific enterprise and shared with and benefited from the colonial structures in science, the third category (Native scientists and their associates) struggled against these structures. The term 'struggle' acquired an important place in the scientific discourse although its meaning and implication for action differed from one individual to another. At the national level, however, these scientists widely shared the national obligation to transform colonial structures of science and create alternative support structures with the necessary autonomy to embark on an independent scientific status.

---

[10] Biographical details on some of these 'soldiers' are available in the excellent work of Armytage (1961). It may be noted that my usage of 'scientific soldiers' is somewhat different from MacLeod's (1975), although I owe the term to him.

[11] In contrast to 'scientific soldiers' some foreign scientists, missionaries and Jesuits were attached to India. For example: Father Lafont who joined St. Xavier College in 1865 (and died in Calcutta in 1908) was in fact one of the founder members of the Indian Association for the Cultivation of Science (IACS) established in 1876. David Hare, a watchmaker, came to Calcutta in 1800 at the age of 25. Later founded great Hindu College in 1817 with R.M. Roy and died in Calcutta in 1842. Simonsen and MacMohan helped in the foundation of Indian Science Congress in 1914 and shared the secretarial work of the ISCA up to 1921 (Simonsen 1928: 11–29).

10.4 Support Structures and Nationalist Orientation in Science

Indian scientists with the support of a small group of missionaries and British scientists embarked on a programme to professionalize science in India within a national perspective. An important objective of the programme was the constitution of specialist groups and small communities in various scientific disciplines. The Indian intelligentsia realized that the success of this objective greatly depended on a series of institutional support structures.

10.5 National Awakening in Science and Education

The first organized effort in this direction was the creation of the IACS on 15 January 1876. The person behind this venture was Mahender Lal Sircar (1833–1904), a man trained in modern science but a staunch advocate of homeopathy. Sircar stated that 'the object of the Association is to enable natives of India to cultivate science in all its departments with a view to its advancement by original research, and (as it will necessarily follow) with a view to its varied applications to the arts and comforts of life'.[12] Independent of the government and with a modest collection of Rs. 61,000, Sircar pleaded that 'we should endeavour to carry on the work with our own efforts, unaided by the government. I want it to be entirely under our management and control. I want it to be solely native and purely national' (IACS 1976: 9). Within a few years of its establishment, seven sections in general physics, chemistry, astronomy, systematic botany, systematic zoology, physiology and geology were organized. Wealthy landlords and other influential Indians generously donated funds for the day to day running of the institution. Until the turn of the present century the greatest contribution of the IACS was the propagation of nationalism in the cultivation of science. A direct spin-off from IACS was the creation of at least four institutions to promote technical education with a national perspective.[13] Satishchandra Mukherjee, a leading educationist of Bengal, launched the Dawn Society in 1902 to promote the idea of national education. In fact, a report of the Indian Universities Commission in 1902 recommended more power within colonial settlers rather than the local intellectuals

---

[12] Mahender Lal Sircar's address to the provisional committee of the LACS. The Committee consisted of twenty-five members with Father Lafont as Chairman and Sircar as Member-Secretary. See LACS (1976: 9).

[13] P. N. Bose, member of the IACS, launched the Indian Industrial Association in 1891 which organized popular lectures on coal and fibres. The Association for the Advancement of Scientific and Industrial Education was founded by J. C. Ghosh in 1904; it sent many students abroad for higher education in science. See also Sarkar (1977).

and educators.[14] The society's magazine, *The Dawn*, provided an important forum for Indian scientists to promote science and literature and popularize science.[15]

In 1903 and 1905 Curzon's attempt to control technical education and exclude advanced research from its definition evoked sharp reactions from the protagonists of national education. The Dawn Society transformed itself into the National Council of Education (NCE) in 1906 with a membership of ninety-six intellectuals to organize a parallel structure of scientific and technical education 'on national lines under national control' (National Council of Education 1956: 3). Two schools of thought emerged in the NCE over the emphasis to be placed on scientific and technical education. Tarak Nath Palit and others launched the Society for the Promotion of Technical Education in 1906 which established the Bengal Technical Institute to promote technical education. The other group of the NCE involving Satish Mukherjee and others established the Bengal National College and School in the same year to promote science along with literary courses both at school and university. In 1907 there were a total of 270 students, out of which 223 were in the school section, ninety-eight at the intermediate level and the rest at the degree and diploma levels. There were eleven national schools in different districts of Bengal with a total enrolment of 731 students. In 1910 the rival camps joined hands again giving birth to the nucleus of the present day Jadavpur University and the University College of Science of the Calcutta University. This research centre received Rs. 24 lakhs from Taraknath Palit and Rash Behari Ghosh and the assets of the Bengal Technical Institute, which, by the turn of the century, immensely contributed to the advancement of science.[16]

The national education movement was however not confined to the Bengal intelligentsia. The Poona Sarvajanik Sabha's demand of 1882 to strengthen higher technical education was taken up by the Indian National Congress after 1885. The Congress passed a Resolution at its third session in Madras in 1887 stating that 'it is desirable that the government be moved to elaborate a system of technical education', which was repeated in different forms in the succeeding years. In the princely state of Baroda, Maharaja Sayaji Rao Gaikwad III established Kala Bhavan in the 1880s, the biggest technical institute established by native Indian states at that time. The significance of Kala Bhavan is that the present-day technology and engineering faculties of M. S. University, Baroda, owe their origin to it.[17]

A major break with colonial science teaching set in with the efforts of M. L. Sircar, Nilratan Sircar and J. C. Bose which resulted in the setting up of the Science Degree

---

[14] The members of the society included well known personalities of Bengal. They included Rabindranath Tagore, Aurobindo Ghosh, Rajendra Prasad, Raja Subodh Chandra Mullick, Radha Kumud Mukherjee and Brajendra Kishore Roychowdhury and others.

[15] Sircar, J. C. Bose and many leading Indian scientists contributed articles on physical and The Emergence of the Indian Scientific Community 105 biological sciences in The Dawn. The magazine launched in 1897 continued till 1913. In 1905 The Dawn had about sixty members.

[16] Some eminent scientists were on the faculty of the University College of Science. C. V. Raman and P. C. Ray were the first Palit Professors. Ashutosh Mukherjee invited a group of brilliant scientists like J. C. Ghosh, M. N. Saha, J. N. Mukherjee and S. N. Bose as lecturers.

[17] For further details on Kala Bhavan see *The Dawn*, Calcutta, from September 1910 to February 1911.

Commission in 1898.[18] This Commission recommended the introduction of separate science courses. When Ashutosh Mukherjee took over as the Vice-Chancellor of Calcutta University in 1912, he sought to give a new lease of life to post-graduate science teaching and research. The colonial government however refused to finance post-graduate research in science at the Calcutta University. The donation of Rs. 24 lakhs by Taraknath Palit and Rash Behari Ghosh made it possible to establish the University College of Science at Calcutta University. This initiative, the establishment of the Indian Institute of Science (1909) through the efforts of Jamsetji Tata and the princely state of Mysore, the efforts of Father Lafont at St. Xavier's College, Calcutta, of P. C. Ray and J. C. Bose at the Presidency College (after 1885) and of J. C. Bose at the Bose Research Institute (1917), laid firm institutional foundations for systematic science by 1920. As the encouragement from the government in the form of scholarships to train Indian students in India and abroad was not forthcoming, a number of scholarships and endowments were instituted by wealthy Indians.[19]

10.6 Scholarships, Science Societies in Translation and Popularization of Science:

Parallel to the establishment of scientific institutions and scholarships for advanced research, popularization of modern science and translation of science literature into local languages received attention from the Indian intelligentsia for the first time. Following the initial efforts of the Serampore missionaries in the 1820s and the Delhi College in the 1830s, Bengal provided the lead in the late nineteenth century for vernacular publications of magazines and books in science. Between 1868 and 1910, ten journals and magazines in science alone and forty-seven in technology were reported from Bengal.[20] Efforts invested in creating a base for modern science

[18] From the communication of D. Bose, Member, Governing Body of the Bose Research Institute and member of J. C. Bose Trust. See Science Today (1983: 21).

[19] The famous Premchand Roychand offered five scholarships of Rs. 1400 per year from an endowment of Rs. 2 lakhs in 1879 to Calcutta University. The Rajabhai Tower and Library at Bombay University were established by the generous grant of Rs. 400,000 given by Roychand. Dadabhai Naoroji offered Rs. 50,000 and collected Rs. 1.75 lakhs for Canning fellowships at Bombay University. J. N. Tata's contribution of Rs. 30 lakhs and SheshadriIyer, the Mysore Dewan's offer of 300 acres plus Rs. 5 lakhs for the establishment of the Indian Institute of Science had no parallel. The Association set up by J. C. Ghosh in 1910 raised Rs. 1 lakh per year to provide scholarships for higher education in science and engineering. See also Visvanathan (1985) who traces at length the professionalization of science accomplished by a group of Indian scientists.

[20] Vigyan Rahasya (1871). Vigyan Vikas (1873), Vigyan Darpan (1876), Sachitra Vigyan Darpan (1882), Chikitsa Darshan (1887), Tatwabodhini Patrika and Bengal Spectator are some of the important periodicals in Bengali dealing with science and technology.

Table 10.2 The Indian School of Chemistry in the 1920s: main researchers

Main researcher	Period of activity	No. of publications	No. of collaborating researchers
P.C. Ray	1894–1920	107	37
E. R. Watson	1910–1924	25	12
P. Neogi	1907–1917	14	3
J. J. Sudborough	1912–1925	30	24
R. L. Dutta	1912–1918	28	14
J. N. Rakshit	1913–1917	12	3
B.K. Singh	1913–1926	17	12
S. C. Jana	1914–n.a.	1	2
H. K. Sen	1914–1915	5	2
J. L. Simonsen	1915–1926	27	4
P. C. Ghosh	1917–1920	4	2
J. C. Ghosh	1914–1918	7	4
P. C. Mitter	1918–1926	4	4
B. N. Ghosh	1918–1920	7	3
B. B. Dey	1911–1918	5	3

Source Ray (1918, 1958) and Guay (1986: 82)

in Indian languages were however not confined to Bengal. These activities extended to other parts of India, as is evident from Table 10.2.[21]

In 1875 the Calcutta Book Society (formed in 1817) contained 1544 titles out of which 333 were its publications on science and technology. Between 1868 and 1900, ten journals and magazines in science alone and 47 in technology were reported from Bengal (Bhattacharya et al 1989). A total of 2124 science publications were translated into Indian languages from five Indian provinces. One can see that much of the translation activities were in Bengal, Bombay and Punjab. With the exception of Bengali translations in the Eastern India, Hindi was a popular language for translation of science books in the Northern India during the late 19th Century.

Between the University College and specialized institutions, there were half a dozen societies whose main objective was to popularize science and create a base for modern science among Indians. As noted earlier, the IACS continued its activity of promotion of sciences in different parts of the country.[22] Mention may be made

[21] An excellent survey of writings in science and technology between 1800 and 1950 is reported by Bhattacharya et al. (1989). This Table and information on sciences in Bengali literature is taken from this source, and Ray (1918).

[22] . The Association, as early as 1911, made explicit reference to the equality principle in its statement of the objective the proposed Association sought: 'to give a strong impulse and more systematic direction to scientific enquiry, to promote the intercourse of societies and individuals interested in science in different parts of the country, to obtain a more general attention to the objects of pure and applied science and the removal of any disadvantages of a public kind which may impede progress'. See *Science and Culture*, December 1937, HI (6): 307.

of one of the earliest societies created in 1851, the Agriculture and Horticulture Society of Punjab in Lahore. Agriculture improvements in the region was the main objective of this society Besides the Dawn Society (1904), there were the Scientific Society of Aligarh founded by Sir Syed Ahmad Khan in 1864 in Aligarh.[23] Besides translation of modern science books into local languages, the Society sought to promote liberal, modern education and modern scientific knowledge among the Muslim community. The other important association was the Bihar Scientific Society, Muzaffarpur, founded by Syed Imdad Ali Khan in 1868. The Bihar society also had the main objective to popularize science through translation and lectures. In the West, Bombay Natural Society was founded in 1883 by eight professionals including D. D. MacDonald, J. C. Anderson, H. M. Phipson and E William. India's well-known ornithologist Salim Ali assumed an important role in this Society beginning 1947. It had conservation as the main objective and evolved as the leading institution in conservation and biodiversity research. *The Journal of the Bombay Natural History Society* was launched in 1886. Punjab Science Institute, Lahore, was established in 1886. The main thrust of the activities of these societies was in creating a base for modern science in the vernacular language, i.e., Urdu.[24] In the South, Tamil Scientific Terms Society came into existence in 1916 at Salem by C. Rajagopalachari. In 1918 the Karnataka Vijnana Pracarni Samithi was established for the popularization of science in Kannada language. It published magazine called *Vijnana* from 1918.[25]

10.7 Construction of Specialist Groups: Genesis of Indian Science Community:

Between 1870 and 1920 the native Indian and missionary contribution to the establishment of colleges and initiation of science teaching exceeded British efforts. In the nine universities established between 1857 and 1918—Bombay, Madras and Calcutta (1857), Allahabad (1887), Punjab (1882), Banaras (1916), Mysore (1916), Patna (1917) and Hyderabad (1918)—the Indian contribution was substantial. Wealthy Indians, mainly landlords and political figures from Princely States, donated substantial amounts in these universities. By 1907, forty-five affiliated colleges were established in the three presidency regions where ninety-one lecturers, most of them of

[23] The main objectives of the Aligarh Society included translation of modern science into local languages, publication of journals and rare and valuable oriental books and holding lectures on scientific and related themes. It also undertook to popularize science through scientific experiments. The society disseminated knowledge on the introduction of technology in agriculture and steam for industrial purposes. The aims of the Society was secular in the sense that it totally kept religion out of its purview.

[24] See Report on Industrial Education, Pan II, National Archives of India, New Delhi, 1903. This report covers some details on the Punjab Science Institute, Lahore. For details on the Aligarh and Bihar Scientific Societies, see Habib (1985).

[25] Material for this section on scientific societies was also drawn from Sen (2017). See https://insa.nic.in/writereaddata/UpLoadedFiles/IJHS/Vol52_2_2017__Art09.pdf.

Indian origin, taught B. A. and M. A. subjects in science and engineering. Between 1910 and the 1920s, Indian universities including those in the presidency towns awarded 2134 degrees in all science fields (Mahalanobis 1971: 221). During 1876 and 1920s a series of infrastructure and institutional efforts were made relatively outside colonial science enterprises. All these intellectual efforts resulted in creating a base for advancement of knowledge in science in some critical disciplines such as physics, chemistry, plant physiology, mathematics, Electromagnetic Waves and to some extent in astronomy. We will turn attention to some leading contributions of Indian scientists.

With J. C. Bose and P. C. Ray joining the Presidency College in 1885 and C. V. Raman joining the IACS a little later on a part-time basis, the 'cultivation' of science was transformed towards the 'advancement of science'. Father Lafont at St. Xavier's College, Calcutta, established an excellent observatory for spectro-telescopic investigations. Together with these centres, the Indian Institute of Science (1909), University College of Science, Calcutta University (1913) and Bose Research Institute (1917) constituted research programmes to give a new 'identity' to Indian science. In advancing modern science, Indian scientists resolved to struggle on two fronts. Whilst the existential circumstances compelled them to struggle against colonial structures, the goal of advancing science was also to revive the rational and experimental tradition. The assertion that the method of science is Western and hence alien to our Indian tradition was rejected as baseless by J. C. Bose and others. P. C. Ray's two volumes on the *History of Hindu Chemistry* (Calcutta 1896) and Binoy Sarkar's *Hindu Achievements in Exact Science: A Study in the History of Scientific Development* (New York 1918) are some examples of this orientation. Thus, advancing science meant giving a new status both to the self and to the national prestige.

J. C. Bose's[26] work on micro-wave (1895) and plant physiology (1900) earned him world-wide recognition and he was elected to the Royal Society. On radio receivers, Patrick Geddes, biographer of J. C. Bose, accords him priority over Marconi who patented it. P. C. Ray in 1896 discovered mercurous nitrite and C. V. Raman who entered the IACS in 1907 published about fifty-eight papers by 1920. The research programmes initiated by J C. Bose, C. V. Raman, P. C. Ray, Father Lafont and others were not individual-based programmes. They constituted, for the first time, the embryo of the Indian scientific community. At St. Xavier's College, Father Lafont was instrumental in organizing a research group on spectro-telescopic investigations and in contrast to the data-supply scheme of colonial science, he set up facilities for undertaking basic studies. During the transit of Venus, Lafont collaborated with the famous Italian astronomer P. Tacchini in the astronomical investigations in Madhupur, Bihar.

Four miniature observatories with revolving cupolas were constructed and Lafont recorded the total time of transit. Impressed by the value of solar observations in the cloudless Indian sky, Tacchini persuaded Lafont to erect a spectro-telescope at St. Xavier's College. Lafont brought a number of instruments from Germany and

[26] Bose was a student at St Xavier's College, Calcutta which evolved as one of the leading centres in India in astronomical research. It is said that Father Laffont, founder of St Xavier College, inspired young Bose to draw into physics.

France by raising private donations (Biswas 1989; Kochhar 1991). Mathematician-cum-astronomer, Father Alphonse de Penaranda, joined Lafont in 1876 on an astronomical programme until his death in 1896. Father Penaranda regularly contributed to the column on 'astronomical occurrences' in the weekly periodical Indo-*European Correspondence,* launched in 1865 by the Catholics of Calcutta. Father V. de Campigneulles joined Lafont in 1882–83 and continued the work on spectro-telescopy. He published two books based on studies of the famous total solar eclipse of 1898 by a team of Jesuit scientists of St. Xavier's College, Calcutta.[27] Several international teams too came to India for the study of this famous total eclipse.

Another major institution that came into existence in 1909 is the Indian Institute of Science at Bangalore. One of the planners was the Nobel Laureate Ramsay. The institute began with the institutionalization of physical chemistry, organic and inorganic chemistry, applied chemistry and electro-technology. School of experimental biology and experimental physics was later added to this list. The period between 1911 and 1920 witnessed the focus on applied research areas relevant to industry and society. In 1914 Mahenderlal Sircar's dream of combining teaching and original research was realized by the establishment of the University College of Science (UCS). This had its precursor in the Bengal Technical Institute, which evolved through the National Education Movement of 1905, which demanded boycott of British goods and support for indigenous industry. This was the period when Asutosh Mukherjee, lawyer and a mathematician, became Vice Chancellor of Calcutta University. It was at this time that Tarak Nath Palit handed over the assets of the Bengal Technical Institute to the Calcutta University. In 1912, Mukherjee had received two donations, one for Rs. 14 lakh from Tarak Nath Palit and one for Rs. 10 lakh from Rash Behary Ghosh. Raja of Khaira Estate, which provided funds for construction of buildings and creation of several faculty positions. With this, he was able to establish the University College of Science (UCS) in 1914. In 1917 all postgraduate studies were centralized in Calcutta. The establishment of UCS was unique in the sense that for the first time since the establishment of universities in the fifties of nineteenth century, it took up the higher science teaching in all its departments and the advancement of scientific knowledge through original research as a prime duty of the faculty.

It was not until 1933, that frontiers in physics research received a boost with Sir C.V. Raman taking over as director of IISc. Several eminent European scientists were invited at IISc during the tenure of Raman. C.V. Raman, after his Nobel Prize winning theme on Raman Effect, developed research interest in the optical properties of diamonds at IISC and later at Raman Research Institute. Diamonds were the source of understanding the properties of crystalline structure. Together with Banaras Hindu University and Jadavpur University's chemical technology departments, the University Department of Chemical Technology (UDCT) Bombay was established in 1934 by Indian industrialists and philanthropists. UDCT in the 1940s emerged as

[27] These books are: V. de Campigneulles and H Josson. 1898. The Total Solar Eclipse—January 22, 1898. Calcutta: Thacker, Spincie and Co., 1898, and V. de Campigneulles, Observations Taken at Dumroan, Bihar India During the Eclipse of the 22nd January 1898. London: Longmans Green and Co. (Biswas 1989).

India's most eminent research and educational centre for chemical technology with particular focus on textile engineering.[28]

10.8 Specialist Groups in Chemistry

After serving Presidency College for thirty-eight years, P. C. Ray joined the University College of Science in 1916. For the first time, what is known as the Indian School of Chemistry emerged by the 1920s. Referring to the 126 papers contributed to various societies such as the *Chemical Society* (London), *Journal of the American Chemical Society* and others, *Nature* in its 23 March 1916 issue observed: 'some of these papers are of very considerable value and interest, and indicate enthusiastic work on the part of this newly created school' (Ray 1958: 150). Rasik Lal Datta, Nilratan Dhar, Jnanendra Chandra Ghosh, Jnanendra Nath Mukherjee, Pulin Behari Sarkar, A. C. Ghosh, P. C. Bose, G. C. Chakravorti, to name a few, were part of the school constituted by P. C. Ray as shown in Table 10.2. By 1920 the active publishing community of chemistry was around fifty and about 160 papers were published. The credit for the first advance in research in physical chemistry goes to N. R. Dhar who also made original contributions to electro-chemistry. J. C. Ghosh's theory (1918) on the abnormality of strong electrolytes created a stir in the international community when it was first published. Similarly, the credit for initiating advanced research on colloidal chemistry in India goes to J. N. Mukherjee. The school of chemistry under Ray contributed immensely to the development of chemistry departments in Indian universities. This school contributed to at least four generations of chemists. The Indian Chemical Society was established on 9 May 1924 with P. C. Ray as the first President and his colleagues and students as members. As soon as the Society came into existence, the first quarterly journal of the Society appeared in November 1924. The research papers published in this first journal found peer recognition from the British chemical metropolis. Professor Wynne, the then President of the *Chemical Society*, London, cabled to Sir P. C. Ray congratulating the *Indian Chemical Society*.

10.9 Physics and Specialist Research Groups

In physics, C. V. Raman, J. C. Bose, S. N. Bose and M. N. Saha constituted the Indian School of Physics, but it was C. V. Raman who gave the lead during the first quarter of the present century. The centenary volume of the IACS recognizes this

[28] One can count some of the most eminent business leaders who have come out of UDCT, Bombay and went on to create leading chemical related firms in India. Dr Keki Gharda of Gharda Chemicals, Dr Anji Reddy of Dr Reddy's Labs, Mukesh Ambani of Reliance, I.A. Modi of Cadilla, the three Parekh brothers of Pidilite, R. M. Kedia of Kedia Chemicals, Dr K.V. Mariwala of Bombay Oil Industries, Nikhil Meswani of Reliance Petro, Dr Atma Gupta of Armour Polymers, V.G. Rajadhyaksha of Hindustan Lever, and J.M. Nadkarni of Bombay Paints are just a few names.

as the 'School of Raman'. A. Dey, S. K. Banerjee, S. Appasamyar, S. K. Mitra, D. N. Ghosh, D. Banerjee, T. J. Chinmayanandan and K. S. Rao are some of the scientists who constituted Raman's School of Physics.[29] Under the leadership of Raman, for the first-time physics acquired a professional status at the IACS. Raman and his associates published in reputed foreign journals like *Nature* and *Philosophical Magazine*, but soon the IACS launched its own *Bulletin of the Association* from 1909, which became a vehicle for publishing original Indian contributions. With the coming of Raman and the increase in research activities, the IACS held regular scientific meetings around three sections viz., physico-mathematical, chemical and biological. Geology was added in 1916. The scientific meetings graduated into the Annual Science Convention, the first of which was held in 1917 in which nine papers in physics, four in chemistry and seven in the biological sciences were presented. Surveying the work of physics in Calcutta from 1907 to 1917, Raman observed, 'a real school of physics has grown up in Calcutta the like of which does not exist in any other Indian university and which even now will not compare very unfavorably with those in European and American universities' (IACS 1976: 30). Raman in this meeting also gave a list of twenty-five original papers from the School of Physics in Calcutta which included the works of S. K. Banerjee, S. K. Mitra and M. N. Saha. Around 1918, the *Calcutta Physical Society* was established under the auspices of Calcutta University. To provide a publication outlet of the annual meetings of the IACS, the *Proceedings of the IACS* was launched from 1917. These efforts led to the launching of the *Indian Journal of Physics.*

K. S. Krishnan, the first Director of the National Physical Laboratory, Delhi, joined Raman after 1920. The most spectacular advances in optics were carried out by the Raman School which later won world recognition for the Raman Effect and the Nobel Prize in physics in 1930. It is reported that in the first seven years after this discovery, the Raman Effect was the main subject of more than 700 papers in the scientific literature at that time.[30] In theoretical astrophysics, Saha's theories of thermal ionization and radiation led to the physical theory of stellar spectra by the 1920s. Saha produced thirteen papers between 1917 and 1920, including the work on 'ionization in the solar chromosphere' (1920). Saha provided the base for the ionospheric school at Allahabad University in which he spent seventeen years of his life. Saha's basic work was further advanced by S. Chandrashekar, D. S. Kothari and Majumdar who studied problems connected with the atmosphere of stars, application of Fermi–Dirac statistics to elucidate the internal structure of stars and Kothari's theory of pressure of ionization (Mahanti 1990; Prasad 1938; Sen 1954). S. K. Mitra's research programmes in the 1930s on radio science, wireless research and chemical physics devoted to the interpretation of absorption spectra owe much to the initial impetus in physical sciences given at the turn of the nineteenth century. The number of publications of Indian physicists in Indian journals after 1925 can be

---

[29] C. V. Raman himself refers to the work on physics as the 'school of physics' at Calcutta: see IACS (1976) and Venkataraman (1988).

[30] See https://www.acs.org/content/acs/en/education/whatischemistry/landmarks/ramaneffect.html.

Table 10.3 Number of papers published in Physics from India between 1901 and 1950 in India

Years	Total papers	Single-author	Collaboration. coefficient
1901–1905	29	29	0.0
1906–1910	59	59	0.0
1911–1915	85	72	0.078
1916–1920	155	120	0.113
1921–1925	356	267	0.128
1926–1930	626	431	0.163
1931–1935	1224	820	0.174
1936–1940	1239	827	0.173
1941–1945	1034	727	0.156
1946–1950	1144	789	0.165

Source Raina Dhruv and Gupta (1998)

seen in Table 10.3. This era can be also seen as the rapid phase of professionalization of physics in India.

10.10 Electro-Magnetic Waves and Plant Physiology

Another group which became active between 1900 and 1920 was the group on plant physiology under J. C. Bose. Following his paper in 1900 on the 'Generality of Molecular Phenomena Produced Electrically in Living and Non-living Matter', Bose published four monographs through the publisher, Orient Longman on 'Response in the Living and Non-living' (1902); 'Plant Response as a Means of Physiological Investigation' (1906) with 315 experiments; 'Comparative Electro-physiology' (1907) with 321 experiments; and 'Researches on Irritability of Plants' (1913). With the basic research base in physics and plant physiology, J. C. Bose organized research groups at his Bose Research Institute established on 30 November 1917.[31] Bose did not favor too much specialization that was evolving in the western science. His views at the time of the creation of Institute reveals that he really intended to realize the regeneration Indian traditions in the pursuit of modern science. N. N. Neogi, S. C. Das, Gurupadaswamy Das, Jyotiprakash Sircar, S. C. Guha and Lalit Mohan Mukherji worked with J. C. Bose and published about twenty papers on various facets of plant physiology. In all, J. C. Bose published ninety-seven papers from 1895 to

[31] As reported in Frontline, A journal of 'The Hindu' group of newspapers. 'The institute was set up in buildings designed in the Mauryan style, on a four-acre plot complete with gardens, libraries and lecture halls, in North Kolkata. During J.C. Bose's lifetime, it received an annual grant of Rs. 1 lakh from the government. The institute today is a grants-in-aid autonomous institution under the Department of Science and Technology, Government of India. Its affairs are managed by a Council comprising representatives of the apex Governing Body'. See https://frontline.thehindu.com/other/advertorial/article30225627.ece.

1920 and collaborated with nine colleagues in a quarter of these publications. From 1917 the Bose Research Institute launched its own journal called the *Transactions of the Bose Research Institute*. All the twenty papers in plant physiology referred to above appeared in this journal (*Science Today* 1983; Sen and Chakraborty 1986).

10.11 Indian School of Mathematics

In mathematics, one of the first steps to institutionalize and professionalize mathematics was taken by Ashutosh Mukherjee. He was a Judge at the Calcutta High Court in 1904 and became the first Indian Vice Chancellor of Calcutta University in 1906 for four consecutive terms till 1914. At Calcutta University, Ashutosh contributed to developing courses in mathematics and lectures for post-graduate students; university chairs in mathematics and encourage students to take up mathematics. During 1906 and 1912, mathematics faculty were appointed or recommended in various institutions including four at the university colleges. The most notable appointment in 1906 was the Chair of RashbeharyGhosh at the Calcutta College of Science and Technology. The other important Chair created at Calcutta University was the prestigious Hardinge Professorship of Mathematics. Between 1913 and 1948 four eminent faculty occupied this Chair, namely, W. H. Young, C.E. Cullis, Ganesh Prasad and Friedrich William Levi (Chaudhuri 2019).[32] Little known about Ashutosh was his original contribution to differential equations, known as 'Mukherjee theorems'. Ashutosh became a member of the London Mathematical Association and the University of Cambridge honoured him by including his theorems in their curriculum.

Through the efforts of V. Ramaswami Iyer, the 'Analytical Club' at Fergusson College, Poona, was upgraded as the Indian Mathematical Society in 1911. In Bombay mention should be also made of the entry of D. D. Kosambi at the Tata Indistutute of Fundamental Research around 1944 who initiated group on mathematics. Through Ramaswami Iyer's effort, mathematics found institutional basis in Pune, Bombay and Madras. No account of mathematics in Madras will be completed without mentioning Srinivasa Aaiyangar Ramanujam. Sociologically speaking, he was a deviant case in the sense that he never went to a university or trained or educated in higher educational institution. He was in a way gifted mathematician who worked in Madras Port Trust as Clerk and interacted with G.H Hardy and gained a scholarship to go to University of Cambridge. This happened mainly due to his sheer mathematical brilliance that attracted Hardy. He was elected as the youngest Fellow of the Royal Society on 28 February 1918 at the age of 31 and died at the age of 32.

In 1914, the Rash Behari Ghosh Chair of Applied Mathematics was created at the University College of Science, Calcutta, and Ganesh Prasad the first D. Se. of Allahabad University was appointed to it. After the establishment of Benaras Hindu University in 1918 by M. M. Malaviya, Ganesh Prasad founded the *Benaras*

---

[32] See Chaudhuri (2019) https://insa.nic.in/writereaddata/UpLoadedFiles/IJHS/Vol54_3_2019__Art11.pdf.

Mathematical Society. Prasad's main contribution was in applied mathematics. His discourse with professionals in this area appeared as a memoir entitled, 'Constitution of Matter and Analytical Theories of Heat', published by the Royal Society of Sciences of Gottingen (1903). The other area of his interest was in the theory of real variables mainly on Fourier Series published in the late 1920s. After Ashutosh Mukherjee, he became the President of Calcutta Mathematical Society. Prasad's main contribution was in applied mathematics. Notable among Ganesh Prasad students who achieved reputation as mathematicians in India is B. N. Prasad who after independence founded the Allahabad Mathematical Society.

10.12 Towards Professionalization: Scientific Associations, Societies and Journals

10.12.1 Indian Science Congress Association

The constitution of specialist groups in the university centres and specialized institutions such as the IACS and the Bose Research Institute enabled Indian scientists to assign a distinct national identity to science in India by the 1940s. These achievements, which remained a dream during M. L. Sircar's lifetime, were a significant departure from the era of colonial science. Indians could hardly publish eighteen papers in the journal of the Asiatic Society during 1836 and 1895. The European (mostly British) scientists on the other hand accounted for 1021 papers (Visvanathan 1985: 27). In the next twenty-five years research output from the Indians alone accounted for over 350 papers, the bulk of it concerning original investigations.[33] Another major step in the professionalization of Indian science during early twentieth century was the creation of a common forum for scientists in different parts of the country through the establishment of the Indian Science Congress Association (ISCA) in 1914. The lead actors who initiated this intellectual move were two chemistry professors, J. L. Simonsen and P. S. MacMohan.

The ISCA served as an important platform to catalyze community consciousness as well as unify the scattered specialist groups on a national scale during its annual conventions which took place in different parts of the country.[34] Beginning with a membership of sixty scientists in 1914 the ISCA quickly expanded to 716 members in 1926, 900 in 1934 and 2500 in 1947. In 1914, thirty-five papers were presented in

[33] This figure is only the conservative estimate made from different sources. The actual figure could cross 600.

[34] The Association, as early as 1911, made explicit reference to the equality principle in its statement of the objective the proposed Association sought: 'to give a strong impulse and more systematic direction to scientific enquiry, to promote the intercourse of societies and individuals interested in science in different parts of the country, to obtain a more general attention to the objects of pure and applied science and the removal of any disadvantages of a public kind which may impede progress'. See *Science and Culture*, December 1937, HI (6): 30.

Table 10.4 Total number of papers presented during ISCA sessions, 1914–1941

S. No	Year of the annual meeting of ISCA	Place	Number of papers presented
1	1916	Lucknow	65
2	1917	Bangalore	74
3	1920	Nagpur	100
4	1926	Bombay	436
5	1927	Lahore	481
6	1930	Allahabad	727
7	1931	Nagpur	697
8	1933	Bombay	703
9	1935	Calcutta	776
10	1938	Calcutta	800
11	1941	Benaras	775

Source Proceedings of Indian Science Congress Association, 1914–1941 as reported in Sinha (2020)

different scientific disciplines which gradually increased to 100 in 1920, 727 in 1930, 776 in 1935 and 775 in 1941 (See Table 10.4).[35] Beginning with the establishment of the Indian Botanical Society (1920), the professionalization of science entered an important phase. During the next twenty-five years up to 1947, about 30 more professional societies were created in different disciplines. ISCA was directly involved in the formation of these societies which were inaugurated during ISCA's annual sessions. By the 1930s, with the creation of several scientific societies and associations, science academies began to be formed. One of the earliest Indian academies was the Academy of Sciences of the United Provinces of Agra and Oudh created in 1930 by Meghnad Saha. Down in South, Sir C.V. Raman who shifted from IACS to Indian Institute of Science established the Indian Academy of Sciences, Bangalore in 1934. In the next year in 1935, the National Institute of Sciences of India was created in Calcutta which was later in 1970 changed to Indian National Science Academy after it moved to Delhi. The creation of three academies in the 1930s created a distinct identity of Indian science within the international sphere. An important intellectual and professional platform was created to recognize well known Indian scientists and honor them with fellowships. One of the achievements accomplished by the nascent Indian science community from the late nineteenth century, particularly during 1920 and 1940s, is gaining distinct national identity within the global world of science. There were 11 Indian scientists who were awarded Fellowship of the Royal Society, London, from the early to mid-twentieth century.

[35] Quantitative data on ISCA is drawn from Sinha (2020).

10.13 Concluding Remarks

For countries such as India, colonial experience is important in considering the social processes of professionalization of science and the emergence of national scientific communities. Colonial science or scientific work in colonial enterprises had little to do with the emergence of the Indian scientific community in its emergent period (1900 to 1940s). In a large measure, the genesis of Indian science community had taken roots outside the confines of colonial science enterprises as these institutions were not conducive for the professionalization of science and its advancement. The main purpose of this essay was not to analyze the Indian case within the diffusionist model of George Basalla (1967). The first stage of his model deserves to be outrightly rejected as India was seen as a 'non-scientific society' (see also the Introductory chapter). However, given the implications of Basalla's second and third stages for the spread of Western Science from a phase of colonial science to achieve independent scientific tradition, it is important here to point out that this was not the case. First, institutions and actors who laid the foundations for the emergence of Indian science community had a distinct national identity and purpose and they were outside the colonial scientific enterprises. As David Arnold (2008) observes, 'that India's first major hydro-electric scheme was constructed in the princely state of Mysore in 1898 and that ten years later Bangalore became the site for the Indian Institute of Science (in 1909), should alert us to the significance of even the circumscribed power of the princes in providing an alternative (often more adventurous) source of scientific support and technological initiative to that offered by the British'.[36] Second, the struggle to achieve independent or autonomous scientific tradition towards national science community was waged with colonial science administration. Long intellectual struggle for the professionalization of science was precisely waged by Indian intelligentsia because colonial science structures were not conducive.

After the 1870s it becomes sociologically meaningful to speak of three categories of scientific personnel, 'gate keepers' and 'scientific soldiers', who were part and parcel of the colonial scientific enterprises, and native Indian scientists and their missionary supporters who constituted the third group. The major conclusion of this paper is that the third category of scientists for the first time made organized attempts to undertake basic or fundamental research by the 1920s. Specialist groups, schools and institutions were constituted in physics, chemistry, mathematics, biological sciences and astronomy by the early decades of this century. By the early 1930s the Indian scientific community made its intellectual presence felt in the international scientific world. There were at least nine fellows of the Royal Society as well as a Nobel Laureate in physics. An Indian scientific community was created which notwithstanding its limited sphere of influence, regarded advancing the frontiers of knowledge as a means by which an Indian national identity could be established at

[36] See David Arnold, *Science, Technology and Medicine in Colonial India*, Cambridge University Press, 2008. Accessed https://beckassets.blob.core.windows.net/product/readingsample/328725/9780521617185_excerpt_001.pdf.
25 December 2020.

the international level of science. Emerging nationalism after the 1870s and the 'ideological' position of Indian scientists were in no small measure unconnected to their struggle to achieve international recognition. The support structures created through indigenous initiatives from the 1870s onwards gave a new meaning to the career structure in science. These scientists espoused professionalization by remaining relatively outside the colonial enterprises and worked against the prevailing discriminatory practices in recruitment. Even though the scientific community of this era, as with any other international group of scientists, interacted with the Western metropolis, the research agenda and social goals in advancing scientific research came from their nationalist orientation. During 1900 and 1920 the Indian scientific community was by no means large enough to reflect an 'all-India' character. It was mainly concentrated in some pockets of the Indian 'metropolis', particularly in the Bengal Presidency.

It is argued in this paper that the Indian scientific community was certainly committed to the Indianization of science in the country but this drive did not imply a complete renunciation of its modern intellectual identity. P. C. Ray, J. C. Bose, C. V. Raman and others never resented being part of the modern intellectual world in science. Rather, they struggled to establish an Indian identity in the world of science. Jairath's (1984: 127) characterization of P. C. Ray's monumental work on the History of Hindu Chemistry as a revivalist streak is questionable. P. C. Ray undertook this work in response to the French chemist, Berthelot's similar work, L' Alchimstes Grecs, for the middle ages. In this work Ray tried to trace the rational and experimental tradition in modern chemical sciences in Indian (Hindu) history—signifying a project to link the modern sciences with the relevant domain of Indian tradition. He wrote: 'If the perusal of these lines will have the effect of stimulating my countrymen to strive for regaining their own position in the intellectual hierarchy of nations, I shall not have laboured in vain' (Habib and Raina 1989: 63). As with J. C. Bose, Ray's orientation in the historical examination of Indian science since antiquity was to revitalize our own rational tradition. The experimental orientation of Indian scientists has, of late, been incorrectly underplayed by some scholars. For instance, Raj (1991: 123) holds that Hindu knowledge was clean in contradistinction to Western science which is linked to laboratory and experimentation and which entails soiling one's hands. Referring to the orientation of scientists in mathematics, algebra, astronomy, optics, hydrostatics, etc., Raj concludes that 'it is in the old image of knowledge qua clean knowledge that the *Bhadralok* sought those aspects of western science that would best correspond to it'.

Such an observation certainly cannot be stretched to an extreme 'cultural determination' mould for the simple reason that scientists such as J. C. Bose, C. V. Raman, P. C. Ray and others established their laboratories by constructing much of their own apparatus. Even though C. V. Raman claimed to have spent only about Rs. 200 for the equipment on his Nobel Prize winning work, his work was not theoretical. In 1930, awarding the Hughes Medal of the Royal Society, Lord Rutherford observed that 'Raman's effort must rank among the best three or four discoveries in experimental physics in the last decade' (Bhagvantham 1972: 32). J. C. Bose for instance did not only write books on plant physiology containing descriptions of about 636 experiments, but is also credited with developing 100 'sensitive and experimental instruments for investigating plants'. In the realm of chemistry P. C. Ray notes, 'the fact that Hindus had a very large hand in the cultivation of the experimental sciences is hardly known'. Works on chemistry in the *Rasendra Chintamani* by Ramachandra and *Rasa—Prakasha Sudhakara* by Yasodhara are referred to by Ray as testimony to the tradition of experimentation and observation in India. C. V. Raman in a different vein drew our attention to the ability of the Indian mind in his own words.

> I can assert without fear of contradiction that the quality of the Indian mind is equal to the quality of any Teutonic, Nordic or Anglo-Saxon mind. What we lack is perhaps courage, what we lack is perhaps driving force which takes one anywhere...we need a spirit...that will carry us to our rightful place under the sun...as inheritors of a proud civilization (Venkataraman 1988: 504).

In their efforts to revitalize the tradition of science, Indian scientists did not attempt to selectively glorify all that was 'good' in the past. In his own words, J. C. Bose observed that 'it is a false patriotism to assert that our ancestors knew everything and that we have nothing further to learn…the real golden age is not the past but, in the future, advancement of positive knowledge and the method of experimental verification is most essential' (Sen 1989). Indian scientists denounced the social forces which caused the downfall of experimental and inductive sciences in the middle ages and in doing so cautioned against the prevailing social and political conditions in early twentieth century India. As noted earlier, nationalism provided an ideological base for the emergence of the scientific community. Advancing scientific research formed an important object of their 'ethos'. As Visvanathan (1985: 31) implies, 'ethos' has both the 'cultural' and 'economic' versions. Some connections between science and economic growth through technological progress were apparent both in practice and in intellectual discourse. However, as far as practice was concerned it was in no way comparable to early twentieth century Germany and America or even to later decades of Japan.

References

Adhikari, K. 1987. 'Science, society and the Indian Transformation'. Philosophy and Social Action, XIII (1–4).

Armytage, W. H. G. 1961. A Social History of Engineering. London: Faber and Faber.

Bagal, J. C. 1955. PramathaNath Bose. New Delhi: P N Bose Centenary Committee.

Basalla, G. 1967. 'The spread of Western science', Science, 156 (3775), 5 May.

Ben, David, J. 1971. The Scientists Role in Society. Chicago: The University of Chicago Press.

Bhagvantham, S. 1972. Professor C V Raman: His Life and Work. Hyderabad: The Andhra Pradesh Academy of Sciences.

Bhattacharya, D. P. et al. 1989.'A survey of Bengali writings on science and technology'. Indian Journal of History of Science, 24 (1), 8–66.

Bhatnagar, S. S. 1928. 'Presidential Address (Chemical Section) 15th Indian Science Congress, January, 1928', Proceedings of the Indian Science Congress. Calcutta: Asiatic Society of Bengal.

Biswas, A. K. 1989. 'Reverend Father Eugene Lafont and the Scientific Activity of Saint Xavier's College, Calcutta (1860–1910)'. Paper presented at a seminar on Science and Calcutta, BITM, Calcutta, 21–23 December.

Blume, S. S. 1974. Towards a Political Sociology of Science. New York: The Free Press.
Bourdieu, P. 1975. 'The Specificity of the Scientific Field and the Social Conditions of the Progress of Reason'. Social Science Information (14).
Chakrabarti, P. 2004. *Western Science in Modern India: Metropolitan Methods, Colonial Practices*, Ranikhet: Permanent Black.
Gaillard, J, V. V. Krishna and R. Waast (1997) Scientific Communities in the Developing World, New Delhi: Sage Publications.
Geddes, P. 1920. The life and work of Sir Jagadis C. Bose, London and New York: Longman Green.
Guay, Y. 1986, 'Emergence of Basic Research on the Periphery: Organic Chemistry in India', Scientometrics, 10 (1-2): 77-94.
Habib, Irfan. 1985. 'Institutional Efforts: Popularisation of Science in the mid-19th Century', Fundamenta Scientae, 6 (4): 299-312.
Habib, I. and Raina, D. 1989. 'Copernicus, Columbus, Colonialism and the Role of Science in Nineteenth Century India', Social Scientist, 17 (3-4): 190-91.
IACS. 1976. A Century. Calcutta: IACS (centenary volume).
Jairath, V. K. 1984. 'In Search of Roots — the Indian Scientific Community'. Contributions to Indian Sociology (n. s.) 18 (1).
Kirtikar K. R. 1907. 'The "colonial" model and the Emergence of National Science in India: 1876–1920'. Paper presented at the International Colloquium on Science and Empires, April 2–6, 1990, UNESCO/CNRS: Paris.
Krishna, V. V. 1992. 'The Colonial "Model" and the Emergence of National Science in India, 1876–1920', in P. Petifjean, C. Jami and A. M. Moulin (eds), Science and Empires: Historical Studies About Scientific Development and European Expansion, Netherlands: Kluwer Academic Publishers.
Kuhn, Thomas (1962), *The Structure of Scientific Revolutions*, The University of Chicago Press: USA
Kumar, D. 1983. 'Racial Discrimination and Science in 19th century India', Indian Economic and Social History Review, XIX (1).
Kumar, D. 1995. *Science and the Raj, 1857-1905*, Delhi: Oxford University Press.
Kumar, D. 1986. Science Policy of the Raj: 1857–1905. Ph. D thesis, Delhi University.
Kochhar, R. K. 1991. 'Astronomy in British India: Science in the Service of State'. Current Science, 25, January.
Lucier, P (2009), 'The Professional and the Scientists in Nineteenth Century America', ISIS, 100(4), pp 699-732.
MacLeod, R. M. 1975. 'Scientific Advice for British India: Imperial Perceptions and Administrative Goals', Modern Asian Studies, 9 (3).
MacLeod, R. M. 1987, 'On Visiting the Moving Metropolis: Reflections on the Architecture of Imperial Science' in N Renigold and W Rosenberg (eds.), Scientific Colonialism: A Cross-cultural Comparison, pp. 217–49. Washington: Smithsonian Institution Press.
Mahalanobis, P. C. 1971. 'Recent Developments in the Organisation of Science in India', in A. Merton, Robert K. (1973) [1942], "The Normative Structure of Science", in Merton, Robert K. (ed.), The Sociology of Science: Theoretical and Empirical Investigations, Chicago: University of Chicago Press.
Rahman and K. D. Sharma (eds), Science Policy Studies. Bombay: Somaiya Publications.
Mahanti, S. 1990. 'S K Mitra (1890–1963) - A Pioneer in Radio Physics', Science Reporter, (October).
Mukherjee, A. 1989. 'Indian Botanic Garden, Howrah and the Agri-Horticulture Society of India—A Probe into Colonial Science', Paper presented at a seminar, Calcutta and Science, BITS. Calcutta: December 21–23.
Nandy, A. 1980. Alternative Sciences: Creativity and Authority in Two Indian Scientists. New Delhi: Allied Publishers.
National Council of Education. 1956. National Council of Education, 1906-56. Golden Jubilee Volume, Calcutta.

Petitjean, P. et al (eds.). 1992. *Science and Empires: Historical Studies about Scientific Development and European Expansion*, Dordrecht: Springer.
Prasad, B. (ed.). 1938. The Progress of Science in India During the Past 25 Years. Calcutta: Indian Science Congress Association.
Rai Choudhuri, A. and R. Singh. 2018. 'The FRS nomination of Sir Prafulla C. Ray and the correspondence of NR Dhar', Indian Journal of History of Science, *Notes and Records: the Royal Society Journal of the History of Science*, Vol. 72, No. 1, pp. 57–73.
Raj, Kapil. 1991. 'Knowledge Power and Modern Science: the Brahmins Strike Back', in D. Kumar (ed.), Science and Empire. New Delhi: Anamika Prakashan.
Ramasubban, R. 1977. Science and Society: A Sociological Analysis of the Development of Science and Technology in India. Unpublished Ph. D. dissertation, Bombay University.
Ramasubban, R. and Singh, B. 1987. The Orientation of the Public Sciences in a Post-Colonial Society: the Indian experience', in S S Blume, et. al. (eds.), The Social Direction of the Public Sciences, Sociology of the Sciences Year Book, XI. Dordrecht: Reide l.
Ray, P. C. 1918. Essays and Discourses. Madras: G A Natesan & Co.
Ray, P. C. 1958. Autobiography of a Bengali Chemist. Calcutta: Orient Book Company.
Ray, P. C. 1920. 'Presidential Address of Seventh Indian Science Congress Association, 1920', Proceedings of the Asiatic Society of Bengal, New Series.
Sarkar, S. 1977. The Swadeshi Movement in Bengal 1903-1908. New Delhi: People's Publishing House.
Sarkar, S. 2020. Let there be Light: Engineering, Entrepreneurship and Electricity in Colonial Bengal. Cambridge: Cambridge University Press.
Science Today, 1983. Special issue on J C Bose.17 (9), November.
Schott, Thomas (1991) 'The World Scientific Community: Globality and Globalisation', Minerva, 29(4), PP440–462.
Sen, D. 1989. 'J. C. Bose and Science in India', Paper presented at a seminar on Calcutta and Science, BITS, Calcutta, December, 21–23, 1989.
Sen, D. and A. K. Chakraborty. 1986. /. C. Bose Speaks. Calcutta: Puthipatra.
Sen, S. N. 1954. MeghnadSaha: His Life, Work and Philosophy. Calcutta: Meghnad Saha 60th Birthday Committee.
Stafford, R. A. 1990. 'Annexing the Landscapes of the Past in British Imperial Geology in the Nineteenth Century', in John M. Mackenzie (ed.), Imperialism and the Natural World. Manchester: Manchester University press.
Venkataraman, G. 1988. Journey into Light: Life and Science of C VRaman. Bangalore: Indian Academy of Sciences.
Visvanathan, S. 1985. Organising for Science: The Making of an Industrial Research Laboratory. New Delhi: Oxford University Press.
Whitley, R, 1976. 'Umbrella and Polytheistic Scientific Disciplines and their Elites'. Social Studies of Science, 6.
Whitley, R, 1974. ' Cognitive and Social Institutionalization of Scientific Specialities and Research Areas', in R. Whitley (ed.), Social Process of Scientific Development. London: Routledge and Kegan Paul.

Chapter 11
The Craving for British Honours Among the Scientists of Colonial India: A Case Study of the Centre-Periphery Relationship in History of Science

Arnab Rai Choudhuri

Abstract The relationship between centres of science (or scientific 'metropolises') and the geographical regions lying at the periphery has been a subject of some interest to historians and sociologists of science. The relationship between Britain and India during the heyday of the British Empire was coloured by the complicating fact that India was a colony of Britain. Although Indian scientists were fully aware that some other countries like Germany might have been ahead of Britain in certain branches of science such as physics and chemistry in the early decades of the twentieth century, there was still a special craving for British honours in India, since such honours would bring additional benefits in the colonial setup. Many of the leading Indian scientists of that era openly expressed the view that nationalistic feelings drove their science, but they still wanted British recognitions for their research. Since the Knighthood was an official honour bestowed by the British Crown, many Indian scientists had an ambivalent attitude towards it. However, the Fellowship of the Royal Society was viewed as a recognition by the scientific peers of Britain and it was perhaps the most coveted honour for Indian scientists of that era. It was well known that the FRS election process was not a level playing field for Indian scientists and a few Indian scientists with high international reputation (such as Prafulla Chandra Ray) were never elected FRS even after being nominated repeatedly by their British peers.

11.1 Introduction

At any stage in human history, certain geographical regions emerge as scientific centres or 'metropolises', i.e. regions where scientific research flourishes. Although countries which are economically prosperous and politically strong often turn out to be thriving centres of science, the correlation is not exceptionally strong and one can give some counter-examples. When a scientifically backward region lying at the

---

A. R. Choudhuri (✉)
Department of Physics, Indian Institute of Science, Bangalore, India
e-mail: arnab@iisc.ac.in

S. Ahmed and S. Sarkar (eds.), *Decolonizing Science and Modernity in South Asia*,
https://doi.org/10.1007/978-981-97-1829-0_11

periphery of scientific metropolises strives to catch up, scientists in that region have to struggle against many disadvantages—physical (lack of facilities) and intellectual (lack of tradition). The relation between scientific centres and peripheries has been a subject of interest to historians of science.[1,2]

When pioneers of modern science in India were trying to develop a scientific tradition, India was a colony of Britain. So, the relationship between Britain and India, as far as science was concerned, was not a simple centre-periphery relationship, but was complicated and coloured by the fact that it was also a relationship between a colonial master and the colonized. This becomes clear if we contrast this relationship with the relationship that India had with the other scientific metropolises of that era such as Germany and France. In fact, in some of the branches of science in which many of the Indian pioneers worked—such as physics and chemistry—Germany was arguably the world leader for a few decades. A knowledge of German was essential for a physicist or chemist in the first few years of the twentieth century. In the next section, we shall discuss the attitude of Indian scientists of that era towards Germany (with a few comments about France), since that will help in bringing out the differences with the relationship between India and Britain in sharper focus. In many ways, the Germany–India relationship in science in the early decades of the twentieth century can be taken as an example of the typical centre-periphery relationship. If we compare and contrast the Britain-India relationship in science with the Germany–India relationship, then it becomes clear how the centre-periphery relationship may get modified if the centre is the colonizer and the periphery the colonized.

Although several wealthy Indians (notably Maharajahs, lawyers and businessmen) made handsome donations for the cause of science, still pioneers of Indian science needed the financial and logistical support of British rulers in vital ways. Perhaps, more importantly, British scientists provided a peer group for the early Indian scientists, who lacked peers in their immediate surroundings. Truly important works by pioneering Indian scientists would (and did) get international attention. However, Britain being the ruler of India, British scientists often had to act as advisers and evaluators of various Indian scientific establishments, apart from the fact that many British scientists served in India in various capacities—very often in administrative positions. As a result, the British scientific establishment was more aware of the overall scientific activities taking place in India.

When there was not yet a significant scientific community in India, an appreciation by the British peers often gave an Indian scientist the confidence that his scientific research was on the right track. When the appreciation came in the form of a highly visible honour that would get known in his local community, the Indian scientist might become a local hero. There were two important British honours which a scientist working in a colony could aspire for—the Knighthood and the Fellowship of the Royal Society of London. A few of the pioneering Indian scientists in the colonial era received these honours. Since the Knighthood was conferred by the official government establishment, we shall see that scientists sometimes had an ambivalent

---

[1] Basalla (1967).

[2] Choudhuri (1985).

feeling towards this honour—especially during the era of the Indian freedom movement. However, the election as a Fellow of the Royal Society (FRS) was viewed as a recognition by the British scientific community. Indian scientists realized that it was not a level playing field for them, and the bar for the election of an Indian to the Royal Society was considerably higher than that for a British scientist who might belong to an "old boys' network", whereas an Indian was an outsider. Still, it was the most coveted recognition for an Indian scientist in the early decades of the twentieth century. It was clear that the Nobel Prize, which was the highest form of scientific recognition, could be expected only for works of extraordinary quality. Only two Indian scientists of the colonial era—C.V. Raman and M.N. Saha—were seriously considered for the Nobel Prize, and Saha eventually did not succeed in winning it.[3] On the other hand, several Indian scientists were nominated for FRS in the colonial era and a few got elected. A major part of this paper will discuss how Indian scientists viewed this honour and what light this throws on the special character of the centre-periphery relationship if the centre happened to be the colonial master of the periphery.

11.2 The Attitude of the Indian Scientific Community Towards Germany

During the heyday of colonialism when the Spanish, the Portuguese, the British, the French and the Dutch were scrambling for colonies, Germany was a politically divided country and could not demand a share of the pie. To provide a moral justification for colonialism, some of the colonial masters would be compelled to argue that the races in their colonies were backward and not capable of ruling themselves in enlightened ways. Not being a colonial power, Germany was often more appreciative of the cultural traditions of the colonized people. Goethe's admiration for *Sakuntala* and Schopenhauer's admiration for *the Upanishads* are well known.

Germany emerged as a powerhouse of science even before becoming politically unified (under the umbrella of the Prussian empire in the late nineteenth century). The pioneers of Indian science like P.C. Ray, who was trained in Edinburgh, were very much aware of this. In a letter to J.C. Ghosh and M.N. Saha dated 21.4. (19)21, he wrote (my translation from Bengali):

> My opinion is that you should try to spend some time in Berlin. Nowhere else you will come across savants of such calibre. Actually, barring a handful, men of science in England are mostly mediocre. As we belong to their subject race, they can never appreciate our works with an open mind.[4]

Although an ability to read scientific papers in German was a requirement for a serious scientist towards the end of the nineteenth century and the beginning of the twentieth

[3] Singh and Riess (1999).

[4] Chakrabarti (2009), p. 270.

century, carrying out academic work in Germany required a higher proficiency in German (an ability to speak and write German fluently). In spite of the language barrier, a few Indians ventured to Germany for higher studies. Perhaps Radhikalal Dhekial Phookan, hailing from Assam, was the first Indian to get a doctoral degree from a German university. He got a PhD degree in chemistry from Heidelberg University in 1890 working under the supervision of Robert Bunsen and Victor Meyer.[5] Since P.C. Ray obtained DSc in chemistry from Edinburgh University in 1887 (only three years earlier), probably Phookan was the second Indian to obtain a doctoral degree in chemistry from anywhere in the world. In the next generation, going to Germany for a doctoral degree became much more common. We list two well-known scientists of the next generation with doctoral degrees from Germany who worked in physics,

(i) D.M. Bose got a doctoral degree in 1919 from Berlin University, where he had the opportunity of attending courses taught by Planck and Einstein.[6] Due to the First World War, Bose was forced to stay in Germany for five years even after his work was over and could take up the Ghosh Professorship of Physics offered to him by Calcutta University only after that.
(ii) N.R. Sen got his PhD degree from Humboldt University (Berlin) in 1923 working under the supervision of Max von Laue.[7]

While Phookan could not play an important role in Indian science due to his untimely death in 1896, both D.M. Bose and N.R. Sen were important figures in Indian science of that era.

Indian scientists in the early decades of the twentieth century realized that they have to read German books if they wanted to remain abreast with the latest scientific trends, and many of them learnt German. However, it was not easy to procure German books in India in that era. Some of the leading British publishers of scientific books had branches in India, through which important English scientific books could be obtained. But that was not the case with German scientific books. When Ashutosh Mookerjee offered jobs to S.N. Bose and M.N. Saha in 1916 for teaching new topics of physics at Calcutta University, they had to look for German books. S.N. Bose wrote:

> Meghnad was assigned to study quantum theory and I had to learn Einstein's relativity theory. We came away committing ourselves to being prepared to teach within a year. But where were we to get the books from? There were some books in English on relativity; we got hold of them. But where could we get hold of the writings of Boltzmann, Kirchhoff, Planck? Suddenly, I thought of an idea. Dr. Brühl was the answer.[8]

Brühl, an expatriate German who taught physics in Bengal Engineering College, had a good collection of German scientific books and could help. When D.M. Bose

---

[5] Deka et al. (2019).

[6] Roy and Singh (2017).

[7] Singh (2020).

[8] Bose (1994).

joined Calcutta University after his PhD in Berlin, he also brought a collection of German books with him. He wrote:

> I joined the Physics Department of the University of College of Science in 1919 after my enforced five years' stay in Berlin. Two of my young colleagues in the Physics Department, Meghnad Saha and S.N. Bose, had been starved of information about the advancement of Physics which took place in Germany, especially in Berlin… I gave S.N. Bose two books written by Planck, *Themodynamik* and *Warmestrahlung*, to read, as they were not available then in this country.[9]

It may be mentioned that M.N. Saha and S.N. Bose were to prepare the first English translation of the papers on relativity by Einstein and Minkowski.[10]

It was rarer for Indians to go for doctoral work to France. However, the pioneering chemist N.R. Dhar was one Indian of that era who got a doctoral degree from France. In fact, Dhar first worked for two years (1915–17) in London, obtaining DSc from London University, and then he spent two years in Paris (1917–19) obtaining from Sorbonne the State Doctorate, the highest form of doctoral degree given in France. Dhar has described the reaction of a British education official, who interviewed Dhar in London for a possible job in India, on hearing of this State Doctorate from France:

> At this news, Mr. Twentyman was taken aback. He stated that State Degrees were not awarded [in France] to foreigners who were given University degrees which were more readily available. But I told him that I was lucky enough to obtain the State Degree in the 'first class'.[11]

Two or three days later, Dhar came to know that he was recommended for the Indian Educational Service, in which Indians were rarely put (J.C. Bose taught at Presidency College for three years without salary before he was placed in this service). Dhar's mentor P.C. Ray with a DSc degree from Edinburgh was never placed in this Service. The British education officials clearly rated a State Doctorate from Sorbonne higher than DSc from a British university! Dhar maintained his contacts with the French chemistry community. Perhaps this helped him to win an extremely rare distinction in later life: he was elected to the Academy de Sciences of France in 1961. The only other scientist of colonial India to receive this distinction was C.V. Raman, who was elected to this Academy in 1949. Although it was much rarer for an Indian to be elected to this French Academy compared to being elected to the Royal Society of London, the significance of this was not realized outside a very small academic circle, and the election to the French Academy did not carry the glamour of being elected FRS in the perception of the Indian public. Another pioneer of Indian science to obtain a second doctoral degree from France was S.K. Mitra, who had a DSc from Calcutta University working under Raman and then went to work under Charles Fabry at Sorbonne to get a PhD in 1923.[12]

Apart from Indians who went to European countries for a degree, there were others who would go to Europe when they got opportunities of study visits. Some of them

---

[9] Bose (1969).

[10] Saha and Bose (1920).

[11] Dhar (1974).

[12] Bhar (1963).

preferred to go to Germany and France rather than to Britain. While working on the theory of thermal ionization, M.N. Saha was selected in 1920 for the Premschand Roychand scholarship which enabled him first to spend a year at Imperial College, London. However, he realized that the best facilities for experimental study of thermal ionization existed in the Laboratory of Walther Nernst in Berlin and could arrange to spend his second year in Berlin.[13] When S.N. Bose got an opportunity of spending two years in Europe, he decided not to go to Britain at all. He was in Paris for one year and Berlin for one year.

After discussing how Indian scientists viewed Germany, let us say a few words about how one leading German physicist viewed the scientific developments in India. Arnold Sommerfeld, one of the greatest theoretical physicists of that era, took particularly keen interest in physics research that was being done in India.[14] Less than four months after the discovery of the Raman effect, he wrote in a letter to Raman dated 16th June 1928:

> Your papers ... formed the subject of a lecture which I gave in our colloquium in Munich. ... We are all of us full of enthusiasm for this new discovery. I congratulate you most heartily on this great result. Japan, that has for a much longer time worked in Modern Physics, has nothing to show comparable to this discovery.[15]

Sommerfeld accepted an invitation to visit India in the later part of 1928 and delivered a series of lectures on the newly developing subject of wave mechanics at Calcutta University. In his own words, Sommerfeld was interested in visiting India because

> it was in this ancient land of civilization that, during the last years, strong shoots of modern physics had grown, by which India suddenly emerged in competition of research as an equal partner with her European and American sisters.[16]

Perhaps the most important contribution of a German scientist for Indian physics was Einstein's translation of S.N. Bose's pioneering paper on what came to be known as the Bose–Einstein statistics. It is well known that, when this paper was rejected in the leading British science journal *Philosophical Magazine*, Bose sent it to Einstein enquiring whether it may be possible to publish that paper in a German journal. Einstein was so impressed by this short paper that he himself translated it and arranged for its publication in *Zeitschrift für Physik*. Some scholars have argued that Germany, which was not a colonial power, had a political interest in exerting an indirect influence on India, which was the crown jewel of the British Empire.[17]

---

[13] Choudhuri (2018).

[14] Singh (2001).

[15] Sinha (2007).

[16] Venkataraman (1988).

[17] Manjapra (2014).

11.3 The Knighthood as an Imperial Recognition

It should be clear from our discussion that professional scientists of colonial India were fully aware that Germany was ahead of Britain at that time in physics and chemistry. However, Britain being the colonial master of India, there was a special kind of relationship between Britain and India, which was something more than a typical centre-periphery relationship. Higher education in India was imparted through English, and Indians could more easily access scientific works in English than in other European languages. Any British recognition had a special kind of significance in the colonial setup. Knowledgeable persons might have been aware that the Fellowship of the French Academy was a rarer honour than the Fellowship of the Royal Society. But Indians certainly did not know as much about the French Academy as the Royal Society. It will be extremely interesting to study if scientific communities in French colonies in that era had a similar kind of relationship with France. We are not aware of such studies.

Perhaps the most visible imperial honour in colonial India was the Knighthood. The Knighthood was, of course, conferred for achievements in diverse fields—from public service and law to literature and science. Presumably, outstanding contributions in a branch of science would be the primary criterion for the selection of a scientist for Knighthood—basically the same criterion on the basis of which Fellowship election to the Royal Society was done. However, we shall see that these two processes were fairly independent of each other. The list of Indian scientists selected for Knighthood and the list of Indian scientists selected for the Fellowship of the Royal Society did not have as big an overlap as one might have expected. As we shall point out later, there have been studies of the election process of Indian scientists to the Royal Society. The Society has preserved the nomination certificates with statements on the basis of which elections to the Royal Society were made. Apart from the Indian scientists who were elected to the Royal Society, we also know which other Indians were nominated for the Royal Society but not elected. We are not aware of similar studies for Indian scientists on whom Knighthood was conferred. We do not know the basis for the selection for those Indian scientists. We are also not aware whether other scientists were considered for Knighthood and not selected. In the absence of such information, we can only list the names of the Indian scientists who were Knighted and provide some anecdotal accounts of how Indian scientists viewed this imperial honour.

Here is the list of Indians Knighted for their important contributions to science—J.C. Bose, P.C. Ray, C.V. Raman, J.C. Ghosh, U.N. Brahmachari, S.S. Bhatnagar. Among these six, only three were elected to the Royal Society (Bose, Raman, Bhatnagar). As we shall discuss later, Ray and Brahmachari were nominated for the FRS but not elected, whereas Ghosh was never even nominated. It is clear that the election to Knighthood and the election to the Royal Society were two rather independent processes—the outcome of one not dictating the outcome of the other, although the British scientific community would take note of Indian scientists being Knighted.

For example, when P.C. Ray was Knighted, a congratulatory letter dated 11 February 1919 was sent to him by William J. Pope, President of the Chemical Society (London):

> The Council of the Chemical Society has noted with much pleasure that His Majesty has been pleased to confer the honour of Knighthood on you and has requested me to offer you its hearty congratulations on this well-merited mark of distinction.[18]

N.R. Dhar, who had been a student of Ray, wrote to him from Paris on 8 January 1919 and expressed the hope that the FRS would follow now (my translation from Bengali):

> The Government has done a very wise thing by conferring Knighthood on you. I was delighted to read about this in the 'Times'. Now the Royal Society will find it very difficult not to elect you this year.[19]

However, the Royal Society did not elect Ray.

As the freedom movement started gathering momentum, the Indian scientists who had been Knighted—all of whom without exception were deeply patriotic—often felt ambivalent towards this official honour. It is well known that Rabindranath Tagore gave up his Knighthood after the Jalianwallabagh massacre. On hearing this, his close friend J.C. Bose, who was already Knighted, sent a letter to Tagore with just one line: "Bandhu tumi dhanya" (the closest English translation would be "Friend, you are blessed").[20] It appears that C.V. Raman did not show the insignia of Knighthood even to his wife. His wife Lokam (short form of 'Lokasundari') came to know of this only during a sea voyage when they were having a meal with Sir C.P. Ramaswami Iyer. According to a biographer of Raman:

> It was at their first meal together that CP asked whether Raman had the insignia of knighthood with him. Raman did not. And it was only through this conversation that Lokam came to know what they were—the medals and robe that CP carried with him, which he showed to Lokam.
>
> Raman, though proud of the honour, had never bothered to show these tokens to anyone, even to his wife. They lay untouched in an almirah at home.[21]

[18] Chakrabarti (2009), p. 303.

[19] Chakrabarti (2009), p. 324.

[20] Bose (1994).

[21] Parameswaran (2011).

11.4 The Royal Society of London and the Fellow Election Process

In contrast to the Knighthood which was an official honour from the British Government, the other important British honour—the Fellowship of the Royal Society—was a purely academic honour bestowed by the scientific community of Britain. Nominations from colonial India for the Royal Society have recently been studied by Choudhuri.[22,23] The present discussion will be very much based on these papers.

On the basis of their interactions with their British counterparts, Indian scientists regarded the majority of the British scientists to be fair-minded enough to appreciate a good scientific work from a colony. As we shall point out, Indian scientists found that election to FRS was not a level playing field for them, and the hurdles for an Indian to be elected were much more than for an insider of the British scientific establishment. Occasionally, additional non-academic considerations (whether a scientist was involved in political activities) also played a role in the election process. Still, by and large, Indian scientists did not have the feeling of ambivalence towards the Fellowship of the Royal Society which they might have towards Knighthood. When C.V. Raman and his student K.S. Krishnan were sure that they were on the verge of a major discovery, the first possible honour that came to their mind was that this discovery may fetch a Fellowship of the Royal Society for Krishnan (Raman was already a Fellow). Krishnan wrote in his diary on 9 February 1928:

> he [Raman] remarked that nothing was beyond my reach referring to the Fellowship of the Roy[al] Soc[iety] and that I might probably get the Fellowship before I got my Doctorate.[24]

While Raman might have mentioned the possibility of Krishnan's being elected to Royal Society before his doctorate on the spur of the moment, he actually decided to wait for a few years for Krishnan's scientific reputation to build up before nominating him. Krishnan was nominated for the Royal Society in 1936 and elected in 1941. Apart from the prestige, an election to the Royal Society often brought other benefits as well. In a 1947 letter to H.H. Plaskett, M.N. Saha described the hardships he faced after joining Allahabad University and then wrote:

> Things began to improve when ... I was elected to the Royal Society in 1927. This was regarded in India as a rare distinction and I was congratulated by the Governor of the Province, Sir Wm Marris. ... Taking advantage of the favourable change in the atmosphere I approached the Governor for an annual research grant to enable me in my research work and through his efforts, annually a sum of Rs. 5000 was granted to me.[25]

The Royal Society of London was established in 1660—in an era of political turbulence in England barely a decade after the execution of King Charles I (in 1649). According to a contemporary writer:

---

[22] Choudhuri (2018).

[23] Choudhuri (2021).

[24] Mallik (2000).

[25] 'Saha-Plaskett correspondence', *Science and Culture,* Vol. 84, 2018, pp. 285–292.

> Their first purpose was no more, then only the satisfaction of breathing a freer air, and of conversing in quiet one with another, without being ingag'd in the passions, and the madness of that dismal Age.[26]

Until the Royal Society adopted a set of new *Statutes* in 1847, the Society was not too unlike a gentlemen's club. Robert Clive and Warren Hastings, two builders of the British Empire in India, were FRS! Since a newly inducted Fellow had to pay a fee and it meant an income for the Royal Society, the Society was initially quite eager to admit any foreign scholar who wanted to be a Fellow. According to Heilbron[27], "Voltaire had the idea that anyone who declared his love of science, and deposited his fee, was immediately received a member, and d'Alembert is said to have boasted that he could arrange the election of any traveller bound for England, 'should he think it an honor'." The only ethnic Indian to be elected FRS in that era before 1847 was Ardaseer Cursetjee, elected in 1841. Although not a scientist in the usual sense of the term, he was nevertheless a remarkable man—a shipbuilder of Bombay who had also "introduced Lighting by Gas into Bombay".

Only after the Royal Society adopted new Statutes in 1847 (available at the Royal Society Archives), clear guidelines for the election of Fellows were laid down and accomplishment in a branch of science became the primary consideration for election to the Royal Society. Although a provision was made for electing Foreign Members, a person had to be "one of Her Majesty's subjects" in order to be considered for election as Ordinary Fellow. The procedure for nominating persons for the Fellowship of the Society and then the procedure for election were spelt out clearly in the Statutes of 1847. This procedure has been described in detail by Choudhuri[28] with appropriate quotations from the Statutes. Here we provide a summary of the procedure.

A scientist had to be nominated for the Fellowship of the Royal Society by at least six Fellows of the Society in writing. This nomination would remain valid for five years. The most important part of the nomination certificate was a statement explaining why the person nominated for the Fellowship deserved to be a Fellow. During the years when the nomination remained valid, this statement in the nomination certificate (along with the list of nominators) would be included in a list prepared for circulation among the Fellows. Figure 1 of Choudhuri[29] shows a portion of a page from the list of 1922 in which the nomination statements for C.V. Raman and P.C. Ray appear one after the other in alphabetical order. The Royal Society would elect only 15 new Fellows per year in the early decades of the twentieth century till 1931, though this number was gradually increased in later years.[30] The selection would be primarily made by the Council of the Royal Society, although there would a formal voting among the Fellows. It is clear that many persons who were nominated for FRS would not get elected.

---

[26] Sprat (1667).

[27] Heilbron (1982).

[28] Choudhuri (2021).

[29] Choudhuri (2018).

[30] Andrade (1960).

Before coming to a discussion of the Indians who were nominated during the colonial era, let me say a few words how we can obtain the necessary information. Many important old documents of the Royal Society have been preserved in the magnificent archive of the Royal Society. I had the opportunity of visiting this archive during 24–25 July 2017 to check the various materials connected with Indian scientists of the colonial era kept in the archive. Some of the most important documents have been put online. For example, nomination certificates of the Fellows elected from the beginning of the nineteenth century can be seen online. However, materials available online so far constitute only a small fraction of the materials preserved in the archive. Most of the materials have been catalogued, and this catalogue is available online. A scholar visiting the archive of the Royal Society can request to see the catalogued documents not available online. Only after 1940, the full nomination certificates of even the unsuccessful nominations have been preserved. For earlier times, the only available materials about the unsuccessful nominations are the nomination statements which can be found in the lists of nominations compiled together every year (for the nominations valid in that year) for circulation among the Fellows. These nomination lists for most of years (but not for all the years) have been preserved, but they are not available online. These yearly nomination lists are the only source of information about successful nominations in the early decades of the twentieth century.

Apart from the nomination certificates of the elected Fellows and the statements about unsuccessful nominations given in the yearly lists (providing the names of the persons who signed the nomination certificate), very little materials have been preserved regarding the decision-making process of the Fellow election. There would have been oral discussions within the Council of the Royal Society regarding all the nominations which had to be considered in a year. The records of these oral discussions have not been kept. However, on a few occasions, written enquiries were made about a nomination and some of the written correspondence have been preserved in the archive.

11.5 Indian Nominations for the Fellowship of the Royal Society

From the procedure for the nomination and election outlined above, it should be clear that it was very challenging for an Indian to be elected FRS. The first difficulty was the nomination, which had to be done by at least six Fellows of the Royal Society. In the early decades of the twentieth century when there were very few Indian Fellows, the nomination process had to be initiated by the British peers. Clearly, only an Indian with sufficiently high international scientific reputation to be known among his British peers had a chance of being nominated. Even during the selection by the Council of the Royal Society, an Indian had obvious disadvantages compared to a British scientist. A British nominee for the Fellowship was likely to belong to an “old boys’ network”, knowing some of the Council members personally. This would not be the case for Indian nominees. Even if we assume that most to the Council members were fair-minded persons who were free from racial prejudice, it was natural for them to be favourably disposed towards persons with whom they were personally acquainted and with whose scientific research they might have been familiar. Additionally, it is common for scientists to be sceptical about important scientific works from a place without a well-known earlier tradition. Most of the members of the Council would not be experts in the research field of an Indian scientist. It would not be unnatural for them to be doubtful about the credibility of scientific works done in India. When C.V. Raman was nominated for FRS, the statement in his nomination certificate began with the assertion: “Although trained entirely in India has made considerable addition to our knowledge of sound and light.”

It may be mentioned that, throughout the colonial era, many British scientists working in India were elected FRS. According to Home, who had done an extensive study of the nominations from the colonies of the British Empire:

> Many were clearly admitted on the basis of the senior positions that they held in government scientific agencies rather than the originality or the intrinsic merit of their research.[31]

We are primarily interested in ethnic Indians who were nominated for or elected to the Royal Society. Certainly, the bar for an ethnic Indian was much higher.

Since P.C. Ray and J.C. Bose were the first Indian scientists to achieve high international reputation, it is no wonder that they were the first Indians of the modern era to be nominated for FRS (in 1913 and 1915 respectively). However, Srinivasa Ramanujan became the first Indian of this era to be elected FRS in 1918—G.H. Hardy playing a key role in the election process by stressing Ramanujan’s deteriorating health condition[32], Table 11.1 lists all the Indians who had been nominated for FRS from that time till the Indian independence. The research field of the scientist is indicated inside brackets after his name. The range of years during which the nomination remained valid is given. If the scientist was elected, then his year of election is indicated in bold. Perhaps a comment is appropriate about the use of pronouns ‘he’ and ‘his’ in this context. Regrettably, all the persons appearing in Table 11.1 were men. In fact, even no British woman was elected FRS before 1945.

We note down here the names of those Indians who were elected during the colonial era for their works done in India (with the election years)—J.C. Bose (1920), C.V. Raman (1924), M.N. Saha (1927), Birbal Sahni (1936), K.S. Krishnan (1940), S.S. Bhatnagar (1943), P.C. Mahalanobis (1945). Although Ramanujan was a self-trained genius who picked up mathematics on his own, his professional papers were written only after Hardy brought him to Cambridge. Certainly, S. Chandrasekhar and H.J. Bhabha should not be included in this list, since they were elected FRS for their works done outside India. This list of seven names given above is truly an exclusive list. All the persons included in this list are regarded as heroes of Indian

[31] Home (2003).

[32] Kanigel (1991).

Table 11.1 Ethnic Indian scientists nominated for FRS during the colonial era

Names	Nomination validity years
Prafulla Chandra Ray (Chemistry)	1913–17, 1918–22, 1934–38
Jagadis Chandra Bose (Physics/plant physiology)	1915–1919, **1920**
Srinivasa Ramanujan (Mathematics)	**1918**
Devendra Nath Mallik (Physics/applied mathematics)	1919–23
Chandrasekhar Venkata Raman (Physics)	1922–**1924**
Meghnad Saha (Physics)	1925–**1927**
Raghavendra Row (Medical research)	1926–30
Nil Ratan Dhar (Chemistry/soil science)	1927–31, 1932–36
Birbal Sahni (Paleobotany)	1932–**1936**
Sudhansu Kumar Banerji (Geophysics)	1933–37
L.K. Ananthakrishna Iyer (Anthropology)	1935–38
Kariamanikkam Srivivasa Krishnan (Physics)	1936–**1940**
Shanti Swarup Bhatnagar (Chemistry)	1937–41, 1942–**1943**
Homi Jehangir Bhabha (Physics)	**1941**
Upendra Nath Brahmachari (Medical research)	1942–46
Subramanian Chandrasekhar (Astrophysics)	1942–**1944**
Prasanta Chandra Mahalanobis (Statistics)	1942–**1945**
Karm Narayan Bahl (Zoology)	1943–47
Daulat Singh Kothari (Physics)	1944–48
Suri Bhagavantham (Physics)	1945–49
Rappal Sangameswara Krishnan (Physics)	1945–49
Sisir Kumar Mitra (Physics/ionospheric research)	1945–49 (elected later in **1958**)
Jnanendra Nath Mukherjee (Chemistry)	1945–49
Rames Chandra Ray (Chemistry)	1945–49

For those who were elected, the year of election is indicated in bold

science in the colonial era. Any person interested in the history of science in colonial India is expected to have some familiarity with these persons and their scientific contributions. However, in order to understand how modern science took roots in India, we need to go beyond the small circle of scientific heroes. It is in this regard that this list of Indian scientists nominated for FRS gives us a good clue of who else had high international reputation and what areas of science they pursued. It is obvious that physics is the branch of science which dominated the nominations from colonial India. If we keep in mind that even some of the Indians who later distinguished themselves in other fields—such as S.K. Banerji (geophysics) and P.C. Mahalanobis (statistics)—were originally trained in physics, then the dominance of physics becomes even more overwhelming. The distant second field in which many of the nominees worked was chemistry. This is not a surprise. Physics and chemistry were the two branches of science in which well-defined course curricula existed in India even at the beginning of the twentieth century. It is no wonder that young Indians of that era aspiring for a scientific career would primarily take up physics or chemistry.

The list given in Table 11.1 must not be taken as a list of all Indians who achieved high international reputation. There are some glaring omissions. Although S.N. Bose did his famous work on Bose(-Einstein) statistics in 1924, he was never nominated for FRS during the colonial era. Much later in 1958, after Dirac found out during his trip of India that S.N. Bose was not an FRS, Dirac took the initiative of nominating him and he was elected in the first year of his nomination (along with S.K. Mitra, whose earlier nomination was unsuccessful, as indicated in Table 11.1). Jnan Chandra Ghosh, who developed a theory of electrolytes published in a series of papers during 1918, is the other important Indian scientist of that era who seems never to have been nominated for FRS, although he received the other important honour of the British Empire—the Knighthood. On knowing about the immediate impact which Ghosh's work made, P.C. Ray wrote to him (my translation form Bengali):

> Einstein, Laue, Planck and Nernst are scientists of such calibre that merely describing them as Nobel Laureates would not add anything to their stature. I am delighted to know the opinions they expressed on Jnan's 'bahnbrechende Werke'. I have never experienced such pure joy in my life.[33]

It may be noted that the four scientists whom Ray mentioned for appreciating Ghosh's work were all German. Perhaps one possible explanation of Ghosh never being nominated to the Royal Society is that there were not any important British scientists working in that field who could have been interested in nominating Ghosh. One chemist important in the British scientific establishment, J.R. Partington, worked on a rival theory of electrolytes at the same time and got involved in a controversy with Ghosh.[34,35] These examples show that whether an important Indian scientist got nominated for FRS often involved accidental factors and the list in Table 11.1 should not be taken as a comprehensive list of all Indians who achieved international eminence in science in that era.

Occasionally, the election to the Royal Society might have been influenced by non-academic factors. The Royal Society might carry out a background check for those nominated Indian scientists who were known to be politically active. The correspondence concerning the nomination of M.N. Saha has been preserved in the Royal Society archive and studied.[36] T.H. Holland, the Vice-President of the Royal Society at that time, was a geologist who had worked in India. Holland wrote to A.P. Muddiman, Home Member of the Viceroy's Executive Council, enquiring 'whether Saha's political record is likely to be embarrassing afterwards to the Royal Society'. On receiving a negative report from Muddiman, Saha's nominators, A. Fowler and G.T. Walker, were pressurized to withdraw the nomination. When they refused to

---

[33] Chakrabarti (2009), p. 270.

[34] Mukherjee (1959).

[35] Ghosh (2021).

[36] DeVorkin (1993).

do that, and a majority of the Council members gave their verdict in Saha's favour, he was elected FRS. This incident dramatically shows that the Council of the Royal Society had full academic independence and could ignore opinions of government officials.

We now turn our attention to the FRS nominations from colonial India which were unsuccessful. As already mentioned, their nomination statements can be seen at the archive of the Royal Society. Choudhuri[37] has reproduced all the unsuccessful nomination statements before 1940. Official reasons for non-election are not available. Only in the case of U.N. Brahmachari, the correspondence making enquiries about his scientific reputation has been preserved in the Royal Society archive. Rajinder Singh[38], who studied this correspondence, believes that Brahmachari might have been elected if he lived longer (Brahmachari died before the expiry of his nomination).

Perhaps the most puzzling case among the unsuccessful FRS nominations is that of P.C. Ray. He was a very important public figure in India and was Knighted. One wonders whether his non-election could be due to the fact that his scientific works were not known outside India. Choudhuri and Singh[39] have argued that this was not the case. Ray was held in very high esteem by many British chemists who strongly felt that he deserved to be elected FRS. Although the requirement was that six Fellows should sign a nomination statement, his first nomination statement in 1914 was signed by the unusually large number of 17 British scientists. Since this was the first nomination of an Indian scientist for FRS in modern times, perhaps his nominators wanted to show that Ray had the support from many within the British scientific community. Ray's favourite student N.R. Dhar was carrying on his doctoral research in London when Ray's nomination was being considered by the Royal Society and campaigned within the British chemistry community for Ray's election, keeping Ray informed in a series of fascinating letters written in Bengali. Unfortunately, Ray's letters to Dhar are not available and we do not know Ray's view of the matter. However, Dhar's letters, of which English translations have been given by Choudhuri and Singh[40], vividly convey what election to the Royal Society might have meant for Indian scientists of that era. Although the nomination for FRS was a confidential procedure and nobody other than the Fellows was even supposed to know who the nominators were, it is remarkable that a 24-year-old student like N.R. Dhar could get accurate information about who had nominated Ray and most of them were willing to discuss this matter openly with Dhar. After the death of Sir Henry Roscoe, a well-known British chemist, Dhar wrote in a letter dated 25 September 1917:

> Sir Henry had a generous mind. His daughter Miss Dora Roscoe told me that her father had the earnest desire of seeing Professor Ray elected in near future.[41]

---

[37] Choudhuri (2021).

[38] Singh (2014).

[39] Choudhuri and Singh (2018).

[40] Choudhuri and Singh (2018).

[41] Choudhuri and Singh (2018).

Roscoe must have felt sufficiently strongly about Ray's case to mention it even to family members.

In another letter dated 17 February 1919, Dhar wrote:

> I had gone to see [Dr. Forster]. Without my bringing up the subject, he told me on his own that it is urgent to admit Sir Prafulla Chandra to the Royal Society. The Council should take special care…
>
> All the leading chemists here are aware that you have done very good and solid work and that you are a pioneer who built up a School. Walker, Frankland, Thoran, Pope, Scott, Baker, Forster, Thorpe, Philip—all have told me these things. But their words do not get translated into action. Perhaps they are not sufficiently influential and have their own candidates to push.[42]

Ray was close to Gandhi and a strong champion of the freedom movement. While the Council of the Royal Society might have overlooked Saha's political connections, that might not have always been the case, and Choudhuri and Singh[43] have speculated that Ray's political activities might be a factor behind his non-election. A few years later, N.R. Dhar himself was also nominated for FRS, as we can see in Table 11.1. He was also not elected, although he was later to receive the rarer honour of being elected to the French Academy. This raises the question whether there was a stronger resistance against Indian chemists being elected to the Royal Society compared to Indian physicists. S.S. Bhatnagar was the first Indian chemist to be elected FRS in 1943.

All the unsuccessful Indian candidates for the Royal Society in that era were primarily nominated by their British peers—except S.K. Banerji, whose first two nominators in 1933 were Raman and Saha, who had become FRS by that time. Since only an Indian scientist with high international reputation had a chance of being nominated for FRS, the unsuccessful Indian nominees constitute a remarkable group of pioneer scientists in colonial India. They should be key figures in a comprehensive history of Indian science in that era. Unfortunately, many of them are completely forgotten today and the nomination statements for the Royal Society may provide the best concise account of their scientific accomplishments. D.N. Mallik, one of the earliest unsuccessful nominees, was probably the first Indian to do a doctorate in physics outside India (from Dublin) and was nominated by such luminaries of physics as Rutherford, Larmor and Lodge[44]. Although he taught in the mathematics department of Presidency College and influenced S.N. Bose and M.N. Saha, I have not been able to find any detailed account of his life or a photograph. As research in other areas of science besides physics and chemistry got initiated in India, three of the early unsuccessful nominees, Raghavendra Row, S.K. Banerji and L.K. Ananthakrishna Iyer, were pioneers in three newly developing fields respectively—medical research, geophysics and anthropology. When the National Institute of Sciences in India (NISI), later renamed Indian National Science Academy, was established in

[42] Choudhuri and Singh (2018).

[43] Choudhuri and Singh (2018).

[44] Choudhuri (2021).

1935 in the premises of the Asiatic Society in Kolkata, all three of them were among the Foundation Fellows.[45] The obituaries of Banerji[46] and Ananthakrishna Iyer[47] appeared in the *Biographical Memoirs of the Fellows* of this Academy. However, no such obituary was published for Raghavendra Row. For him, as well as for D.N. Mallik, the nomination statements for FRS give short accounts of their careers with the titles of their most important research papers, giving us an idea of the kind of research they undertook.

It is estimated that nearly 50% of the nominations for the Royal Society used to be successful in the early years of the twentieth century[48] (although this fraction became much smaller in later years due to the increase in the number of nominations). A group of unsuccessful British nominees for FRS would hardly constitute a group worthy of attention from a historian of science. However, the group of unsuccessful FRS nominations from colonial India make up a remarkable group of men. It is quite likely that a British scientist with achievements comparable to some of these Indian nominees would be elected to the Royal Society. That Indians did not have an even playing field must have been clear to all the players. P.C. Ray wrote in a letter to M.N. Saha and J.C. Ghosh (my translation form Bengali):

> What you have written about the Royal Society of England is literally true. On hearing about this year's election, a very even-tempered person like Nilratan also exclaimed that these people are *haramzadas*.[49]

(The closest approximation of haramzada is 'son of a bitch'). In spite of being aware of this not being a level playing field, Indian scientists of that era still craved to be FRS because that was viewed by them and by the Indian society at large as the most important recognition for scientific research. The way Saha was elected in spite of a negative report from a government official shows that there was a certain kind of academic sanctity in the election process, although the odds were heavy against Indians.

11.6 Conclusion

For the pioneers of science in colonial India, British honours such as the Knighthood and the Fellowship of the Royal Society were the most prized recognitions for their scientific achievements. The craving for these honours in colonial India shows that the relation between the colonizer and the colonized could be somewhat different from a typical centre-periphery relationship in science. It was realized within the Indian scientific community that Germany was ahead of Britain in certain areas of science in

[45] Sahni and Tandon (2017).

[46] Rao (1966).

[47] Ramdas (1937).

[48] Choudhuri (2021).

[49] Choudhuri and Singh (2018).

that era. The scientific relationship which India had with Germany at that time can be taken as a typical centre-periphery relationship. However, Britain being the colonial master of India and having the overall responsibility for education and research in India, the relationship between India and Britain was somewhat different. Indian scientists were trained in English by using English books and the general public were also more aware of the British honours compared to other similar honours. Winning of a British honour would often result in various advantages for Indian scientists in the colonial setup. Since the Knighthood was viewed as more of an imperial honour, even Indian scientists who won this honour often had an ambivalent attitude towards it—especially during the era of freedom struggle. On the other hand, the FRS election was viewed as an endorsement by the scientific peer group in Britain. Since the nomination process for FRS required the support of at least six Fellows of the Royal Society, it is clear that only Indian scientists with high international reputation would be nominated. The persons nominated for FRS from colonial India make up a group truly remarkable in many ways. However, only a few of this group got elected. While most scientists in the British establishment might have been sufficiently open-minded and free from racial prejudice to appreciate works done by Indians, not belonging to the "old boys' network" must have posed an additional hurdle for Indian scientists in the election process. It was clearly recognized that the FRS election process was not an even playing field for Indians. The non-election of stalwarts like P.C. Ray remains puzzling. Still, scientists of colonial India aspired for FRS because it was the honour which provided the best combination of such factors as personal gratification, social visibility and enhancement in professional standing.

Acknowledgements I am grateful to Sahara Ahmed for inviting me to write this article. I thank Deepanwita Dasgupta, Deepak Kumar, Suvobrata Sarkar and Rajinder Singh for valuable comments on an earlier version of the manuscript.

References

E.N. da C. Andrade, *A Brief History of the Royal Society*, London: Royal Society, 1960, pp.12–13.

George Basalla, 'The spread of Western Science', *Science*, Vol. 156, 1967, pp. 611–622.

J.N. Bhar, 'Sisir Kumar Mitra', *Biographical Memoirs of Fellows of the Indian National Science Academy,* Vol. 1, 1963, pp. 107–127.

D.M. Bose, 'Foreword', in *Collected Scientific Papers of Meghnad Saha*, New Delhi: Council of Scientific and Industrial Research, 1969, pp. v–vii.

S.N. Bose, 'The early phase of my explorations in science', in *S.N. Bose: The Man and his Work, Part II: Life, Lectures and Addresses,* Calcutta: S.N. Bose National Centre for Basic Sciences, 1994, pp. 244–249.

Jagadis Chandra Bose, *Patrabali,* Kolkata: Bose Institute, 1994, p. 157.

Syamal Chakrabarti, *Oitihya Uttaradhikar O Bijnani Prafullachandra*, Kolkata: Sahitya Sansad, 2009.

Arnab Rai Choudhuri and Rajinder Singh, 'The FRS nomination of Prafulla C. Ray and the correspondence of N.R. Dhar', *Notes and Records: the Royal Society Journal of the History of Science*, Vol. 72, 2018, pp. 57–73.

Arnab Rai Choudhuri, 'Practising Western Science outside the West: Personal observations on the Indian scene', *Social Studies of Science*, Vol. 15, 1985, pp. 475–505.

Arnab Rai Choudhuri, 'How the Saha ionization equation was discovered', *Physics News*, Vol. 48, 2018, pp. 27–44.

Arnab Rai Choudhuri, 'FRS election as a recognition for scientists of colonial India', *Indian Journal of History of Science*, Vol. 54, 2018, pp. T223–228.

Arnab Rai Choudhuri, 'Unsuccessful FRS nominations from colonial India', *Indian Journal of History of Science*, Vol. 56, 2021, pp. 131–148.

Ramesh Chandra Deka, Gaurangi Maitra and Ranjit Kumar Dev Goswami, 'Dr. Radhikaram Dhekial Phookan: A chemist from Assam', *Indian Journal of History of Science*, Vol. 54, 2019, pp. 93–103

D.H. DeVorkin, 'Saha's influence in the West: A preliminary account', in S.B. Karmohapatra (ed), *Meghnad Saha Birth Centenary Commemoration Volume*, Calcutta: Saha Institute of Nuclear Physics, 1993, pp. 154–202.

N.R. Dhar, *Reflections on Chemical Education*, Calcutta: Indian Chemical Society, 1974, pp. 30–31.

Rajarshi Ghosh, 'Ghosh's law: a retrospection', *Resonance*, Vol. 26, 2021, pp. 515-522.

J.L. Heilbron, *Elements of Early Modern Physics*, Berkeley: University of California Press, 1982, pp. 115-116.

R.W. Home, 'The Royal Society and the Empire: The colonial and Commonwealth Fellowship. Part 2. After 1847', *Notes and Records of the Royal Society of London*, Vol. 57, 2003, pp. 47–84.

Robert Kanigel, *The Man who knew Infinity*, New York: Washington Square Press, 1991, pp, 291–295.

D.C.V. Mallik, 'The Raman Effect and Krishnan's Diary', *Notes and Records of the Royal Society of London*, Vol. 54, 2000, pp. 67–83.

Kris Manjapra, *Age of Entanglement: German and Indian Intellectuals across Empire*, Cambridge: Harvard University Press, 2014.

J.N. Mukherjee, 'Jnan Chandra Ghosh', *Biographical Memoirs of Fellows of the Indian National Science Academy,* Vol. 1, 1959, pp. 31–43.

Uma Parameswaran, *C.V. Raman: A Biography,* Gurgaon: Penguin Books India, 2011, p. 122.

L.A. Ramdas, 'L.K. Ananthakrishna Iyer', *Biographical Memoirs of Fellows of the Indian National Science Academy*, Vol. 3, 1937, pp. 17–26.

Rao, 'Sudangsu Kumar Banerji', *Biographical Memoirs of Fellows of the Indian National Science Academy*, Vol. 13, 1966, pp. 47–54.

Suprakash C. Roy and Rajinder Singh, *D.M. Bose: A Scientist Incognito*, Kolkata: Bose Institute, 2017, pp. 30–32.

M.N. Saha and S.N. Bose, *The Principle of Relativity, Original Papers of A. Einstein and H. Minkowski,* Calcutta: University of Calcutta, 1920.

S.K. Sahni and P.N. Tandon, *Indian National Science Academy's Inspirational Journey in 75 Years*, New Delhi: Indian National Science Academy, 2017, pp. 243–244.

Rajinder Singh,'Sir Upendranath Brahmachari and the Fellowship of the Royal Society of London', *Science and Culture,* Vol. 80, 2014, pp. 29–37.

Rajinder Singh, 'Nikhil Ranjan Sen – The formative years', *Indian Journal of History of Science*, Vol. 55, 2020, pp. 269–278.

Rajinder Singh and Falk Riess, 'C.V. Raman, M.N. Saha and the Nobel Prize for the year 1930', *Indian Journal of History of Science,* Vol. 34, 1999, pp. 61–75.

Rajinder Singh, 'Arnold Sommerfeld—the supporter of Indian physics in Germany', *Current Science*, Vol. 81, 2001, pp. 1489–94.

Dinesh Chandra Sinha, *Prasanga: Kolkata Bishwabidyalay,* Calcutta: University of Calcutta, 2007, p. 349.

Thomas Sprat, *The History of the Royal-Society of London*, London, 1667, p. 53.

G. Venkataraman, *Journey into Light: Life and Science of C.V. Raman*, Bangalore, Indian Academy of Sciences, 1988, p. 54.

Chapter 12
An Argument for Scientific Creativity in Peripheral Contexts: The Case of Satyendra Nath Bose

Deepanwita Dasgupta

12.1 The Problematic World of Peripheral Science

While telling the story of the emergence of a modern scientific community in India, there are some well-known narratives that one generally falls back on while explaining its trajectory. Common to all such narratives is the idea that the modern sciences were born in India as a form of a Western transplant—much like how a plant or an organism is transferred to a new location. The transplantation narrative then takes usually the following form. The Western sciences were incorporated in India first in the shape of various field enterprises but then soon took root in the form of an education system. From these two loci, there arose on the sub-continent a group of people that modeled itself on their counterparts in Western Europe, following their footsteps in solving various problems about nature. This, more or less, still is the self-image of the Indian scientific community, and it continues to be repeated endlessly in textbooks and in scholarly journals. My essay will seek to provide an alternative way of thinking about this issue by introducing the idea of peripheral creativity, and in what way such creativity can give rise to research interfaces between two unequal groups in the task of making new knowledge. Such an interface can technically be called a trading zone, so a new scientific community is often born in the intense effort of creating trading zones. If successful, such zones can become more deeply rooted over time, forming the nucleus of a future scientific community.

The phenomenon of peripheral creativity is familiar enough in the domain of art[1] but it has never been used to understand peripheral science; instead, the default model has always been to invoke the idea of a transplant. Departing from this well-known

[1] Think for example of Islamic art which uses abstract geometric forms for the artists were forbidden to portray human or animal figures.

D. Dasgupta (✉)
Department of Philosophy, The University of Texas at El Paso, El Paso, USA
e-mail: ddasgupta@utep.edu

S. Ahmed and S. Sarkar (eds.), *Decolonizing Science and Modernity in South Asia*,
https://doi.org/10.1007/978-981-97-1829-0_12

tradition, in the passages below I shall try to sketch an agent-based cognitive picture of peripheral creativity and argue that once we use this new approach, it can allow us to see Satyendra Nath Bose's story—as well as other similar stories—in a new light.

Bose's story indeed is a problematic one, and a quick perusal of his story, which today we find described in the histories of quantum mechanics raises some very deep problems. The first problem of course is Bose's overall trajectory. Having made a brilliant beginning in radiation theory with his 1924 paper, why did he not immerse himself further in quantum mechanics by means of more publications? After all, he was physically present in Europe and in direct touch with many of its contributors when the quantum theory was being born. And yet, by all accounts, his later life took an altogether different form—he dabbled in various projects that were often not really connected to one another and ended by dropping out almost completely from international science. The suspicion that Bose did not realize his full potential because of his own lack of efforts is present both in the Western and the Indian sources. On the Western side, this takes some times an extreme form, claiming that Bose did not understand his own statistics, the so-called Bose statistics being an elementary sort of mistake, the significance of which was grasped only by Einstein.[2] On the Indian side, hero-worship of Bose continues of course but it is always mixed with a deep disappointment that with him things never really went any further.

With this problem of Bose's personal trajectory, there arises the overall vexed problem of a scientific community that engages in the practice of science from a periphery. Should peripheral communities at all devote their energies to projects of this sort? What can they possibly gain out of it? Or perhaps they should abandon all such projects of modernity, and simply try to keep alive their own pre-colonial traditions? Since many scientific communities in Asia and Latin America began their journeys in science via an initial peripheral phase, this question becomes an important one. The post-modern arguments against such modernizing projects are well-known and well-rehearsed.[3] Yet, in spite of the colonial wrapping in which the sciences arrived, it seems that Bose and his contemporaries always saw in them a new set of of promises, and wanted to incorporate it into their society, albeit under their own directions. Was Bose at all successful in this attempt? If yes, what was his idea of science, and what kind of epistemic and scientific contributions did he manage to leave behind?

Within these two sets of questions looming over our horizon, I now approach the story of Satyendra Nath Bose, and consider his proof of Planck's law that famously gave rise to the idea of a class of elementary particles, now known as Bosons, and the first quantum statistics of indistinguishable particles. My goal is to show that Bose's path through radiation theory and beyond cannot be understood by calling it a case of transplant. Instead, we must perceive in his efforts the outlines of an agent-based

[2] "My guess is that he simply made a mistake, the kind of mistake that is very easy to make in any statistical undertaking and especially in the present case when the whole business of cutting up the phase space individual particles into cells of size h^3 is obviously a makeshift affair…". For the full article, see Delbruck (1980).

[3] For detailed examples of this sort, see Arunachalam (2015).

example of peripheral creativity, which shows us what such people can achieve with their very modest resources. This notion of peripheral creativity also reveals how a strong division of labor can arise between a center and its peripheries, and how this dependency might adversely affect the career of a budding peripheral scientist.

12.2 On the Idea of Peripheral Trading Zones

Before we go any further with these ideas, let me quickly explain what I mean by "peripheral," and when exactly one community becomes peripheral to another.

Generally, by the term "peripheral" I mean something that is dependent on another group and needs the acceptance and the consensus of that bigger community. Such dependence might arise because of the lack of a suitable community at home, or a certain shortage of resources, training, and available mentors. Because of these difficulties, such people often have to pick up their skills in the absence of recognized mentors or institutions. And perhaps because of this epistemic dependence, the work that they produce often receives less attention, most of the limelight falling upon the creativity of their corresponding centers.

Now, all this might suggest that nothing very interesting could ever happen in peripheral science except perhaps certain imitative projects, and this view has indeed been held by a number of thinkers. And yet, contrary to such expectations, I shall try to show that peripheral researchers can, and often do, display a great deal of creativity—initiating shifts of thought that do arise from their very marginal circumstances. Indeed, the very marginality of a peripheral situation often becomes a strong incentive for overcoming it—pushing someone to embrace a line of thought ahead of all the others. Out of such episodes arise a new interface or a trading zone, via which the newcomer secures a foothold in the research process of science. He or she then naturally takes up the task of training others to follow them in their footsteps. Contrary to our usual expectation, it is indeed the peripheral person who often initiates such trades. If successful, this produces an iterative cycle that may hopefully continue over time. The success of a peripheral community thus depends upon the ability of its protagonist to sustain those iterations over as many cycles as possible. As one can clearly see from the diagram below, this requires drive, energy, initiative, a profound ability to take a risk, and thus, it is quite different in structure from the passive mindset associated with a transplant (Fig. 12.1).

But there are risks inherent in this peripheral trade, which can be seen from the diagram above. There is clearly a power asymmetry between the two sides in the trading zone, for the peripheral person needs the assent and the consensus from the bigger metropolitan community but not vice versa. This opens up the possibility of failure, should such assent fail to materialize. Because of this differential epistemic authority, a controversy can cause a peripheral agent to fade out of the picture after making a few initial contributions. Peripheral trading zones are thus fraught with risk, but they also carry a strong potential for creativity. In studying the

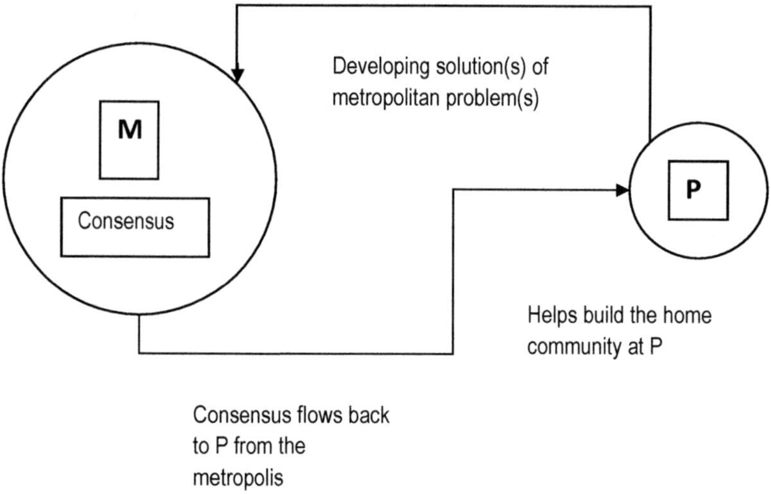

Fig. 12.1 A peripheral community and its trading zone link with metropolis

structure of such trading zones, one naturally focuses on their cognitive structure—the grasping of new representations, its changes over time etc. All this demonstrate perhaps a surprising property of the human mind—*that sometimes a deficit or a challenge can turn out to be the greatest creative opportunity.*

Peripheries like this can of course arise in many situations even when the two sides share the same social and geographical space but hold asymmetric distributions regarding epistemic authority. In what follows however, I shall explore a full-fledged colonial situation in early twentieth-century India where the peripherality arose because both from the lack of political sovereignty as well as from the loss of traditional knowledge.

12.3 Entry in Science from a Periphery

With this description of peripherality in hand, let me now turn to the story of Bose and see how he first found his foothold in the scientific community. In June 1924, Satyendra Nath Bose, then a young unknown Indian mathematician, sent Einstein an unsolicited letter in which he enclosed a short paper, and requested Einstein its publication in a German scientific journal. In this paper, Bose did two novel things—first, he derived the coefficient $8\pi\nu^2/c^3$ by using a particle picture of radiation based entirely on Einstein's theory of light quanta without using any tools from the classical wave theory of light. Second, he introduced a new way of counting those light quanta treating them essentially as indistinguishable—now formally known as the Bose-Einstein statistics. With Einstein's quick endorsement of these two moves, the paper won Bose—until then completely unknown to the European scientific community—a lasting place in quantum theory and in its histories of quantum mechanics. Bose's results were then immediately picked up and extended by Einstein to monatomic gases—predicting the novel phenomenon of Bose-Einstein condensation. Those two events opened a floodgate of theoretical activity among the European physicists, giving birth to what we now know as the new quantum theory. With this came the recognition of a new kind of elementary particle in nature now known as Bosons, which not surprisingly, obeys Bose-Einstein statistics. By initiating this unforeseen chain of events, Bose thus became forever a part of the story of quantum mechanics and that of Western normal science.

The extraordinary thing was that until that moment Bose was a complete outsider to the European physicists and their discussions on the problem of blackbody radiation. He had, until that moment, no direct contact or acquaintance with the physicists on whose work he was trying to launch his contribution, neither had he been trained by them in any shape or form. His connection to them was entirely indirect, as only a reader of their published works in German scientific journals. This extraordinary story of a newcomer walking into a problem situation with a readymade proof in hand so intrigued Bose's European contemporaries that Bose's story had always been told in the West with a mixture of fascination, disbelief, and wonder.[4] It seemed that Bose had somehow created a fundamental concept in quantum mechanics even before quantum mechanics was officially born. This unanticipated entry of a newcomer into a specialized knowledge community raises the question why Bose was so drawn to the Blackbody problem, and via what paths did he finally produce a derivation of Planck's law that was different from the previous derivations of Planck, Debye, Pauli, and Einstein?

To answer these questions, we will now have to turn to the details of Bose's derivation of Planck's law. My analysis will be concerned with three aspects of this process—the chain of the events by which Bose stumbled upon this problem, the representations that he thereafter inherited and manipulated to arrive at his result, and finally, the controversy that made him leave radiation theory altogether. I think it can well be said that Bose's entry into the blackbody radiation problem came from his deep desire to build a tradition of the sciences for India and see himself as a knowledge-maker and not merely a teacher of science. I shall conclude that with some qualifications, Bose was successful in his project, and even though his own international success lasted only for two brief years, he remained as a classic example of the risks and hopes inherent in the life of a peripheral scientist.

4 Similar attitudes also persist regarding Ramanujan. Both are of course brilliant cases of peripheral creativity.

12.4 The Self-training of a Peripheral Scientist

Born in 1894 in the colonial British city of Calcutta to a middle-class family that had a long-standing tradition of education, Bose obtained his Masters in Mixed Mathematics[5] from the Presidency College, Calcutta in 1915. A brilliant student all his life, he was a part of the famous 1909 cohort of the Presidency College, many of whom later took up the business of contributing to the sciences to combat the colonial condition of their country. The way they diagnosed their condition was a mixture of political as well as epistemic factors. The introduction of English education in 1857 had thrown out all the previous organized systems of knowledge, along with their languages of intellect, i.e., Sanskrit and Persian. While the elites of the society welcomed this step into 'modernity', it could hardly be disputed that the society, as a whole, had suffered a massive epistemic loss, for all its previous knowledge systems were now perceived as stagnant and backward, and no longer in step with the new world.[6] Sociologically speaking, this amounts to a severe loss of 'knowledge societies', i.e., "technical and professional expertise that organize large areas of material and social environment".[7] It was therefore necessary to recover this loss and build a new knowledge tradition, and this requires full-fledged teaching universities. Since scientific knowledge was taken as the very hallmark of modernity and nationhood, and since science functioned as the main tool of political power in the colonies, research in science, especially in pure sciences (as opposed to field sciences), was seen by these young men as a part of their project of building a free nation. It is through such contributions that they could hope to reverse the stigma of being a stagnant society, a target to the 'civilizing mission'.

Shortly after Bose completed his Masters, the Calcutta University launched its first initiative of opening a graduate department in the sciences, for until that time, the University had simply been an examining body and not a true teaching institution. But under the leadership of Ashutosh Mukherjee, its dynamic vice-chancellor, the University received generous donations from Taraknath Palit and Rashbehari Ghosh, two well-known lawyers, who stipulated that only Indian professors would be appointed with their endowment. Thus began the efforts to establish a full teaching university at Kolkata. Along with a few others, Mukherjee invited both Bose and M. N. Saha to join the new Department of Physics at the University College of Sciences, Kolkata, and gave them one year to prepare themselves. Raman too joined this effort in 1918 as the Palit Professor. While Raman would stay in Calcutta for a long time, producing his most well-known work from that city, Saha and Bose would soon move to Allahabad and Dhaka respectively, but still remained intent on the task of building a culture in the sciences for India. In trying to negotiate in this way between two different worlds, their minds must have been characterized by what is called

[5] This would now be called Applied Mathematics.

[6] On the phenomenon of the sudden death of Sanskrit language, see Kaviraj (2005).

[7] On the notion of knowledge societies, see Knorr-Cetina (1999).

universalism and cross-culturalism—i.e., an attempt to combine the best parts from different traditions and build a new house of thought.[8]

Their work began by seeking to grasp the 'new' physics that was then sweeping Europe. Together Bose and Saha along with their classmate P. C. Mahalanobis translated Einstein's essays on General Relativity into English, to be used for their university courses. Thus, courses on the new physics were introduced for the first time in the syllabus of the University College of Sciences, Kolkata.

The story of those exciting days is told in detail in Jagdish Mehra's biographical notes on Bose.[9] Devoid of any mentors or an expert community, Bose got down to work with a few texts of Planck, Gibbs, Einstein, and Pauli, mainly obtained from P. J. Bruhl, an Austrian expat living in Kolkata, and his own critical motto of never accepting an idea as long was he was not satisfied with its logical consistency. He also taught himself both German and French in the process. Such isolated heroic efforts are generally destined for failure, but in Bose's case, this produced a remarkably different result. His peripheral isolation shielded him from the collective thinking habits of his European predecessors, making him see the logical gaps in their thinking, and finally enabled him to take step outside of their conceptual box.

Bose's teaching duties, first in Calcutta, and then later in Dhaka, confronted him daily with Planck's derivation, and he felt deeply dissatisfied with it.[10] So far, all derivations had been based on the classical idea of light as a continuous wave, and yet each of them also used Planck's quantum of action, that involves discontinuity. Thus, a logical contradiction was present at the very heart of those proofs, beckoning for someone to resolve it. Seeking to make himself a part of the group of those distant knowledge-makers, Bose threw himself into the task, and after four years of intense effort, came up with a derivation that he thought was free from all classical relations (1924).[11] It was now necessary to publish this result, and it is in this task that he ran into difficulty.

Having published several papers earlier in the *Philosophical Magazine*, a leading physics journal of his day, he sent this paper too to the same journal, but this time the journal showed no sign of accepting his contribution. Bose was certain however, that his work was of fundamental importance, for he had removed the standing logical flaw from all derivations of Planck's law. Given the publication impasse, Bose then took an extraordinary step similar to what Sreenivas Ramanujan had done many years earlier—he sent his work to the highest authority in that area, i.e., to Einstein. But to understand the solution that Bose sent to Einstein, and the problem into which he had thrown himself, we will now have to delve a bit more deeply into the problem that occupied his attention.

---

[8] For extended discussions on universalism etc. see Dasgupta (2007).

[9] See Mehra (1975).

[10] Much later in life, Bose recollects that in Dhaka University he was teaching thermodynamics and Maxwell's electromagnetic theory in his M.Sc. classes. He was also studying, on his own, the theory of relativity and the quantum theory. See, Blanpied (1972).

[11] Bose's 1920 paper on Rydberg's law on spectral emission shows that Bose had already begun thinking about the problem of radiation. See Bose (1920).

12.5 The Problem of Blackbody Radiation

The problem of blackbody radiation arises from thinking about a perfect theoretical body that absorbs—and thereafter emits—all radiation that falls upon it. Such a theoretical body can be taken to be roughly realized in a burning heap of charcoal or in a hot oven cavity, and the radiation emitted from such a hot body can be measured per the classical wave theory of light. According to this theory however, all absorption and emission from such a hot body would be continuous (i.e., across all frequencies), so the energy density inside the blackbody cavity would rise to infinity over time!

Dubbed as the ultraviolet catastrophe, it is precisely this prediction that led to a clash between the well-accepted wave theory of light and the experimental data, for the two could in no way be reconciled. The experimental work of Lummer, Rubens, and Pringsheim had by then revealed the famous blackbody curve, quite inconsistent with all theoretical predictions. Instead of rising to infinity, the energy density inside the cavity fell off sharply at higher frequencies—producing a curve that clearly shows the Wein displacement typical at higher frequencies. Nature was simply refusing to obey the predictions of the classical wave theory.

Seeking desperately to fix this mismatch, in 1900 Max Planck reluctantly proposed a bizarre idea. Suppose, he said, that the molecules embedded within the oven walls *do not emit their energy continuously*—instead, the energy is emitted in discrete bursts or packets. Thus, was born $E = hv$, Planck's quantum of action, and with it a new era was ushered in physics. Assuming this bizarre behavior to be true of his resonators, Planck then wrote down a formula which described the behavior of a blackbody in all experimental situations.

$$p(\nu, T) = \left(\frac{8\pi \nu^3}{c^3}\right) \cdot \left(\frac{h\nu}{\left[\exp^{\frac{h\nu}{kT}} - 1\right]}\right)$$

Here T stands for temperature, v a specific frequency, h is Planck's constant, c is the speed of light, and ρ is the energy density per unit frequency interval. The expression 8π etc. stands for the angular integration of the amplitude.

A new law was thus discovered but this led to an even worse situation, for now no one, Planck included, had any idea if this was just a mathematical fix or an undiscovered aspect of nature, or how to derive this expression from the existing wave theory. Trying to remedy this unsatisfactory condition, Planck then sought to derive his new law in the following manner. Essential to his proof was the idea of a number of material oscillators embedded deep within the walls of oven cavity that transmitted their energies to the stationary radiation field (enclosed within the cavity) in discrete jumps. Several things enter this calculation—a phase space argument describing the discrete behavior of a resonator, the average energy of a resonator of frequency ν, the integration over all the resonators inside the blackbody cavity to capture U, the average energy density. The full entropy of the system is then realized by serially integrating things over the phase spaces of all the oscillators, thus connecting it with a distribution of energy elements among the N oscillators. Each such distribution was technically called a "complexion."

But this idea of an oscillator taking only discrete values spelled an end to the classical picture of things. Essentially, the average energy of an oscillator splits the cavity into a number of discrete cells in phase space, *each cell separated from its neighbor by an interval of h.*[12] In employing this phase space argument, Planck was thus led to view *h*, Planck's constant, as a finite extension of the elementary area in phase space. In other words, Planck became persuaded that classical electrodynamics is essentially incompatible with his new radiation law, but he refrained from speculating further on this. This message was correctly grasped by Bose, who 24 years later took this idea to its logical conclusion.

Peter Debye then sought to offer a new proof in 1910 without bringing in any resonators. Deeming all such devices to be ad hoc and artificial, Debye calculated the probability of a given complexion and the total entropy of the system by counting the normal modes of radiation which a blackbody cavity can sustain. Since all normal modes and their behaviors could be deemed as independent of one another, this strategy too is somewhat similar to Planck's resonators. In trying to preserve the classical theory—and perhaps avoid any Planck-like 'inconsistencies'—Debye calculated the number $N_\nu d\nu$ vibrational modes standing inside the cavity and thus obtained $8\pi\nu^2 V d\nu_s/c^3$—the average energy inside the cavity. Assuming that a quantity of energy $h\nu$ gets distributed over each of these standing modes, he then calculated the probability of distributing $N_\nu f_\nu d\nu$ (f_ν being an arbitrary distribution function) quanta of energy hν among these modes by means of the following relation:

$$W_\nu = (N_\nu d\nu + N_\nu f_\nu d\nu)!/(N_\nu d\nu)!(N_\nu f_\nu d\nu)!$$

This reasoning implies that there can only be a discrete set of allowed modes inside the cavity, each mode of oscillation working like a harmonic oscillator. It also turns out that we do not need any detailed knowledge of the properties of those resonators—simply thinking about the exchange of energy between matter and the radiant energy is enough.

A very interesting proof was then offered by Einstein, who used an altogether different approach. Instead of bringing in any resonators or normal modes, Einstein modeled the energy inside the blackbody cavity per his own light quantum hypothesis and used Bohr's notion of an atom with its discrete energy levels. To calculate the probability *W*ν for an atom to be in a stationary state, he assumed that a stationary state *m* will pass on to another stationary state of higher energy *n* by absorbing a light quantum. However, the reverse process by which an atom shifts down to a lower energy state (by emitting a light quantum) could happen in two different ways. Either the atoms will emit radiation independently of the external field, such as in

---

[12] In Planck's own language, "It is now a matter of finding the probability *W* so that the *N* resonators together possess the vibrational energy U_N. Moreover, it is necessary to interpret U_N not as a continuous, infinitely divisible quantity, but as a discrete quantity composed of an integral number of finite equal parts." See Plank (1920).

radioactivity, or they may be 'persuaded' to emit a quantum by the radiation field. This *induced transition*, dependent on the radiation field, was called 'negative radiation'. In a thermal equilibrium, the upward and the downward transitions would be held in a perfect balance. Thus, he saw the equilibrium state within the radiant cavity as a function of the numerous up and down transitions that are held in balance, as opposed to seeing it as the most probable mathematical arrangement of the system. Finally, he took guidance from Bohr's correspondence principle, which says that classical theory must be the limit of the quantum theory, used Wien's displacement law, and thereby derived Planck's law.

This, very roughly, was the structure that Bose inherited from his European predecessors.[13] From Planck, he inherited the idea of combinatorial arguments, and the message that there must be only one oscillator per cell. From Debye, he picked up the idea of equations of state as well as his use of statistical mechanics. Finally, from Einstein he inherited his main framework—the notion of light quanta. As for the elementary processes of absorption and emission, Bose thought of them in a statistical manner, as the most probable *crowd behavior* of a system, and not as its causal and dynamical transitions. As we shall shortly see, this led to a serious disagreement with Einstein. Overall, his project was to find a clean logical derivation free from the inconsistency of his predecessors who used both the classical and quantum relations in their proofs. By training a mathematician, and by his readings, a budding amateur physicist of very wide interests, Bose stepped into the blackbody radiation problem driven by its intrinsic interest, and his own program of critical reading. One must also remember that he was also playing his favorite game, i.e., finding different solutions to given problems, something that he had often done during his student days.

12.6 A Clean Derivation of Planck's Law

There are two ways to unpack Bose' reasoning—first, by analyzing each of the two factors in Planck's law, or by tracking his reasoning during the stages of derivation. In John Stachel's paper on Bose, we see the first approach, whereas in Partha Ghose' s approach we see a form of the second.[14] Indeed, in his communication to the Prussian Academy of Science (July 10, 1924) Einstein himself presents Bose's reasoning in the form of four steps. Since my concern here is to explore the evolution of Bose's thought, and the emergence of a new cognitive representations its final result, I shall stick to the well-known second approach, showing how Bose manipulated his representations at each step, eventually coming up with a new picture of reality.

Recall that Planck's law is often written in the following form:

---

[13] The literature on Blackbody radiation is enormous, and it is impossible to sum it up in a few paragraphs. My idea has been to stick to the few people that Bose mentioned in his first paper, thus showing his direct line of descent.

[14] See Stachel (1992). See also Ghose (2009), pp. 296–329.

$$p(\nu, T) = \left(\frac{8\pi \nu^3}{c^3}\right) \cdot \left(\frac{h\nu}{\left[\exp^{\frac{h\nu}{kT}} - 1\right]}\right)$$

To arrive at his derivation, Bose then proceeded in the four following steps.

Step 1: His starting point was to see the blackbody radiation as a gas of particles enclosed in volume V. Using Einstein's theory of light quanta from the very beginning, he began his proof by using a phase space argument inherited from Planck that modeled the position and the momentum of those light quanta. Since each light quantum has a location in space, and also a directed momentum, its phase space coordinates would be (x, y, z), and the momenta along those coordinates would be (p_x, p_y, p_z). This simple reasoning splits up the radiation field into elementary cells of volume h^3 each, and $4\pi v^2/c^3$, which had until then been modeled only as a collection of normal modes inside the blackbody cavity, now receives a new interpretation. Throwing in a factor of 2 to account for what he called the 'two states of polarization', Bose easily derived $8\pi v^2/c^3$, the co-efficient in Planck's law without even mentioning the classical wave picture.
Step2: The next task naturally would be to compute A_s or the number of cells available for occupation by light quanta. Now, since each cell can only contain one quantum state, and since permutations of light-quanta within a cell do not give rise to a new macro-state, the problem of counting the number of quantum states reduces to the problem of counting those cells. An elementary cell of volume h^3 would contain at most one such quantum state.
Step 3: Having modeled the free radiation in this new way, Bose then proceeded to determine the thermodynamic probability W of each micro-state that together would make a macro-state. Here he proposed an amendment of the older Boltzmann statistics, demanded by his (modified) phase space argument. Since the permutations of light-quanta within a cell can no longer produce a new macro-state, *particles inside a cell might be treated as indistinguishable.* It is thus enough to note the occupation number of those cells and not which particles were going into which cell. In doing this however he never explicitly remarked he was switching over to a new statistic.
Step 4: Having performed all those crucial moves, Bose then quickly concluded his proof by following the standard statistical mechanical procedures, holding everything subject to the constraints on E and N, and thereby rigorously deriving Planck's law.

12.7 A Case of Peripheral Creativity or a Shot in the Dark?

Two great shifts of thought occur in this innocent-looking proof that occupy only a modest 4-pages and contains no mention of the novelty of his ideas.

Let us follow Bose's own language a bit further to see the stage-wise progress of his thought that used the works of his predecessors and yet led to the emergence of a new ontological picture.[15]

Bose's opening statement noted his dissatisfaction with the inconsistencies of his predecessors who used both the classical theory as well as discontinuous relations. Noting down this inconsistency, Bose then proposes the following clean-up. Assuming from the very beginning that the radiation inside the cavity is a gas of light particles and not a collection of waves, Bose says that the energy E can be calculated by finding the sum of the energies of light quanta over all frequencies, in other words,

$$E = \sum_s N_s h\nu_s = v \int \rho_\nu d\nu$$

The important first step in his proof thus began by seriously modeling blackbody radiation as a collection of light quanta, enclosed in volume V. Much later in life, in a conversation with Jagdish Mehra, Bose recalled his entry into the blackbody problem in the following manner: "I had studied Planck's derivation of his radiation formula and also his *Theorie der Wärmestrahlung*. I knew about phase space and Boltzmann statistics from Gibb's book. Also, I had Boltzmann's *Vorlesungen über Gastheorie* from Brühl. I knew about Einstein's derivation of Planck's formula and how he had never been satisfied with the derivation of this quantum law; he always kept on coming back to it. Planck's condition ran counter to the classical ideas. Planck was also aware of the difficulties, but he had never been able to resolve them. He had relations which were derived from Maxwell's electromagnetic theory and others in which discontinuities appeared. He wanted to reconcile his theory with classical theory."[16] But unlike his European predecessors, Bose applied his phase space arguments *consistently*, contemplating throughout a cavity filled with particles. Debye for example, had occupied himself with a cavity that was filled with the standing waves of radiation, and Einstein, in his March 1916 paper, talked about the causal processes of emission and transmission by the atoms in the blackbody cavity. Recall that even with Planck, the classical wave theory remained the established picture of things, and the quantum processes showed up only in the peculiar interaction between radiation and the elementary resonators in the cavity walls. Thus, Bose's unique contribution to this problem lay in drawing the discussion away from those processes of interaction, and to associate the light-quanta first, with a new interpretation, and then, with a new statistic.

The state of a single light quantum is now characterized, as per this new view, by a point in 6-D phase space lying on the surface of a cylinder.

To compute the number of states of a light quantum whose frequency lies between ν_s and $(\nu_s + d\nu_s)$, the phase space must now be divided into tiny cells of volume h^3 each, thus extending Planck's quantizing of the action of the oscillators to the radiation field itself. Here, in accordance with his professed motto of consistency

[15] See Bose (1924).

[16] See Mehra (1975).

and coherence, Bose abandoned the oscillator picture altogether, shifting wholly over to a new particle model. Multiplying by V to get the total volume of the spherical shell of the phase space, dividing it again by h^3, and multiplying it again by 2, Bose quickly obtained:

$$A_s = 8\pi V \nu_s^2 d\nu_s / c^3$$

The co-efficient $8\pi\nu^2/c^3$ now receives a completely new interpretation, as Bose had mentioned in his first letter to Einstein. Bose's consistent use of a phase space argument in conjunction with the light quantum hypothesis had modified the older picture into a new one—*measuring the density of radiation is no longer a matter of counting the waves within a cavity.*

His second great shift of thought occurs in rigorously holding on to Planck's hidden message regarding the counting procedure for complexions. Since the total number of cells *is* all the possible arrangements (or states) of a quantum in a given volume, the next step clearly is computing the number A_ν, i.e., the number of cells available for occupation by the light quanta of frequency ν. Classically, there would be infinitely many such arrangements, but on the particle picture that Bose was working with, those possible quantum states, following Planck, would have to be separated by the finite amount of energy $h\nu$.

An elementary cell of volume h^3 would thus contain at most one such quantum state, and the whole problem of counting the number of quantum states now reduces to the somewhat different problem of counting those cells.

If all cells were distinct, the number of such cells would clearly be A_s! But since each cell can contain only one quantum state, the permutation of light quanta within a cell can no longer produce any new state, i.e., the contents of those cells are not available for inspection. All cells thus can be partitioned into classes *by simply writing down their occupation numbers*, i.e., by noting how many quanta are sitting within each cell. This proposal indicates a radical break with the traditional method of calculating a macro-state by looking up which particle goes into which cell. So, in place of the older Boltzmann statistics used by Planck[17]:

$$W = N!/N_0!N_1!N_2!\ldots \tag{12.1}$$

We now have:

$$W_s = As_0!/p_0^s!p_1^s!p_2^s!\ldots \tag{12.2}$$

(p_0 denotes states that are empty, p_1 denotes states with one quantum, and so on…)

The proposed change simply *assumes* that light quanta within the cells are indistinguishable in the following manner (Fig. 12.2):

[17] Planck represents this as a case of N gas molecules being distributed among the elementary phase-space regions of 0, 1, 2 and so on. See Mehra (1982), p. 567.

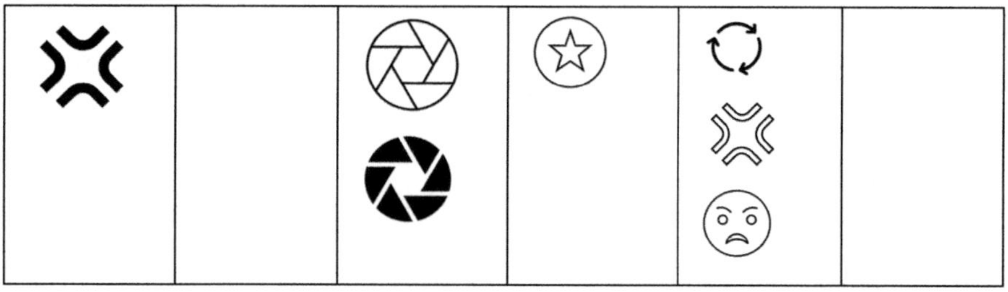

(a)

(b)

Fig. 12.2 **a** Maxwell-Boltzmann statistics. **b** The result of Bose's amended statistics. Note that in (**b**) the particles have now become indistinguishable, making them capable of a herd behavior

Finally, since the radiation inside the cavity must be brought to an equilibrium, W must be maximized. This Bose did via the standard manipulations of statistical mechanics, also used by Debye before him, holding the total energy E and the total number of A constant. Planck's law was thereby easily derived as a logical result.

Departing from the long tradition of his European predecessors who counted oscillators, standing waves, or causal transitions, Bose simply thought of black-body radiation as a gas of free particles from the beginning, in accordance with the light-quantum hypothesis of Einstein. Housing each such particle into one cell, he then simply asked how such a radiation field would achieve its equilibrium, i.e., its maximum probable state. Under Bose's coarse-grained counting, which dropped the individuality assumption from the particles, the quanta became capable of a new 'herd instinct' at very low temperatures, when the available number of energy states shrinks to a minimum. This consequence of Bose's work was immediately grasped by Einstein, who applied this method to the case of monatomic gases, predicting the phenomenon of Bose-Einstein condensation. If in the first half of his proof, Bose's substantial contribution lay in introducing a phase-space argument that allowed him to re-arrange the old problem consistent with the particle model—his most decisive contribution in the second half lay in dropping the individuality assumption that was weighing down all his European predecessors. With these two moves, Bose completed the arguments began by Planck twenty-four years ago, ushered in a new ontological concept of a type of elementary particles, now known as *bosons*, and provided the statistics for those indistinguishable particles.

12.8 A Controversy Over Radiative Processes

Having finished this first paper, Bose then almost immediately wrote a next one, and sent it to Einstein on 15th June 1924. This second paper was longer, more ambitious, and dealt with the problem of interaction between matter and radiation. By 1924, Bose was familiar with the Compton effect, and also with Einstein's own photo-electric effect.

Since in his first paper he was concerned with radiation only, the goal of this second paper was to derive the general conditions for the statistical equilibrium of a system that consists of both matter and radiation *but is independent of any special assumptions about the mechanisms of energy transfer.* Hoping again to build upon Einstein's work, but remove all ad hoc assumptions from that work, Bose now rejected Einstein's special assumptions about there being two kinds of radiative processes—spontaneous and induced transitions—by means of which an atom of higher energy shifts down to a lower energy level. Instead, he proposed that the transition from the higher to lower state can be explained without Einstein's additional hypothesis of an induced transition. The spontaneous downward transition (such as in radioactivity), Bose believed, could be explained as a property arising from the statistical nature of the radiation field itself, consistent with all its equilibrium conditions. "The problem of thermodynamic equilibrium of radiation in the presence of material particles can however be studied using the methods of statistical mechanics, independently of any special assumption…in this way we obtain a general relation which is valid for all special assumptions…".[18] The thermodynamic probability of this bigger system, Bose argued, would simply be a product of their two probabilities. In the rest of this paper, Bose provided a derivation for this more generalized formula.

This time Einstein had strong objections. Even though he communicated Bose's paper to *Zeitschrift für Physic*, he added two critical footnotes to this paper regarding elementary radiative processes.[19] To respond to these objections, Bose wrote a third paper while he was in Europe. This third paper too was duly sent to Einstein from Paris, but the outcome this time was quite different. It was never published, suggesting that Bose had abandoned his paper after hearing Einstein's critical response. Since the paper does not appear in the Einstein archives, and since, astonishingly, Bose did not keep a copy of it for himself, we cannot know what the lost paper had contained, except via the only thing that still exists—Bose's third letter to Einstein. On 27th January 1925, Bose thus writes to Einstein from Paris:

> I have tried to look at the radiation field from a new standpoint and have sought to separate the propagation of quantum of energy from the propagation of electromagnetic influence. I seem to feel vaguely that some such separation is necessary if Quantum theory is to be brought in line with Generalized Relativity theory.[20]

---

[18] See Bose (1924). Translated in English, the title of the paper reads: "Thermal Equilibrium in the Radiation Field in the Presence of Matter".

[19] Bose always remained of the opinion that his second paper was just as important as his first. However, this paper was generally ignored, perhaps due to Einstein's critical remark.

[20] See Ghose (2009), p. 327.

Partha Ghose, in his article, provides along speculative reconstruction of this lost paper. Bose's language seems to suggest that he was by then on the track of a notion called an empty wave (an electromagnetic wave carrying no energy-momentum), which subsequently did become an area of theoretical activity.[21] Faced with Einstein's criticism however, Bose apparently decided to abandon all plans of pursuing radiation theory any further and devoted his two years in Europe in picking up crystallography and various other experimental techniques. Thus, his idea of attaining a generalized expression for a system containing both matter and radiation remained unfulfilled, and what we remember today is only Einstein's generalization of Bose's idea in the form of Bose-Einstein condensation but nothing about Bose's own attempts at generalization. Those envisioned projects were shortly completed by Dirac by his idea of second quantization, but by then Bose's proposal lay long forgotten. It is worth remembering that it was Dirac who coined the term *boson* and recommended Bose for the fellowship of the Royal Society.

This complex interaction between two men, which began with an instant acceptance, which gradually dissolved into criticism and the ultimate abandonment of the project, can perhaps be best understood if we take note of a few things from Bose's point of view. The first of course is Bose's manifest hero-worship of Einstein, revealed clearly in his letters that address him as "Master". It seems clear that Bose had projected onto Einstein the long-standing traditional relationship of a master and disciple in the Indian tradition, and perhaps felt bound by it, even when this relationship started going against his interests. The second is Bose's self-perception as a scientist from a distant colony who urgently needed to pick up new skills to enrich his research programs further. Thus, instead of approaching others over the matter of his second and third paper, he simply abandoned the whole project. Einstein most likely remained unaware of this strong emotional investment on Bose's part—his interest in Bose lay in Bose's ideas but not in Bose as such—therefore when Bose's ideas began to go against his deeply-held views on quantum theory, he firmly rejected those proposals.

One perceives, in this interaction, both the strengths and weaknesses of Bose's situation, and the fragility of the trading zone that Bose inhabited before he could secure an entry into scientific practice. Bose's strengths were many, among them his great mathematical competence, his insistence on clean logical derivation of everything from its first principles. Generally, his technique was to pick up an existing problem around which a certain amount of controversy has accumulated, and then try to clean up 'the mess'. This signature style pervades all his theoretical work, which we see foreshadowed even in the first five papers that he had published before 1924, also his in later papers on the Unified Field theory.[22] In fact, William Blanpied, who produced perhaps the most sympathetic assessment on Bose, noted that Bose

[21] The opposite of this is called a full wave.

[22] In 1953, Bose applied himself again to Einstein's program of Unified Field theory, providing a solution to Einstein's field equations in a series of five papers. In a letter written on Oct 4th, Einstein thanked him for his interest in the matter but also asked: do these solutions have any physical meaning?

always sought to give everything a mathematical and statistical form, often at the expense of a physical interpretation.[23] This is not surprising, if we remember that Bose was originally trained as a mathematician, and turned himself into a theoretical physicist mainly to meet the unusual demands of his day. Per my view of peripheral creativity, his very isolation from the European scientific community worked in his favor, allowing his thoughts to remain uncluttered by their classical difficulties. But he still needed their consensus and their acceptance in order to be part of the scientific community and build further on his trading zone.

Bose remained ever appreciative and deeply grateful of Einstein's interest and encouragement of his work, for it is thanks to Einstein's personal support that he could visit Europe during 1924–26 and thus become part of the community of theoretical physicists in Europe. Overtly, he always maintained that he had no regrets about losing priority on his ideas about quantum mechanics, and yet, in certain other moments, (especially in the account given by Partha Ghose), it seems that he could never really forget that some of his best ideas were rejected by Einstein, and thus was never seen as his contributions to the quantum theory. Among them would be his ideas of looking at radiation field in interaction with matter by using statistical techniques, and the more speculative idea of an 'empty waves' carrying no quanta of energy. These were areas which shortly became areas of interest in quantum mechanics, over which he could have perhaps made bigger contributions. With hindsight, we now see that Einstein was perhaps not the best person Bose should have approached, given Einstein's lifelong ambivalence about quantum mechanics, and given that Bose's approach in the last two papers was tending towards that very direction. In any case, by 1925 Einstein's attentions had already shifted—radiation theory was no longer the focus of his work.[24]

It often seemed that Bose operated in a style quite different from the self-absorbed focus of a normal scientist. He tended to solve others' problems and then lose interest in those matters after they were solved. Thus, many of Bose's ideas never saw the light of publication under his own name. To this can be added a certain amount of disorderliness and perhaps his excessive hero worship of Einstein. Bose was certainly guilty of not keeping copies of his own papers, and perhaps in the end spread himself too thin by attempting to get into too many areas.

But what we see in Bose's story is not just the outcome of his personal habits but rather the complicated dynamic between a peripheral scientist and his/her metropolitan community whose assent and authority he or she absolutely needs to become part of the consensus practice. This one-way transaction makes the relationship between a periphery and its metropolis quite fragile and asymmetric. Indeed, part of a colonial scientist's problem stems from this asymmetric bond by which he or she is linked with his/her metropolitan community. While he or she contributes often to the normal science at the metropolis, they cannot always build further on that work. Instead, their work gets integrated with the metropolitan tradition, which

---

[23] See Blanpied (1972).

[24] See Pais (1982) for Einstein's ambivalence about Bose's work.

it serves and supports, and to which it often gives a new direction, but tends to leave his/her next set of ideas behind.

What then became part of normal science was Einstein's extension of Bose's new counting method to monoatomic gases that predicted the phenomenon of Bose-Einstein condensation, but Bose's own approach of finding a generalized formula for both matter and radiation was completely forgotten. In the subsequent literature on Bose—in spite of the extraordinary praise that he receives both from Abraham Pais and John Stachel—their discussions center more on Einstein. Indeed, Stachel's paper focus almost entirely on why Einstein did not discover Bose statistics, given that he was so well-primed to respond to it. Generally, Bose's contribution tends to be seen as a case of an astounding intuition—in Pais' beautiful language, a shot in the dark. But if we employ the notion of peripheral creativity, we can see that this was truly the case of creating a new representation, standing in the middle of a difficult trading zone. In moving between the arguments of Planck, Debye and Einstein, Bose was able to shed their various constraints, and rise to the level of a new ontological picture. Indeed, towards the end of his life, in an interview with Raman's student Ramasheshan, Bose very modestly rejected the widely prevalent view that he did not really understand his own proof, recalling in mind his four years of intense effort that he then likened to a real physical pain.[25] Cognitively speaking, Bose may have been unaware that a revolution in physics would shortly follow his contributions, but this would have been equally true of Planck himself and many others. Bose certainly knew that he had solved a crucial conceptual problem in radiation theory, as that theory stood at that particular point.

Yet, all this resulted into little personal mileage. After his two years in Europe, Bose returned to his job in Dhaka and applied himself to chemistry and crystallography! The one solid expertise that he had gained from his Europe trip was a working competence in several areas of experimental science, and a deep interest in X-ray spectroscopy and crystallography. In Dhaka University, he tried subsequently to build a well-equipped lab suitable for such research work and training, which eventually became the home for K. S. Krishnan's work. His purpose then had become not to blaze a trail on his own, but rather to build a working tradition in the sciences by increasing the breadth of his various research programs. Simultaneously, he sought to become an effective mentor, and, finally, to create a program of science education in one's own language. Still later in life, he trained a generation of women students, and encouraged them to develop their own brand of hands-on science. Creating a soil map for India, maintaining a deep interest in archelogy, art, music, and literature his was the paradigm case of a Renaissance mind. Indeed, his curiosity and span seemed endless. This decision to switch gears was perhaps self-conscious: "I was a comet", Bose said, "who came once, never to return again…".[26]

I think at this point we are now finally equipped to address the question whether peripheral scientists like Bose should engage at all in trying to build any trading zones. Do they thereby weaken or even completely lose their place in their own traditions?

---

[25] Ramseshan (2000).

[26] See Mehra (1975).

My analysis of Bose shows that no such outcome needs inevitably follow. Indeed, having created trading zones between their two worlds, such people become uniquely capable of creating new research programs that might draw on multiple traditions and can visualize different enriched projects for home. One can here remember Bose's research efforts in creating maps of different clays and soils in India by using the techniques of X-ray diffraction and differential thermal analysis. In this genre also fall his beautiful Bengali essays, written for *Bangiya Bijnana Parishad*, which are almost completely forgotten today—but in them he once again demonstrates that one can very well conduct sophisticated scientific discourses in Bengali language.

12.9 A Failed Peripheral Scientist?

Granting that his was the orientation of a polymath rather than that of a Kuhnian normal scientist, shall we then view Bose's trajectory as a case of failure? After all, apart from his 1924 paper on light quanta and the second paper which contained the failed statistical scheme of how a radiation field interacting with material particles would achieve its equilibrium, Bose left scant imprint upon international science. His entire set of publications amounted to a meagre thirty-two papers, his projects on organic chemistry and crystallography over which he spent many years of his life earned him little credit in publications, and his idea of teaching science in one's native language was simply laughed out of the court.

But if this is the view that we maintain, then perhaps we will be thinking of science too much in the mode of what Kuhn calls normal science. Yet, other modes of scientific practice are surely possible. After all, both Newton and Mendel worked on their celebrated problems for only a certain period of their lives, and never came back to them again. This sort of Renaissance mindset is also a perfectly legitimate way of practicing science, and it is not always necessary that a scientist's ideas appear under his or her name in the shape of a set of publications. One may also leave behind a track record by providing examples and influence and by becoming the nucleus for a future scientific community.

Standing now in the early decades of the twenty-first century, it is now time for us to come to an evaluation about Bose's career. By the word 'scientist' today we mean a person of a rather narrow interest who publishes on a few specific topics and makes sure that his or her profile remains always updated. But perhaps what we need more today are thinkers of another sort—someone who can stimulate others to think, who recognizes and solves problems in multiple domains of intellectual interest—sometimes ushering by that process a few peak moments of creativity.

What was the idea of science that Bose himself operated with? This can be glimpsed from his many speeches and some of his students' memoirs. In his Presidential Address at the Indian Science Congress in 1929, Bose makes the following interesting remark: "The ultimate aim of scientific inquiry is to arrive at a minimum number of hypotheses, which will explain the maximum number of facts."[27] In the classroom, his challenges to his students often took the form of a thought experiment—he challenged his students to imagine that all books and documents on a specific subject had been destroyed, and it was now the students' turn to rebuild that lost knowledge from their own understanding.[28] In his student days he often provided multiple solutions to given problems in examinations, thereby completely dazzling his instructors. Rebuilding solutions by multiple paths was something he loved doing, and it is precisely this sort of mathematical game-playing that enabled him to construct a clean derivation of Planck's law.

Having now come to the end of my story on Bose, I now want to go back to the idea of the transplant model, with which I began my paper. If we take history of science seriously, it becomes impossible to consider Bose's oeuvre as a case of transplant. His iconic research problems did not come to him from his coursework during formal education, rather they came to him via his own critical efforts at self-training. His well-known mindset of building things from scratch via multiple pathways gave him his most famous proof. Contrary to the well-worn transplant model with which we write the history of Indian science today, a new scientific community is not born if we introduce a number of research programs on its horizon. Rather, it is born from the intense efforts of a few people who tinker with those ideas, create new representations, and in that process scales various peaks of creativity that starts a new trading zone. True, such process begins with the reception of some new material but that would be true equally of scientific revolutions elsewhere, such as in Europe, which could not have begun without Greek texts, Arabic numerals, and the Chinese printing press and compass. But it is not in the materials received, but in one's interaction with it, that we should locate the hallmarks of scientific thinking. Mere transplantation produces what Richard Feynman unkindly calls cargo cult science, and which—as Feynman tells us—has little to do with creativity.[29] Peripheral tinkering, on the other hand, is a genuinely creative process—one can here think of C. V. Raman's early acoustic researches on violin on a shoestring budget or Michael Faraday writing down his notes after attending Davy's lectures. Indeed, the human mind can sometimes reach its highest creative moments during such peripheral interactions and come upon a new breakthrough. This idea of a transplant being the main mechanism for scientific thinking in India therefore is a case of bewitchment by language, a la Wittgenstein, and as we develop new cognitive accounts, it is hoped that such accounts would be increasingly set aside in favor of new agent-based understanding of peripheral creativity. Until we do that, Indian contributions to science will have little mention in the mainstream accounts of history and philosophy of science. Far from being a dependent process, peripheral creativity involves a process of enrichment of metropolitan thinking in the same way that human peripheral vision is a case of enrichment of our central vision.

[27] Presidential Address at the Indian Science Congress 1929, held at the Asiatic Society in Calcutta.

[28] Singha (1981).

[29] See Feynman (1985).

References

Subbiah Arunachalam, 'Science on the Periphery: Can it Contribute to Mainstream Science?' *Knowledge and Policy*. Vol 8, 2015, pp 68–87.

William Blanpied, 'Satyendra Nath Bose: Co-Founder of Quantum Statistics', *American Journal of Physics,* Volume 40, Issue 9, 1972, pp. 1212–1220.

William Blanpied, 'Satyendranath Bose: Co-Founder of Quantum Statistics', *American Journal of Physics*, September, 40, 1972, pp. 1212–1220.

Bose, S.N. 'On the Deductions of Rydberg's Law from Quantum Theory of Spectral Emission', *Philosophical Magazine*, Vol. 40, 1920, pp. 619–627.

Bose, S.N., 'Wärmegleichgewicht im Strahlungfeld bei Anweisenheit der Materie, *Zeitschrift für Physik.* Vol 27, 1924, pp 384–393.

S.N Bose., 'Plancks Gesetz und Lichtquantenhypothese', *Zeitschrift für Physik,* Vol 26, 1924, pp. 178–181.

Subrata Dasgupta, *The Bengal Renaissance: Identity and Creativity from Rammohun Roy to Rabindranath Tagore*, New Delhi: Permanent Black, 2007.

Delbruck, Max. 'Was Bose-Einstein Statistics Arrived at by Serendipity?', *Journal of Chemical Education*, Vol.57, 1980, pp. 467–470.

Richard Feynman, 'Surely You're Joking, Mr. Feynman! Adventures of a Curious Character', 1985, W.W Norton and Company.

Ghose, Partha, 'Bose-Statistics: A Historical Perspective', in Kamaleshwar C. Wali (ed.), *Satyendra Nath Bose: His Life and Times*. New Jersey, London, Singapore: World Scientific Publishing, 2009.

Sudipta Kaviraj, 'The Sudden Death of Sanskrit Knowledge', *Journal of Indian Philosophy*, Vol. 33, 2005, pp. 119–142.

Knorr-Cetina, Karin '*Epistemic Cultures*'. Cambridge, Massachusetts, London, England: Harvard University Press, 1999.

Jagdish Mehra, 'Satyendra Nath Bose', *Biographical Memoirs of the Fellows of Royal Society*, 1975, pp 116–154.

Jagdish Mehra, 'Satyendra Nath Bose', *Biographical Memoirs of the Fellows of Royal Society*, 1975, pp 124–125.

Jagadish Mehra, 'Satyendra Nath Bose', *Biographical Memoirs of the Fellows of the Royal Society*, 1975, pp. 21116–21154.

Jagdish Mehra, *The Historical Development of Quantum Theory*, Vol I, Part II, New York, Heidelberg, Berlin, Springer-Verlag 1982.

Pais, Abraham, *Subtle is the Lord: The Science and the Life of Albert Einstein*. Oxford, New York: Oxford University Press, 1982.

M Plank, 'On the Law of Distribution of Energy in the Normal Spectrum', *Annalen der Physik*, Vol. 4, 1920, pp. 553–560.

S. Ramseshan, 'A Conversation with Satyendranath Bose about Five Decades Ago', *Current Science*, Vol. 78, No. 5, 2000, pp. 636–638.

Purnima Singha, *Bijnana Sadhanar Dharay Satyendra Nath Basu*, Visva Bharati Publications, Kolkata, 1981.

Stachel, John. 1992.'Einstein and Bose', in *Einstein from 'B' to 'Z'*, Vol 9, Birkhäuser:Boston, Basel, Berlin, 1992.

Part V
Science and Multiple Modernities

Chapter 13
The Dreams of Reason: Rabindranath Tagore and the Invention of Science in India

Shiv Visvanathan

13.1 The Other Colonialisms

This paper is an attempt to understand Rabindranath Tagore within the debates of science and technology in nationalist India. I use Tagore as an equivalent of linguistic shifter, whose meaning and argument varies with context and use. Tagore speaks in dialects which vary depending on who he is conversing with. The article seeks a synthesis because Tagore's views on science are presented in a fragmented fashion, with a few references to Bose, a deep emphasis on the Gandhi debate and little reference to Geddes. It begins with an articulation of other colonialisms, the dissenting West in the national debates. The other colonialisms become a preclude to highlighting the availability of eccentricity in India.

The next section captures the romance and everydayness of the Tagore-Bose relationship. The third section re-examines the making of Santiniketan as a conversation between Geddes and the poet. This sets the stage for the Tagore-Gandhi debate on khadi and the concluding comments.

I want to begin with a confession which is almost a statement on method. I am not an authority on Tagore. I cannot even claim to be enthusiastic about him. I decided I must leverage my illiteracy to understand Tagore in a different way. I knew little about his poetry and less about his music. But as a historian of science, I did know that Tagore was close to scientific legends like Patrick Geddes, the Scottish biologist and Jagdish Chandra Bose, an extraordinary scientist.

Originally published in the *Economic and Political Weekly*, Vol. 48, No. 47, 2013, pp. 43–49. Reproduced with the permission of the Editor, *Economic and Political Weekly*, for which both the editors of the volume *Decolonizing Science and Modernity* express their gratitude.

S. Visvanathan (✉)
Jindal Global Law School, Jindal University, Sonipat, India
e-mail: svcsds@gmail.com

S. Ahmed and S. Sarkar (eds.), *Decolonizing Science and Modernity in South Asia*,
https://doi.org/10.1007/978-981-97-1829-0_13

I decided to begin not with the great literary texts, but with the letters that Tagore wrote and received from these two intellectuals. Reading them, I was surprised. There was a different smell to these conversations. These men did not come out like immaculate conceptions, but vulnerable people struggling to create a different world. There was a smell of loneliness about them. One sensed the vulnerability. I suddenly realised that nationalism and the nationalist narratives had their limits. They have been unfair to these people. They have created an angular Tagore. In fact, I will go to the extent of saying that historians of nationalism, chroniclers of the swadeshi movement have been illiterate about science and virtually silent about science as a theatre for civilisational debates. I realised that nationalist history was a failure of story-telling; that nationalism, in talking about science, failed miserably. It was not just a failure of literacy; it was a tragedy of the imagination.

Reading these documents, one sensed an excitement; these were remarkable people at a remarkable time. They were sheer theatre, larger than life personalities in an era which was not short of heroes. There was eloquence to their struggle.

But this was not just a fight against colonialism. These men realised that colonialism was a corset that forced you to live within the Victorian, imperial imagination. But these men followed the dictums of Cornelius Castoriadis who made a distinction between the imagination and the imaginary (1998).[1]

The Greek philosopher argued that the imagination is the current ecology of categories around an idea or an intellectual system. The imagination is what an idea conventionally extrapolates as a set of possibilities, but the imaginary is the horizon that goes beyond the current paradigm and talks of visions not yet born, ideas which have not yet been worked out, and worlds which are still to be constructed.

Rabindranath Tagore was a man who dealt in imaginaries. He went beyond the imagination of the colonial era. He was no ordinary nationalist. In fact, he went beyond the tired categories of the modern nation state to dream of a dialogue with the West. Tagore's ideas were like an experimental theatre, tentative, playful, and full of a poignancy which one needs to understand. The narratives of nationalism construct him like a statue, a monument formidable in silence, colossal in achievement. I think it is time we see Tagore like an animated cartoon, a caricature of himself, inventing, planning, complaining, gossiping, creating a world and looking at it simultaneously as a theory of creativity. What made Tagore creative was that he saw nationalism as a theatre of possibilities and he realised that the violence of nationalism emptied it of possibilities that he wanted to explore. Let me elaborate it in terms of two points.

For Rabindranath Tagore, displacing the West in a colonial sense was important. But nationalism that only devoted itself to fighting imperialism was doomed to tiredness. Tagore wanted India to be confident. He wanted India to smell the West, taste it and understand the differences within it. He realised that imperialism is only one phase of the West, that there were other Wests that one could talk to and conspire with. Like Gandhi, he realised that one of the great contributions of the national movement could be the opening of the West to its own possibilities. He made nationalism a playful, inventive exercise of a civilisation literally conversing with another.

---

[1] Castoriadis (1998).

He realised the importance of power and yet, realised that a politics of knowledge must be a playful exercise. It is in this context that I want to introduce the idea of the other colonialisms. I am using the idea of the other colonialisms to talk about the availability of eccentricity, of plurality within the national movement.

Our national movement had a tremendous diversity to it, and it was a diversity anchored on the fact that the English had their own diversities to offer, and that English eccentrics and dissenters had fascinating ideas that we could chew on. A nationalism that only looked on colonialism as a collection of narrow-minded administrators and visionaries was a failure of the imagination.

The British sent many of the great eccentrics to India. The role call is amazing. Let me just cite a few. Think of Allan Octavian Hume. People call him the "Father of the Indian Congress". He was something more. He was actually the greatest ornithologist of the 1gth century. And he began "Stray Feathers", a journal which Salim Ali took over and professionalised. Hume, in fact, wanted to be theosophist, but it was finessed out of its leadership by Helena Blavatsky. In fact, the theosophists provide another example of the Western imagination, where the occult is metaphor and provided a new humus for art and childhood. Out of theosophy emerged the great educational experiments of Maria Montesorri and the great art of Nicholai Roerich. It was Roerich who suggested the idea of the "Green Cross" as the equivalent of the Red Cross. The green cross would be dedicated to the protection of monuments and seek to save the cultural heritage of a country during war. Unfortunately, this, like many ideas of the other colonialisms, belonged to the department of the lost causes.

One can add to this availability of eccentricity, a man like Albert Howard, who marvelled at traditional agriculture and invented the idea of organic farming. Few people know that the concept of organic farming was worked out in Indore and then transferred to the world. To this roll call of dissent, one can add the names of Patrick Geddes, Annie Besant, Charlie Andrews and later, J. B. S. Haldane.

Tagore is a man who thrived on this alternative imagination, seeking to hybridise it with the best of Indian civilisation as he saw it. I am emphasising this roll call of dissent and marginality to emphasise that nationalism was a theatre of ideas, a collection of thought experiments which has tremendous relevance for today when India needs such ideas. One man who understood this, who went beyond his poetry to discuss the dream of science was Tagore. Tagore was not just an educationist; he enjoyed ideas, dreamt ideas, lived ideas and wrote about them, especially in his letters. It is Tagore's dreams of science that I want to make the text, context and pretext here. One is surprised to see that outstanding historians like Sumit Sarkar give a paragraph to Bose in a standard book on the swadeshi movement.[2]

The rest of my story is divided into three parts. In the first, I talk of Bose and Tagore; in the second, of Tagore and Geddes, and finally of the much-analysed encounter between Gandhi and Tagore on the charkha (spinning wheel). Each is an event on its own, but each telescopes into the other to give us an intriguing picture of a multifaceted man.

---

[2] Sarkar (1973).

13.2 The Tagore-Bose Letters

The first thing I read on the Tagore-Bose encounter was Tagore's "Bose lecture of 1938". It was written in a kind of lofty prose, whose language left me cold. One realised that a lecture like this verges between salutation and hagiography.

It is sometimes difficult to talk of the everydayness of friendship in these movements. It is the Tagore-Bose letters that capture the nature of this relationship. Reading the letters which are sometimes less than literary, one senses the nuances of emotion, the desperation to be recognised. One senses the quality of a friendship. Bose was the older man, already a legend in Europe before Tagore became recognised abroad. I am just going to use the letters as a text to talk about these two men, one of whom was a great poet and the other, a great scientist.

Bose was the older man. In fact, when he returns from his triumph in Europe, Tagore goes to congratulate him. Bose is not home and Tagore leaves a collection of magnolia flowers on his work table as a signature of his visit.[3]

Their conversation is intense. Bose talks about his fears to Tagore; the entire struggle of a national scientist struggling to be creative, is caught in the poignancy of these letters. Tagore waits for every episode of science in Bose's life. Bose, in turn, waits excitedly for every story. In fact, when Tagore once invited him to stay with him at his house near the Padma River, Bose accepted with one request. He wanted Tagore to write him a new story everyday and read it to him every evening.

Geddes adds "it is in this fashion that one of the most beautiful series of Tagore's short stories came to be written".[4] Bose, in turn, tried to promote Tagore's stories at every opportunity. In fact, during his second visit to England in 1900, Bose got "The Kabuliwala" translated into English.

Peter Kropotkin, the anarchist and the scientist, was delighted with this story and declared that it reminded him of the great stories of his own countrymen.[5] Jagdish Bose was a man who was fascinated by literature and immersed in mythology. Geddes reports that his Calcutta drawing room contained a striking picture of Mahabharata by Nandlal Bose and the lecture hall of the institute had the same painter's The Quest. Bose was fascinated by Karna whose life he saw "as a continuous struggle between his divine qualities and human triviality".[6] Bose in fact was a life modelled on Karna. If he was born in England, he would have been one of the leading scientists, but since he struggled in India, he had to break the barriers to let his talent be known.

---

[3] Geddes (1920); Meller (1994).

[4] Patrick Geddes, *The Life and Work of Sir Jagadis C Bose*, ibid.

[5] Patrick Geddes, *The Life and Work of Sir Jagadis C Bose*, ibid.; Hellen Meller, *Patrick Geddes: Social Evolutionist and City Planner*, op. cit.

[6] 'A Scientist and a Dreamer', Collection of JC Bose Letters, Vol 5, Bose Institute, Kolkata, 1997.

13.3 Battling for Recognition

The letters capture the poignancy or the struggle and the despair of Bose's battle to be recognised by the Royal Society for his work on the "Physiology of the living and the non-living". Bose realised the dangers of too quick an acceptance; he wanted to be accepted by the mainstream scientists and to be published in The Transactions of the Royal Society. He did not want to be seen as a magician, an oriental creature, an intellectual orphan to be adopted by British eccentrics, especially the Psychic Research Society. He is shrewd enough to understand the dangers of such acceptance. He realised his work belonged to a classic strand, but the letters revealed that he was pragmatic about the divides he had to cross between marginal and mainstream, colonial and imperial, physiology and physics. The letters chronicled his movement across the divide. In his letters, Bose writes about how he had to contend with the scepticism of Oliver Lodge and the plagiarism of scientists like Waller who was fighting for priority over Bose.

One senses the pain and agony of two intellectuals, struggling to be creative in an atmosphere which is hostile. Bose, in fact, received a meagre salary and the letters show that he appealed for a larger stipend. The finance member of the viceroy's council rejected the plea, that "he found it difficult to maintain himself on his small means" with the observation that it was adequate for a native gentleman[7]. In 1901, he wrote to Tagore from London, "You do not know what solace I get from your letter. You cannot even think how many hurdles I face in every step here. Sometimes, I feel totally dejected. I have written to you earlier that the more there will be uniqueness in my work, the more there will be obstacles in accomplishing it", and then adds in the next paragraph, "For all this, I will have to work alone".[8] Bose often consulted Tagore. In fact, during one of his visits, he was offered a professorship abroad and wrote desperately to Tagore asking whether he should accept it.…attempt to create a classic Indian university at Santiniketan. It was the conversation of two eccentrics, each obsessed with his own life, each deeply admiring the other, but refusing to concede an inch to the other. It was an erratic relationship between two temperamental and lonely men, mediated by the generosity of two friends. The Geddes-Tagore relationship was sustained by the patience of Arthur Geddes, Geddes' son who later taught at Santiniketan and the Christian patience of Tagore's friend, Charlie Andrews. The recent edition of the Tagore-Geddes correspondence is full of intermittent silences filled by these two generous men.[9]

[7] Ibid.

[8] Ibid.

[9] Fraser (2005).

13.4 Geddes and Santiniketan

Tagore's friendship with Patrick Geddes was a more tempestuous and erratic one; Geddes was an eccentric, a prophet of marginal ideas. Patrick Geddes spent about a decade in India, served as the first professor of Sociology at Bombay and wrote over thirty town plans on urban India. None of them were ever accepted. But some of them are seen today as ecological and urban classics, especially his two-volume report on town planning in Indore.

Geddes was a great biologist who realised that colonialism was a dumbing down of the Indian mind and what was needed was a new idea of creativity which went beyond the traditional critique of colonialism and the standard critique of the west.

For Geddes, the Bengali babu was a comic monster. And in fact, he felt that the tragedy of swadeshism lay in its mimetic obsession with the west. In one of his letters, he claims that the cricketer, Ranjit Singhji, might be better known than Buddha in England but Ranjit Singhji was the prime example of a snob, of a colonial mind, who set the wrong example of creativity.

Geddes felt that one had to go beyond the Ranjit Singhji of science who had only come off second best and look for an original order of creativity. Geddes argued for J. C. Bose as the exemplar of this alternative model and, in fact, became Bose's first biographer.

Geddes realised that he needed a sparring partner for his ideas and was naturally driven to Tagore. He was fascinated by Tagore's educational ideas and Tagore, in turn, admired him. In fact, Tagore's classic story, The Parrot's Training was dedicated to Geddes. It is a devastating story about a raja who comes across a delightful bird. Watching it singing and hopping, he realises that it is completely ignorant, and money must be spent on educating it. He consults the pandits as resident experts on education. They tell him that the bird must not be allowed to roam free or even live freely in his nest. They insist that the bird be put in a gilded cage which of course required constant scrubbing and cleaning. Meanwhile, the raja summons scribes to copy scholarly books who amass copies and copies of books which turn out to be an expensive enterprise. A whole department of education has sprung around the education of the bird. The raja is deeply pleased and then someone remembers the bird. The little bird lay dead in its nest.

Geddes saw it as a fable of university education, of the conventional minds being created by the colonial model which created in Calcutta and Madras, cheap clones of English education for a clerical life. The Tagore-Geddes relationship is an attempt to find a common site for conversation between two difficult and classically different experiments—between Geddes' attempt to create an Indian college in France and Tagore's attempt to create a classic Indian university at Santiniketan. It was the conversation of two eccentrics, each obsessed with his own life, each deeply admiring the other, but refusing to concede an inch to the other. It was an erratic relationship between two temperamental and lonely men, mediated by the generosity of two friends. The Geddes-Tagore relationship was sustained by the patience of Arthur Geddes, Geddes' son who later taught at Santiniketan and the

Christian patience of Tagore's friend, Charlie Andrews. The recent edition of the Tagore-Geddes correspondence is full of intermittent silences filled by these two generous men.[10]

Geddes, like Tagore, was not sympathetic to Indian nationalism. If Tagore saw in it the beginnings of cog…public in order to win their confidence…I began anyhow. All I could do was to offer the five little boys who were my first students, my company. I talked and sang to them, played with them, recited to them our epics, improvised stories…took them on excursions to neighbouring villages.[11]

Tagore argued with Geddes that he lacked the perspective to make a plan and stick to it. He felt temperamentally that the greatest part of his work must remain "in the subsoil of the obscurity of mind. All my activities have the character of play in them, they are more or less like writing poems only in a different medium of expression".[12] Tagore acknowledged that Geddes' own ideas had the same sense of play but they had a different idiom which he did not have the power to use. Geddes in reply acknowledged that while he worked out the prosody of thought, the equivalent of musical notations, Tagore could compose songs as well as poems. But planning remained a problem of Santiniketan which Arthur Geddes was to work on systematically.

The quarrels between Tagore and Geddes were about the architecture of the university. Geddes objected to the five buildings of Suren Kar whom Geddes called a shocking amateur and his buildings as "five blots of wasteful architecture".[13] Geddes felt that architecturally Kar was a painter who designed Santiniketan from fanciful borrowings from architectural models and then adds that Tagore's translation of the university did not do ample justice to his own vision.

He claimed that the ashram of India is always in the poet's mind and yet the university did not translate into a way of life and a laboratory where people lived out the ideas they thought about. He claims, "I doubt that the poet has done justice to his own idea of the university".[14]

What one notices in these letters are the dreams of a new kind of university. Geddes realised the affinity between science and art and claims that, "The attraction of the physical world from heavens to the landscape has from the first days of science been its beauty, its aesthetic appeal." He observes, "Song and music came early into physical science with Pythagoras, and light and colour with Newton".[15]

Geddes and Tagore quarrelled about the West and Geddes argued that the poet did not know the West. Geddes argued with Tagore that he did not understand the divisions of the West and that Islam had a great slice of the West over centuries and this made a shamble of the Balkans again and again.

[10] Bashabi Fraser, ibid.

[11] Ibid.

[12] Ibid.

[13] Ibid.

[14] Ibid.

[15] Ibid.

The letters move between global interpretations of history and the concreteness of planning hostel rooms and village tanks. Geddes warned Tagore that the Santiniketan he was about to build could not compete with universities like Oxford and Cambridge on a great world stage but what Tagore felt was that the grid he worked on was a different one. What Tagore wanted was not an imitative act but a conversation of differences. In his classic paper, Tagore argued that the university in the West was a microcosm of the city and that the city in the West had always grown by conquering nature; cognitively, whether it was city or university, the Western mind was hostile to nature. In contrast to this the mind of India always built its great intellectual centres in the forest where Indian sages and their ashrams sought to live in harmony with nature. Tagore argued that the coming dialogue of civilisations had to be a dialogue with the universities but this had to be a dialogue of differences where the forest universities of modern India debated with the city universities of the modern west. It was a dialogue of mentalities which had to be a dialogue of civilisations.

The Tagore-Geddes letters are tantalisingly incomplete. It is a story of an unfinished project and an unfulfilled dream. Tagore's idea of Santiniketan was eventually devoted to the Humanities and the Arts. Its sources of creativity were restricted to these imaginations. Tagore failed to create a department of science which could embody his vision of nature. All Santiniketan eventually had was a department of agriculture where Geddes's son Arthur Geddes worked for a while before he returned to Edinburgh to become professor of Geography. In the next section, I hope to discuss the Gandhi-Tagore debates but within the perspective of Tagore's wider view of science.

13.4.1 Jugalbandhi

The Gandhi-Tagore debates have been called one of the great jugalbandis (a duet of two solo musicians) of the Indian national movement. It evoked a polarity and reciprocity of opposite positions. But it is more than a struggle of opposites and reciprocity of ideas. The two actors as two great national figures lay out the possibility of the whole. They create, in fact, a choreography of technology. Imagine it as a performance. At one end, stands a sage, at the other, the Mahatma, the one elaborate in his robe, the other, minimalist in his dress; one semaphoring renunciation, the other, affirming sensuality in the aesthetics of his culture. Together, they were outlining the possibilities of a future. It is a debate crying out for a third term. Nehru is a third term, because he represents the heuristics of the future.

Fundamental to the debate is the metaphor of language and the nature of language as a model for technological change. The debate begins with a prelude about English and the battle about English is not a diversion. It is a debate about civilisation and the question centres around whether one should employ English.[16] Gandhi's answer is a matter of fact. He claims that the English language has turned India into a race

[16] Prabhu (1961).

of imitators and for Gandhi, imitation is impotence. He argues that the genius of creativity requires the power of the vernacular. He confronts the genius of two men, Ram Mohan Roy and Tilak: one is the virtual father of modernisation in India, and the other, the legendary icon of Maharashtrian politics. Gandhi insists that eventually both were secondary figures because they did not operate in the vernacular. Gandhi claims that Kabir, Nanak, Chaitanya and Guru Gobind were greater men than Roy and Tilak. The tragedy was that their English had shrunk their genius.

Tagore creates a parallel world. If Gandhi talks about the power of the vernacular, Tagore warns about the isolation of the provincial. The provincial isolates, the provincial territorialises. In a world where giving is exchanging, ideas demand openness and reciprocity. Tagore contends that Gandhi had misunderstood Roy. He argued that Roy was confronting a period of history which demanded a comprehension of two sets of cultures—one had to be conversant with Hinduism and Islam, and one had to be ready for an engagement with the rest. Roy realised the importance of openness to the West and Tagore argues that by denying the West, swadeshism was turning into an intellectual enclosure movement.

Gandhi in his response, as a matter of fact, claims that English is studied for commercial and political value and he rebuts the charge of provincialism. Gandhi insists "mine is not the religion of the prison house" and demands the culture of all lands blows freely but he insists that he refuses to be blown off his feet. By this time, one realises that both are masters of the symbol and both are performing an exercise in philosophical pedagogy for the future. The reciprocity between the two arguments is fascinating. Gandhi creates a text for technology and Tagore with all his questions, creates a set of pretexts and contexts within which technology has to enact a set of ethical and aesthetic questions.

13.5 A Ganglion of Connections

Gandhi sees khadi as a ganglion of connections. It is khadi as cloth, it is weaving as prayer, but it is also an act of passive resistance. Khadi is resistance, boycott and production but Tagore replies that swadeshism is a costume ball of borrowed terms. In fact, he insists "we have no word for nation in our language". Tagore insists that technology cannot be a mere symbol; it has to speak the language of everydayness and he sees swadeshism as a mirror-image of Macaulayism. If Macaulay denied the East the potency of even five books of any value, swadeshism seems to repay the compliment to the West.

Gandhi challenges the idea of the swadeshism as mimicry. He contends we do not require the English to talk about democracy and that swadeshism can make these concepts live beyond the English idea of democracy. Gandhi wants to outthink the West by rescuing it. Tagore is worried about the coerciveness of demagogic swadeshism which might create a Chinese wall between India and the West. Tagore argues that the charkha is no panacea for independence. He sees it as a list of behaviours, a collection of attitudes.

Tagore then says that the charkha must be both nature and culture and the society that reduces itself to nature is a form of parasitism. The naturalism of the bee in a beehive is not for man because society has to be about truth, desire and dreaming. The charkha becomes a form of technological timidity associated with "lower types of life".

Tagore felt that the charkha was both bad science and bad technology and he summons a homely metaphor to drive home the argument. He calls that piece of heritage we call the pickle. The pickle he said was a form of hygiene. It was built on the rituals of pollution rules. When the women of Bengal prepared a pickle, they took care to keep themselves clean.

Tagore then added that the caste practices of hygiene and pollution that went into the making of the pickle was an early form of pasteurisation. Yet a pickle is no more a science than a kitchen is a laboratory. A successful caste practice cannot be a basis of bacteriology. Tagore felt in a similar way, that the charkha was just an empirical practice. It could not be elevated to a theory of technology or a comprehensive solution to economics.

Tagore argued that if freedom is choice and opportunity, to restrict one's way of life to the charkha was to accept atrophy and entropy. The mix of sensuality, freedom and creativity which constitutes Tagore's aesthetic sensibility found khadi almost philistine, a mere biology which fails to see the erotic, a technology that fails to see the sensual. For Tagore, the charkha like the beehive has neither sensuality, freedom nor a sense of the erotic. One senses that while Gandhi's ethics might be powerful, Tagore's aesthetics of technology which demands that technology return to the erotic is also valid.

Tagore argued that the charkha in its search for economics is not just an insistence of the mediocre but a pursuit of the magical. He equates the charkha to a mantra reminiscent of the fakir with his gold-appearing trick. For him, the charkha is a part, masquerading about the whole. What swadeshism, with its magic formulae of khadi does, is to bundle out economics and replace it with ethical dictums. Tagore calls it a "confusion of levels" and contends that mathematical mistakes cannot be cured by moral maxims. The poverty of khadi becomes a poverty of philosophy.

Gandhi is slow to respond to Tagore's argument which he describes as a series of word pictures, a powerful critique of those mimicking the call for charkha. He volleys back to Tagore a set of questions every Indian must ask, questions about the chronic famines of India, questions about the parasitism of Indian cities. He asks us to question technology as a chain of being. "Trace the course of every piece that finds its way into your pocket, and you will realise the truth of what I write." Gandhi suggests that Tagore's politics of difference might hide a philosophy of indifference to the suffering of modern India. The encounter simmers as swarai becomes a tussle between a poet preaching freedom, and a politician preaching love.

Gandhi describes Tagore's description of the charkha as poetic licence. Tagore sees the charkha as Gandhi's private dream articulated as a private language, claiming to be a public good. He fears that each village would become a collection of isolated units and khadi would create the prison house of technology.

Gandhi argued that the charkha was the language of the core, which linked work, livelihood, technology, dignity and employment in an ethics of scale that moved from the local to the global. It was not a fetish but a sacrament. For him, the charkha represented the essential unity of India. It encodes unity in variety. One talks of the monotony of khadi and sees the charkha as a technology masquerading in a uniform, the other talks of obsolescence, erasure, of communities being displaced by the machine. Gandhi then adds bluntly that if "the poet spends half an hour a day, his poetry would gain in richness". Tagore offers Plunkett's idea of cooperation in agriculture as a model. Gandhi shrugs it aside as an example and offers the charkha as a model with myriad transformations. In one sense, the debate was never over. In fact, I see it as a pedagogic performance where the Gandhi-Tagore debates is being enacted almost everyday in the battles over technology and science in India.

13.6 Concluding Fragment

How does one read the sheer richness of conversation, the celebration of cultures we call the national movement? Tagore stood out as an outlier even in a cascade of eccentrics. He was sceptical of nationalism which he only saw as a legitimation for violence. Nationalism, he felt was an amplified version of the parochial or the vernacular. It over-amplified the cultural ego, made it deaf to arguments of the other. Tagore and Bose both felt that internationalism did not quite solve the problem either. Internationalism was a conversation of nation states. As a result, it failed to create the dialogue of civilisations. Internationalism created greater wholes while emptying the message of content. What one needed was a claim to identify a culture which could still read the other as a possibility, be at home in it and yet be one self. This was the kind of cosmopolitanism, Tagore wished to create. Nationalism and internationalism were emanations of the past. It was a comfort across variety of times. Cosmopolitan provided a sense of locality and the case of nomadism. It had a sense of ease both with oneself and the other. What it refused was easy solutions where the power of an Indian self gets banalised by being exoticised into the other. Tagore and Bose did not debate with marginality. They wanted to look modern paradigms in the eye and then show new possibilities. In that sense the cosmopolitan man going beyond easy oppositions of west/east, accidental, town/country, seeks conviviality in difference. The letters to Bose, the experiment at Santiniketan and the debate with Gandhi become three variants of this search for the cosmopolitan. Tagore went beyond translation. He wanted dialect to be also global while realising it could degenerate to the parochial. The dream of the cosmopolitan anchored in science and poetry is the great message of Tagore.

I must confess that every man reads his biases into a text, and the hermeneutics I have offered is of a man who sees the world of Tagore as a world searching for a poetics of science. Tagore believed that science and technology were prosaic, everyday efforts operating in terms of logic but what Tagore added to them was an imaginary, where science also becomes a dream of possibilities, a series of thought experiments about the nature of freedom and logic of choice. For me, the Indian national movement is a collection of wonderful stories, powerful fables, fragments of a philosophical mind that we need to recover so that we realise that the choices we make with our textbook intellects are only reflexes of the richness of debate from a nearby past. A genealogy of these debates might recreate the genealogy of dissent for an India cascading towards a more homogeneous world.

References

Cornelius Castoriadis, *The Imaginary Institution of Society*, Cambridge: MIT Press, 1998.
Bashabi Fraser, *A Meeting of Two Minds: Geddes-Tagore*, Edinburgh: Wordpower Books, 2005.
Patrick Geddes, *The Life and Work of Sir Jagadis C Bose*, London: Longman, 1920.
Hellen Meller, *Patrick Geddes: Social Evolutionist and City Planner*, London, Routledge, 1994.
R. K., Prabhu (ed.), *Truth Called Them Differently*, Ahmedabad: Navjeevan Press, 1961.
Sumit Sarkar, *The Swadeshi Movement in Bengal, 1903–1908*, New Delhi: People's Publishing House, 1973.

Chapter 14
Understanding Gandhi's Approach to Technology: After the Critique of Modernity and Industrialism

Dhruv Raina

The clamour for decolonizing the academy has more or less acquired global dimensions, and the more specific demand in the world of higher education has been for the decolonization of the sciences and social sciences.[1] This paper raises a couple of questions that obliquely have to do with some of these concerns. While recognizing that the history of territorial decolonization can be traced to over a century, contemporary demands for the 'decolonization of minds' and social science theory are over half a century old. However, I do not think that these two histories can be easily disentangled. So, while this paper relates to some contemporary concerns of decolonization, they have as much to do with an earlier historical period. In a way the paper brings together two distinct historical moments under a genealogical rubric. It is against this backdrop that I we turn to Gandhian reflections on technology.

Gandhi scholarship over the last several decades has portrayed Gandhi as an increasingly complex thinker in contrast to his modernist contemporaries. But this is not really an issue that I intend to engage with as a Gandhi scholar, which I am not. On the other hand, I shall try to show that Gandhi's approach and thinking about technology was neither that of a Luddite as he is often portrayed nor that of a traditional modernist, but founded on circumstantial case-based reasoning keeping in mind the benefits of the concerned social constituencies. Ashis Nandy and several others have elegantly pointed out that Gandhi's intention was not to develop a systematic theory or ideology, but offer an alternate vision in opposition to the hegemonic oppressions of the time.[2] Furthermore, the question of Gandhian thinking about technology will be addressed from two ends: the one historical and the other contemporary.

[1] Bhambra et al. (2018) and Herbjørnsrud (2019).

[2] Nandy (1981).

D. Raina (✉)
Zakir Husain Centre for Educational Studies, School of Social Sciences, Jawaharlal Nehru University, New Delhi, India
e-mail: d_raina@yahoo.com

S. Ahmed and S. Sarkar (eds.), *Decolonizing Science and Modernity in South Asia*,
https://doi.org/10.1007/978-981-97-1829-0_14

14.1 The Historical Context of the fin de siècle

We shall commence with the historical debates towards the end of the nineteenth century and the beginning of the twentieth century that transpired within the circle of intellectuals associated with the *National Council of Education* and *The Dawn Magazine* in Calcutta.[3] The salience of debate within the circle of the National Council of Education and the splinter society called the Society for the Promotion of Technical Education is relevant to the discussion that follows since most of the involved interlocutors were in the vanguard of the Swadeshi movement that erupted in Bengal in the first decades of the twentieth century. Further for all practical purposes until the beginning of the third decade of the twentieth century the city of Calcutta (now Kolkata) was the scientific capital of India and the cauldron of heated discussions both on the nature of science and technology.[4] Within the circle just mentioned discussion on technology and industrialization focused upon two core issues, namely the path of industrialization that was best suited to India's economic realities and the second within this context was the future of the artisanal and crafts traditions in India.[5]

The decades 1890–1920 in late colonial India were years marked by heated debates and detailed reflection on the character of India as a civilization and the future pathways of industrial development and modernization. In fact, it could be considered the period of a reflexive reckoning with modernization and the processes this transformation entailed. *The Dawn Magazine*, a review published from Calcutta, served as a cultural forum for the expression, cultural appropriation and translation of perceived processes and notions associated with modernization during the fin de siècle.[6] In fact, the contributors to the journal during these three decades included the cultural constellation of the Bengali cultural amphibian class of the time. It is possible to identify a progressive metamorphosis in the central themes discussed in the pages of the journal during these years. In the first phase the contributors were preoccupied with the notion of Indian history and Indian civilization, particularly in the light of an emerging national consciousness that was witnessed towards the end of the nineteenth century.[7] Gradually, focus shifted to India's resource endowment, its tradition of artisanal skills and the gradual decline of the handloom and textile industries in the face of the onslaught of mass-produced goods. During this assimilative phase, a core issue was to develop a logic or set of criteria that would enable a choice between retaining that which was considered worthwhile from the past and assimilating only that which was valuable from the West. By this time, it was axiomatically accepted within this circle that technology was an important agent of social transformation. If reservation there was, it related to the Western model of industrialization, which had become the subject of critique as much in the West as amongst the National Council of Education. Elements of this critique overlapped with

[3] Chandra (1977) and Sarkar (1975).

[4] Kumar (1994).

[5] Visvanathan (1985).

[6] Raina and Irfan Habib (1996).

[7] Raina and Irfan Habib (1995).

Victorian anti-industrialism; and this related to the squalor, congestion and pollution characterizing industrial urban conglomerates. In aesthetic terms of course there was a generalized revulsion towards the mass-produced consumer good that had ushered in era of the depersonalized artifact.[8]

The larger concern was the perception that Western industrialism was highly insensitive to the artisanal base of the country, in which the artisan's skills and resources were progressively immiserized, and clearly out of pace with India's civilizational charter. Two alternatives appear to have emerged for the concerned interlocutors, the one institutional and the other paradigmatic. The institutional change desired suggested a greater involvement of the state in the form of protectionist policies, but more importantly the state was required to be more pro-active in the establishment of technical institutes to undertake the task of upgrading the skills of Indian artisans against the backdrop of modern industrial development.[9] This effort, it was believed would draw the artisan from the periphery of the economic life of the nation into the mainstream.

At the paradigmatic level it appears that three alternatives presented themselves to this Bhadralok circle. The first in any case was a non-option, namely the reigning paradigm of Western industrialism from which our interlocutors wished to disidentify themselves. Counterposed to this paradigm was the Tolstoyan vision of a pastoral utopia premised on the idea of autarchic self-sustaining communities. The Gandhian vision was inspired by this frame but was all the same quite different from it. The third alternative considered the possibility of deploying scientific and technical knowledge to work at smaller scales than those at which large scale technologies were seen to be profitable within the then prevalent level of global industrialization. In other words, it was being suggested that there existed lower scales of production where technological knowledge could be deployed to produce economies of scale. This required a different regime of skills to locate the Indian artisan with this upgraded knowledge.[10] Certain historical contingencies accelerated the rate at which our interlocutors were compelled to create the institutions where they would give concrete form to the developmental scenario they had envisioned. It appears now that in this long-drawn revaluation of the so-called founding precepts of Indian civilization, the meaning and relevance of central notions of progress, and industrialization of India were deeply contested.[11] In a sphere distant from that of the NCE Gandhi was to pose a cognate question: "Does economic progress clash with real progress? By economic progress, I take it, we mean material advancement without limit and by real progress we mean moral progress which again is the same thing as progress of the permanent element within us".[12]

A colleague and I had ploughed through the archive of this debate more than a decade ago and had conjectured that Gandhi had been deeply informed by these

---

[8] Raina and Irfan Habib (1993).

[9] Ibid.

[10] Ibid.

[11] Raina and Habib, 'The moral legitimation of modern science', op.cit.

[12] Gandhi (1926).

reflections in the elaboration of his own position on modern technology. We were mistaken in ascribing the influence of the Swadeshi reflections on Gandhi's thinking about Swadeshi and technology. Gandhi was never directly a part of the Swadeshi movement sweeping through the intellectual and scientific spheres of late nineteenth early twentieth century Bengal. But as is evident, he addressed several of the issues that were debated and articulated within that discourse. Furthermore, during the first two decades he did not participate in the debates on the cultural meaning of East and West that appeared in the pages of *The Dawn Magazine*. Nevertheless, with many Western and Indian thinkers alike he more or less had come around to the conclusion that the West was no longer an exemplar of civilization. In this too he shared a great deal with his Bengali colleagues involved in the Swadeshi struggle. The pages of *Hind Swaraj*—a pamphlet that appeared in 1909[13]—were inspired by Indian and Western critics such as Tolstoy and Thoreau of industrialism and materialism, but rejected the founding discourses of Social Darwinism and progress that conflated material and spiritual progress.[14] Consequently, it would be more appropriate and historically tenable to argue that the first two decades of the twentieth century served as a crucible where different ideas, visions and imagined developmental trajectories were hotly contested and debated, and these visions circulated as much within *The Dawn* circle and the evolution of the Gandhian conception of technology.

Gandhi was quite close to most of his scientific and technological colleagues associated with the Swadeshi movement and the founder of modern Indian chemistry Prafulla Chandra Ray himself became a Gandhian.[15] Both the Swadeshi movement and Gandhi interrogated the spread of industrialization and sought to rejuvenate the handicraft tradition, but for Gandhi this revival was to stem from village focused community development. These positions were clearly embedded within his larger conception of political struggle wherein he argued in an article in *Young India* in 1926, that freedom from British rule would be futile if the nationalistic, industrial and civilizational assumptions of the West were not jettisoned as well. In other words, the future of India rested on her ability to resist "the onrush from the West".[16] If on the one hand Nehru and Gandhi are seen as complex conjugates of each other then on the other so are Gandhi and the grand old engineer from Karnataka namely Visvesvaraya.

But during the high tide of the Swadeshi movement the founder of the school of modern chemistry in India P.C. Ray continued in Gandhi's footsteps. Ray's students would not traffic with Gandhian ideas about the charkha. During the fateful 1930s when the National Planning Committee was working towards India's economic, scientific and technological future it was Visvesvaraya's, and Saha's desire that India "choose the cold logic of technology" which won the day. But here too the old inherent tensions did not go away. The Bombay Plan is testimony to the variety of opinions

[13] Gandhi (1938).

[14] See, for example, his address to the YMCA at Colombo in Ceylon (Sri Lanka), which was reprinted in *Young India* (1927).

[15] Raina (1997) and Shiv Visvanathan, *Organizing for Science*, op.cit.

[16] Gandhi (1926).

and negotiation that went into its making.[17] In a very thoughtful but forcefully argued piece Shiv Visvanathan recreated a Gandhi in today's environment albeit in a "social science format". But in doing so he remarks on the historical Gandhi as one of "the great scientists of the Swadeshi era", including him in the pantheon of Ramanujan, J.C. Bose and P.C. Ray.[18] Irrespective of the rationale for Visvanathan's inclusion of Gandhi in the pantheon of Indian science, the larger point is that Gandhi was deeply embedded and, in a way, central to the spirit of Swadeshi.

The second case I wish to discuss on the social relations of technology that has a Gandhian twist to it is that of the programme developed by the scientists Amulya K.N. Reddy in the 1970s. The programme developed at India's premier Indian Institute of Science, Bangalore and was orchestrated by a leading electrochemist, the founder of a programme on technological alternatives, turned energy policy analyst and winner along with three others of the Volvo Prize for their two-volume path-breaking work *Energy for a Sustainable World*.[19] It could easily be argued that between the 1970s and the end of the 1990s he was easily amongst India's scientists who reflected most deeply upon the relation between science and society and influenced in significant measure the efforts of social movements that sought to democratize the culture of science and influence technological choices that impacted on development.

The remarkable feature of the programme developed by Reddy at the Institute of Science was that long before post-Kuhnian philosophy of science provided an intellectual panoply to contest the traditional disciplinary hierarchies of power and influence he emerged within the world of scientific certitudes of the Indian Institute of Science, not just a dissident scientist but an icon of doubt. I had the good fortune to work with him between 1981 and 1984, and our conversations continued long after I left the city of Bangalore. A striking feature about the efforts of Reddy and his associates was the manner in which certain key Gandhian ideas and ways of looking at the world from the other side were woven into their technology development programmes. When are value considerations factored into programmes of technological development? Does it happen before the programme commences or does the re-engineering commence after the programme is completed? These questions while central to the study of technology and culture were of salience to the larger concerns of Gandhi about the place of modern technology in contemporary life. I shall return to the contemporary conjuncture later in this essay and now review basically some of the themes that appear in the scholarship on Gandhi and technology.

[17] Kumar (1995).

[18] Visvanathan (1997).

[19] Johansson et al. (1988).

14.2 The Two Myths Concerning Technology

The historical and contemporary moments provide an avenue for thinking through a variety of concerns that run as constants that animate Gandhi's thinking about industrial civilization and its conjugate modern technology. In order to read through the multifaceted aspects of his utterances on the subject that sometimes appear so contradictory, it would be appropriate to frame these utterances along two axes, in order to transcend some of the essentialist dichotomies customarily employed in framing his thoughts on the subject. These dichotomies include the traditional versus the modern, rural India versus urban India. Amidst these dichotomies, Gandhi is sometimes portrayed as this resolute figure opposed to the votaries of science and industry, namely Jawaharlal Nehru and the engineer-planner Mokshagundam Visvesvaraya.[20] These characterizations are at best caricatures, as Nehru himself was quick to point out to Aldous Huxley who was possibly unnerved by Gandhi's supposedly Luddite streak. In 1933, while in prison Nehru had read *Science in a Changing World*, a volume edited by Thomas Holland containing a number of articles on science in its relation to society, some of them authored by left wing scientists from Cambridge. The volume also contained an article by literateur Aldous Huxley entitled "Economists, Scientists and Humanists", where he pointed out that:

> Tolstoyans and Gandhi-ites tell us that we must 'return to Nature' – in other words, abandon science altogether and live like primitives or, at best, in the style of our medieval ancestors.[21]

Nehru's letter to Huxley is a long detailed one that merits a separate discussion. But I shall quote this letter at a little length not just to clarify the point that Nehru never saw Gandhi as a Luddite and the deep respect he had for the latter's views even though they thought differently. Nehru writes:

> I am not competent to say anything about the Tolstoyans but I have been closely associated with Mr. Gandhi and the "Gandhi-ites" for the last fourteen years and can thus presume to know something about them. Indeed, in a political sense, I have myself been a "Gandhi-ite" for all these years. I feel that, quite unintentionally, you have done them a grave wrong in the reference you have made…I do not propose to discuss Mr. Gandhi's personality in this letter… Essentially he is a man of religion and essentially the typical modern is not… His attitude to science is very far from being hostile. He welcomes it and takes advantage of it in a variety of ways, and often people, willfully misunderstanding him, accuse him of inconsistency because he does so… Large-scale industrial production he does not like but he has never suggested to my knowledge, that it should be scrapped. He wants,…to decentralize industry…and for this purpose he wants to take the fullest advantage of the scientific method.[22]

Between 1925 and 1927 Gandhi had occasion to address students of science in different parts of the country. In 1927 while talking to students of India's premier Indian Institute of Science he clarified that he was not an opponent of science but

---

[20] Zachariah (2005).

[21] Huxley (1933).

[22] Jawaharlal Nehru, Letter to Aldous Huxley (1933), in Singh (1988).

wished to ensure that scientific research was constrained by non-violent, humanistic and moral considerations. He exhorted these students further: "How will you infect the people of the villages with your scientific knowledge? Are you then learning science in terms of the villages and will you be so handy and so practical that the knowledge that you derive in a college so magnificently put and I believe equally magnificently equipped—you will be able to use for the benefit of the villager".[23] The criticism of the cult of modern technological progress was premised on the idea that human inventions were probably inferior aesthetically to the products of nature, and that the existence of complex technological systems encourages the belief in technological utopia. In this Utopia inner spiritual development is traded for gross national product, which is the deliverance of complex machine technology. When upbraided for rejecting technology's potentialities for alleviating the toil of labour Gandhi maintained:

> I do want growth, I do want self-determination, I do want freedom, but I want all these for the soul. I doubt if the steel age is an advance upon the flint age. I am indifferent. It is the evolution of the soul to which the intellect and all our faculties have to be devoted.[24]

14.3 Interdisciplinary Engagements with Technology in Society

However, the sociology of science and technology offers a cognitive device to map the inherent ambivalences in the discussion. The political dimensions of science and technology are approachable from a variety of perspectives by reflexively turning upon the different disciplines, frames and approaches. Bruno Latour for one has pointed out that discussions on the relationship between science, technology and society are frequently torn between the myths of technological autonomy and neutrality.[25] Within disciplinary academic communities studies of this complex relationship have been institutionalized under various rubrics depending upon the focus of concerned investigative communities or social movements either contesting particular policy legislations or very specific scientific and/or technological interventions within the public sphere.

These academic approaches are interdisciplinary in nature though some have argued and not without reason that they are post disciplinary, drawing upon developments in a variety of disciplines and discursive fields. These include the economic analyses of Smith and Marx, social realist literature, political reform movements, disciplinary specializations such as history and philosophy of science, technology and medicine, the new social movements that include the environmental movement as well as cognitive movements such as STS.[26] The canvas is spread over macro studies

[23] M. K. Gandhi, *The Hindu*, 19 March 1925, quoted in Iyer (1986).

[24] M. K. Gandhi, 'The Great Sentinel', *Young India*, October 13, 1921; in Gandhi (1923).

[25] Latour (1994).

[26] Fudpucker (2001).

of science and technology often rooted in the myth of technological autonomy, to micro studies that are of a social determinist nature.

The central metaphor proposed by the social studies of technology to define its object of investigation is that of the seamless web.[27] Though presently articulated in the context of regnant technoscientific worldview, and as part of a particular theoretical vocabulary or discourse, the idea itself has earlier precedents in reflections about science, technology and society. Some of the STS approaches abandon the myth of autonomy and the use-abuse model, centering instead a conception of science and technology as value-laden, historically contingent, shaped by and shaping socio-cultural and political institutions. This perspective of the discipline enables it to overcome both unabashed Baconian optimism and Heidegerrian technological nihilism. The discipline also entertains the possibility, given its historical proximity to science activism, for structuring mechanisms for exercising democratic control over contemporary technoscience.[28]

Consequently, one important aspect of social studies of science has been the elaboration of a critique of technological determinism. Two essential postulates of technological determinism are the development thesis and the primacy thesis. According to the latter, technological development is driven by purely internal technological factors. And the development thesis suggests that technological development results inevitably in societal development. Social constructivist research has challenged these founding postulates of technological determinism; though there have been those who have argued for a softer technological determinism. In this view a diversity of cultural and social factors play a central role in influencing the construction of technological systems, and this development does have a positive impact on society.

The works of scholars such as Langdon Winner reveals the politics encoded within technological systems and makes the case for searching out technological alternatives a precondition for democratization within society. For some time now the strong version of technological determinism has been out of fashion and Winner has stepped down from his earlier views encountered in *Autonomous Technology* to a modified determinist thesis,[29] recognizing that there exists a societal tendency to adopt some version of technological determinism.[30]

If technological determinism was ideologically located in the triumphalism of the 1960s; other perspectives that emerged came in from the opposite end and sought to understand how the social impacted on the face of science of science and technology. Thus, Thomas Misa has pointed out how technologists and philosophers are often

[27] Bijker et al. (1987).

[28] Lars Fugslang, 'Three Perspectives in STS in the Policy Context', in Stephen H. Cutcliffe and Carl Mitcham (eds.), op.cit., pp. 35-50.

[29] Winner (1977).

[30] Langdon Winner, 'Where Technological Determinism Went', in Stephen H. Cutcliffe and Carl Mitcham (eds.), op.cit., pp. 11-18.

committed to the frame of technological autonomy/determinism. The frame developed by the social determinists occupies the opposite end of the spectrum and envisions technology as largely socially determined.[31] The social constructivist frames attempt to navigate between determinism and being either promotional or apologetic about science and technology. Furthermore, the constructivist perspective contrary to conventional thinking, unveils the fuzzy and convoluted boundaries that separated the scientific from the political; between scientific neutrality and value-ladenness, and between perspectives highlighting the one-way impact of science on society and ones that illustrate their mutual shaping.

However, it must be acknowledged that the technologically oriented frame has revealed the variety of ways in which technology has progressed; the STS perspective has demonstrated that this technological progress is not identical with general progress in society. In that sense STS is a critical enterprise that provides a "comprehensive evaluation of the nature of scientific and technological change and the forces that impel it". In the STS perspective the understanding that technology lacks autonomy, implies that human choices and intervention could render us active agents that control and shape technological futures.[32]

In addition to the technological determinist perspective where science and technology shape society, and the constructivist perspective where society shapes science and technology; there are other perspectives such as the interactionist approach that emerged during the post World War II years within very specific policy contexts. These perspectives have had their adherents and could be seen as competing conceptual frameworks for looking at the relationship between science, technology and society. The interactionist approach shares with constructivism the hope of tangibly contributing to the democratization of technological culture and decision making on technological matters that are of global dimensions. The social shaping perspective seeks to empower actors and networks, and the determinist version either popularizes science and technology as a panacea for developmental problems or rejects it as a diabolical force that seeks to enslave society to the inevitable march of technological development. These perspectives are often portrayed to be in conflict and upholding incongruous standards of theory and methodology.[33]

Approach	Context
Science and technology shapes society	Determinist/ technocratic
Society shapes science and technology	Radical/business targeted
Interactive approaches	Voluntaristic/relativist

In summary it could be suggested that the myth of technological autonomy embraces an entire range of technological determinist positions. Furthermore, it could be established that the philosophical critiques of technology that argue that

---

[31] Misa (1988).

[32] Susan E. Cozzens, 'Making disciplines Disappear in STS', in Stephen H. Cutcliffe and Carl Mitcham (eds.), op.cit., pp. 51-66.

[33] Cutcliffe (2000).

technology has a logic of its own that would subjugate us as human beings are premised on one version of this myth. Similarly, the myth of neutrality envelopes again an entire range of social determinist positions; where people kill people and not guns that kill people.[34] Into this genre would conform a number of social histories of trade unions, business histories and the like.[35] In the subsequent discussion I shall argue that Gandhian ideas on the subject of Western industrialism, science and technology begin to make sense when seen within this frame of technological neutrality and social determinism.

A subject of Gandhi scholarship since the 1970s has been the civilizational critique of Western industrialism and modernity that is seen to be the characteristic of the Gandhian corpus. But equally it could be suggested that there is more than a hint in several of his writings that other starting points could lead to other trajectories and possibilities. In other words that alternate perspectives of the social could produce other industrial and technological schemas. In other words Gandhi's thought was never limited by the inevitability of a technological frame or a civlizational schema.

The more contemporary scholarship on Gandhi and Nehru appears to be revealing a new dimension of their thought and the relationship between their thinking. Khilnani once pointed out that an essential asymmetry characterizes the more recent scholarship on Nehru and Gandhi. Gandhi for one has with the passage of time been portrayed as this increasingly complex figure, while Nehru is portrayed in more reductive terms as the uncritical devotee of science and technology.[36] Evidently, these dichotomies have not been very helpful and more than anything else have been historically untenable. This polarization of positions foregoes recognizing that both Gandhi and Nehru worked as colleagues, and that it was Gandhi who chose Nehru as his successor. In an attempt to depart from this portrayal of Gandhi and Nehru as antithesis of each other Ram Guha has in an interesting paper attempted to bring out their commonalities rather than their differences. Guha begins by pointing out the considerable influence Gandhi's thought and writings have had on the environmental movements in India, across the political spectrum from the 1970s and commencing with the Chipko movement. These movements have drawn on two Gandhian resources. The first has to do with Gandhian techniques of non-violent protest and the second draws upon the Gandhian polemic against industrialization.[37] Guha argues that the environmental movement has reclaimed Gandhi by demonizing Nehru as this rapacious, consumerist, profligate modernist. The construction itself is a product of unsuccessful attempts to resolve a puzzle namely that:"…the development experience in India has been marked by a profound insensitivity to ecological considerations". It is argued that the puzzle is conveniently explained by proposing

[34] Bruno Latour, 'On Technical Mediation – Philosophy, Sociology, Genealogy', op.cit.

[35] Thomas J. Misa, 'How Machines Make History, and How Historians (and Others) Help them to do so', op.cit.

[36] Khilnani (2002).

[37] Guha (2007).

a conspiracy theory involving the ultra-modernist Nehru and Guha challenges this construction.[38]

Mahadev Desai for one informs us that Gandhi was once asked whether he was against all machinery. Gandhi said,

> How can I be when I know that even this body is a most delicate piece of machinery? The spinning wheel is a machine; a little toothpick is a machine. What I object to is the craze for machinery, not machinery as such.... The supreme consideration is man. I would make intelligent exceptions. Take the case of the Singer's sewing machine. It is one of the few useful things ever invented.
>
> But, "said the questioner," if you make an exception of the Singer's sewing machine and your spindle, where would these exceptions end?
>
> Just where they cease to help the individual and encroach upon his individuality. The machine should not be allowed to cripple the limbs of man.[39]

The point made above is central to Gandhi's thinking about technology. There is a consensus amongst most contemporary scholars that Gandhi was certainly not a nostalgic anti-modernist. Akeel Bilgrami has departed from the neo-Gandhian understanding and argued instead that in Gandhian thought it is possible to find affinities with the radical dissenting traditions of seventeenth century England. Thus, the central question Gandhi would have asked about a technology would relate to considerations of whether it was or was not controlled by ordinary people. According to Bilgrami in the Gandhian scheme from the seventeenth century onwards a new alignment between commercial and mercantile interests combined to generate a predatory attitude towards nature. In fact, Bilgrami reconstructs affinities between Gandhi's ideas about science and the cultural anxieties that surfaced in the West as the dominion of the new science expanded. This he does through a selected reading of Gandhi's writings from *Hind Swaraj* and *Young India*.

In his view a probable reconstruction of Gandhi's reasoning would proceed on the following lines. The West embarked on a path wherein nature was demystified in the seventeenth century. Since this occurred at the same historical juncture as that of the rise of the revolutionary sciences, Bilgrami suggests that Gandhi equated the process of demystification with the institutionalization of the new experimental methods: "...he saw in it a conception of nature whose pursuit left us disengaged from nature as a habitus, and which... engendered a zeal to control it rather than live with it".[40] This form of reasoning has its antecedents in late seventeenth century England and even the continent. The critique does not deny or reject the achievements of the new science. Consequently, Bilgrami argues, it cannot be construed as a Luddite reaction.[41] I do not proceed to follow-up on Bilgrami's argument, but propose that inasmuch as the Gandhian critique was more in the vein of culturally appropriating a new set of practices rather than rejecting them in toto. The larger point was that this attitude to nature not only resulted in ecological disaster, but endangered human relations as

---

[38] Ibid, p. 122-23.

[39] Desai (1938).

[40] Akeel Bilgrami, 'Gandhi, Newton and the Enlightenment', in Karawan et al. (2008).

[41] Ibid.

well. In the 1970s this understanding began to echo in the environmental critiques and social movements that more or less erupted globally that were speaking of the "...dehumanizing effects of rampant globalization and destruction of the earth...".[42] As mentioned above, that is the conjuncture wherein the ecological movements begin to reclaim Gandhi.

Pushing the argument further it is seen from this perspective that the complexity of technological systems sets up imperatives that even its admitted economic benefits cannot always outweigh its negative effects. Complex machinery requires a measure of centralization of control and coordination. The alienation of coordinators from those being coordinated creates conditions which breed violence. Hence centralization was seen, in addition as inimical to community ties eroding the "... the natural and personal relations characteristic of the small group", and in the long run work against the realization of genuine democracy.[43] Clearly this view of technology and industrial culture is hedged in and nuanced and when read within context and the parentheses within which Gandhi expressed these views, the Luddite ascription begins to make less and less sense.

14.4 The Impact of Gandhi on Thinking About Technological Alternatives:

If the 1970s was the period of a crisis in thinking about development and modernization,[44] in India too it was manifest as a crisis of the Nehruvian mandate that prompted several responses at the level of institutions and political economy. While the fuel crisis was the most startling event of the times with a global impact, the environmental and ecological crises, in their relation to the former prompted a rethinking about development, modernity and industrialism. In this environment the mid-1970s in India was also a witness to a growing perception that there was a radical disconnect between science, technology and industry. This prompted policy makers to critically rethink the Nehruvian mandate and the centrality of science and technology within that landscape.[45] In the newly emerging landscape, the buzz words were "development" and "development policy", "social relations of science" and "science policy", and in the interstices between "science and development", "development and science" policy constituted the terrain of the new discourse.

Clearly given the widespread nature of poverty in India trickle down had not worked, development strategies and technology transfer had not dispelled poverty.

---

[42] Bilgrami (2007).

[43] Sibley (2021).

[44] Elzinga and Jamison (1981).

[45] Raina (2003).

This structural crisis manifested itself as one of organizational and institutional legitimacy in the world of science and academic research.[46] The rise of new social movements was reflected in the migration of a few students from elite centres of science to alternate technology movements, people science movements etc. These were new blips on the screen of Indian science. I take up the discussion of one such programme from the world of science, namely the founding of the Centre for the Application of Science and Technology to Rural Areas (ASTRA) at the premium Indian Institute of Science (IIS) to locate where and how Gandhian thinking about technology structured its subsequent efforts. A number of professional engineer-scientists, chemists, biotechnologists at IIS with "conventional" intellectual commitments within the world of science, and notable reputations to match constituted themselves into a Cell called ASTRA to undertake research on the developmental problems of the state with special reference to those of the rural areas of Karnataka. The activity was spearheaded by the then leading electrochemist, and later energy policy analyst and winner of the Volvo Prize, Amulya Reddy, who initially provided the group with both intellectual and organizational leadership.[47]

Obviously, such a cell within the confines of the institute would have had problems justifying itself—since the mandate of the institute was to pursue research at the frontiers of science. Located within the institute, the scientists who formed ASTRA had to have a reputation within their own scientific disciplines in order to survive the possible ridicule of their colleagues who may have accused them of finding an easy way out. Two circumstances weighed in their favour: (1) most of them had reputations in their respective fields—some of them were leaders, others were making their reputation in their respective disciplines, (2) the director, Satish Dhawan, a leading space scientist had conferred his imprimatur on the programme. However, the lower standing of "rural development" in the disciplinary hierarchy of the sciences would continue to haunt those associated with ASTRA. This did not stop it from existing in the institute as what I have called an island of doubt amidst an ocean of certitude.[48]

The moon-ghetto metaphor, a post Apollo metaphor, was often employed to dispel such criticism. But unless one actually worked on rural technologies it would be difficult for a practicing scientist to be convinced that problems of rural technology could be as intractable as those, they were accustomed to working on. In this case, ignorance was truly bliss. Within the confines of the institute ASTRA in a more clandestine way and Reddy more vocally remained the voices of dissident science.[49] While I argue that Gandhi's was a social determinist vision of technology, I now try to demonstrate how social determinism operated within its very opposite; namely that the members of ASTRA entertained an understanding of technology that was technological determinist. Nevertheless, contrary to their understanding there were values and ideologies that underpinned their endeavours that were sometimes explicitly articulated, but more often than not hardly ever rose to the surface of conscious

---

[46] Ibid.

[47] Reddy (1993).

[48] Raina (1993).

[49] Reddy (1978a, b).

expression. The inability to articulate these values could arise from the sway of the culture of technocracy that would have disdainfully looked upon utterances that diverged from the ideological commitment to technological neutrality.

The critique of industrial civilization that resurfaced in the 1970s was deeply indebted to the writings of Gandhi—in particular the British economist E.F. Schumacher's *Small is Beautiful* harked back to an idea that large scale industrial technology does not necessarily operate with economies of scale or lowers per unit costs.[50] Gandhi's writing centre stages the concerns of rural India, Gandhian economics and incest taboos certainly shaped the priorities and set the horizons for the explorations of the group. This would be emphatically stated in a classic paper called the Nairobi paper that Reddy authored on the choice of technology. These criteria served as the touchstone for the research programmes of his group as they broke out of the idea of "technological fixes" and as they moved out of the laboratories of the Indian Institute of Science into the villages of Karnataka.[51] In a way what this understanding shared with some of the Swadeshi thinkers discussed above was the rejection of the idea of "economies of scale justified mass production". Secondly, the Gandhian caution against consumerism was now reformulated in their work in terms of the possibility of envisioning and creating a sustainable future.[52]

It is not very difficult to find the roots of Gandhian thought in the criteria proposed by Reddy for the choice of technology. These were as follows: (1) does the technology satisfy the needs of the neediest, (2) does the technology mobilize local human resources, (3) is the technology environmentally friendly.[53] A fourth criterion was introduced in the late 1980s that had to do with the empowerment of women at the local level. It was hoped that these criteria would guide the development of technologies that would revitalize the village economy or so it was hoped. In any case, a number of movements and NGOs in and around the state of Karnataka were inspired by these ideas. In other words, the focus on quotidian rural life resulted in a number of programmes on energy and habitat with a deep sensitivity to environmental factors that was factored into their research methods—this sensitivity was a constituent element of research during this period.[54]

In the career of several of these movements for technological alternatives and addressing the ecological crisis we find sometimes apparently strange appropriations

[50] Schumacher (1975).

[51] A. K. N. Reddy, 'Technologies Appropriate to Rural Development', op.cit.

[52] On innumerable occasions Reddy prefaced his academic and public lectures with Gandhi's remark that there was enough to satisfy human need but not enough to satisfy human greed. The point was elaborated upon in Gandhi's own words: "We should still have, as we have always had, in our midst people who make the pursuit of wealth their aim in life. But we have always recognized that it is a fall from the ideal.... I have heard many of our countrymen say that we will gain American wealth but avoid its methods. I venture to suggest that such an attempt, if it were made, is foredoomed to failure. We cannot be "wise, temperate and furious" in a moment" [Gandhi, based on a lecture delivered in Allahabad, December 22, 1916]; Gandhi (1922).

[53] Reddy (1978a, b).

[54] A. K. N. Reddy, 'Technologies Appropriate to Rural Development', op.cit; for a more recent account of this work see Raina and Hoysala (2017).

of Gandhi and Mao: the exhortation to return to the village or the people to learn from them.[55] Between Gandhi and Mao, a certain idea had acquired currency within this community of scientists. There were two aspects to this realization. The first expressed the idea that so-called traditional technologies had evolved over centuries and had undergone a great deal of adaptation in local contexts and environments; and this technological knowledge and the concomitant skills and know-how were transmitted like genetic material from generation to generation.[56] A related idea was to undertake the task of comprehending this universe of local knowledge and skills, and use this as a platform, whenever possible, to generate new technologies and systems, since that would possibly minimize the problems of adaptation and assimilation. This created an internal pressure to desist from arriving at generalizable technological fixes to the technological problems encountered as scientists were vulnerable to doing, since this was part of their cognitive socialization.[57] The additional factor that possibly prevented this from happening was the realization that the problems of these areas were caused by the uncritical imposition of techno-fixes to contexts where they were not relevant.

14.5 Gandhi's Contemporary and Future Relevance

In attempting to reconstruct Gandhi's ideas about technology it is difficult to find in his collected writings the terms "science" and "technology" employed in ways that would enable a reconstruction within the S&T discourse. However, what one would encounter are discussions about civilization, industrialization and mechanization; and when civilizations is qualified as Western civilization it becomes the source for Gandhi to mount a civilizational critique not just of the West but the cultures of mechanization and industrialism.[58] I have tried to argue that Gandhi's ideas about industrialism and technology were structured by a version of some social interest hypothesis, particularly from the perspective of rural India and the disempowered. Furthermore, his utterances and writings on technology surface incidentally in discussions on salient political or socially urgent issues and themata. To the extent that these utterances have a structure, it is proposed that their coherence resides neither in pragmatism nor are there any intimations of a dystopian conception of technology.

If this essay has worked against the grain of the standard understanding of Gandhi's approach to technology, two further points are in need of clarification. Standard Marxist and modernist readings of Gandhi have portrayed him as a Luddite. It could as well be argued that there is a conjuncture of one version of Marxism and Gandhian thought on the question of technology and this need not come to us as a surprise. The

[55] Dhruv Raina, 'Technological Determinism embodied in an Appro-tech Programme', op.cit.; Dhruv Raina, *Images and Contexts: The Historiography of Science and Modernity in India*, op.cit.

[56] A. K. N. Reddy, 'What is Appropriate Technology', op.cit.

[57] A. K. N. Reddy, 'Technologies Appropriate to Rural Development', op.cit.

[58] Deepak Kumar, *Science and the Raj*, op.cit., p. 35.

rootedness of Gandhian thought and Marxism in a social determinist understanding of the relationship between technology and society cannot be overlooked. And yet as we know there are important distinctions between the two understandings in so much as Gandhi and Gandhians would offer and foreground another vision of modernity that the Marxism of Gandhi's era would not have entertained anchored as it was in the notion of progress. Secondly, even on Gandhi's relation to Enlightenment thought, while remaining within the broader frame of Bilgrami's argument, it has been pointed out that while Gandhi valorized rationale debate, he differed from the Enlightenment philosophes' belief in 'indivisible' reason. On the contrary, Gandhi's notion of the rational reckoned that it varied with peoples and cultures. This made it all the more urgent to be receptive to the 'rational' of the adversary. Gandhi differed from the Enlightenment in that his notion of rationality was more dialogical.[59]

From the perspective of a social interest theory Gandhi was thus predisposed, as a lawyer would, towards case based reasoning rather than a fully developed philosophy of technology.[60] This entailed carefully evaluating the benefits and losses to the involved social constituencies and the long-term consequences of the chosen technological options. On situating each of these utterances within the context of their articulation and the frame of social interest or social determinism the supposed contradictions that characterize his "technological" worldview become non-consequential. The fundamental issue in this approach was the cultural significance and appropriateness of different technological regimes to the concerned social constituencies. In the 1970s within the world of science policy making, the crisis of legitimacy was handled by rearticulating this kind of social interest theory in terms of the "relevance of technology" to the lives of the disempowered. This idea had a significant half-life in several appropriate and alternate technology movements of the 1970s and 1980s.[61] In a modified form then, Gandhian ideas were reconfigured and stretched to become relevant to the new social contexts of technology and development. These movements have furthermore drawn on the democratizing possibilities that underpin his thinking on industrialization. This thinking is then elaborated upon to extend it into the domain of science and technology.

Finally, these movements have ignored the larger political framework of his thinking and understandably so, in addition to which other aspects of his life considered quaint by today's standard have been more or less blotted out of the reckoning of these movements. Perhaps that is the destiny of appropriations of the ideas of most thinkers. Ideas that are shelved, we know reappear in the most unlikely contexts and are refashioned in ways over which Gandhi or Gandhians could have no control.

---

[59] The point has been elaborated upon by Ronald Terchek and discussed in Hardiman (2003)

[60] Gandhi remarked that he was not too perplexed with the question of the consistency of his views. In a remarkable passage he offers a clue as to how he would like to be read: "I would like to say to the diligent reader of my writings and to others who are interested in them that I am not at all concerned with appearing to be consistent. In my search after Truth, I have discarded many ideas and learnt many new things…and, therefore, when anybody finds any inconsistency between any two writings of mine, if he has still faith in my sanity, he would do well to choose the later of the two on the same subject". Gandhi (1933).

[61] Dhruv Raina, 'Technological Determinism embodied in an Appro-tech Programme'. op.cit.

The question remaining is what will the Gandhian legacy hold for the world of post-industrial technology? The scientists of the 1970s to whom Gandhi was important were ensconced in a worldview where the social mattered; and hence during the crisis of dysfunction years they appealed to the notion of the "social relevance" of their world and research. But the science and technology in the neo-liberal world is marked according to Ziman by a new set of norms that have replaced the old Mertonian norms. Under these circumstances Gandhi will continue to remain an ethical shadow surfacing in future discourses.

As pointed out in the introductory remarks, the movements for decolonization date back to the last decades of the nineteenth century. As I have argued elsewhere, the issues and concerns of decolonization that have been articulated over the last century and a half and the modalities and practices of decolonization have ranged across a wide spectrum. The paper highlighted some of these concerns from the beginning of the twentieth century when the discourse on technology and industrialism was inaugurated as a disidentification with Western industrialism. This process had commenced during an era of decolonization dominated by a territorial imaginary.[62] Gandhi had nudged the discourse towards a critique of modern civilization and industrialism. The paper chronologically concludes when the discourse of decolonization goes beyond the territorial and the struggle to decolonize theory is commences. Gandhi recognized the significance of the critique of modernity as part of the cognitive struggle for independence. In the 1970s this critique was expanded into a critique of modernization theory that was followed by major innovations in science studies, history proper and social theory.

References

Gurminder K. Bhambra, Dalia Gebrial and Kerem Nişanoğlu (eds.), *Decolonising the University*, Pluto Press, 2018.

Dag Herbjørnsrud, "Beyond Decolonising: global intellectual history and reconstruction of a comparative method", *Global Intellectual History*, https://doi.org/10.1080/23801883.2019.1616310; 2019, pp. 1–27.

Ashis Nandy, 'From Outside the Imperium: Gandhi's Cultural Critique of the "West"', *Alternatives*, Vol. 7, No. 2, 1981, p. 173.

Bipan Chandra, *The Rise and Growth of Economic Nationalism in India*, New Delhi: People's Publishing House, 1977

Sumit Sarkar, *The Swadeshi Movement in Bengal*, New Delhi: People's Publishing House, 1975.

Deepak Kumar, 'Calcutta: The Emergence of a Science City', *Indian Journal of the History of Science*, Vol. 29, No. 1, 1994, pp. 1-7.

Shiv Visvanathan, *Organizing for Science*, New Delhi: Oxford University Press, 1985.

Dhruv Raina, and S. Irfan Habib, 'The moral legitimation of modern science: Bhadralok reflections on theories of evolution', *Social Studies of Science*, Vol. 26, No. 1, 1996, pp. 9–42.

Dhruv Raina, and S. Irfan Habib, 'Bhadralok perceptions of science, technology and cultural nationalism', *Indian Economic and Social History Review*, Vol. 32, No. 1, 1995, pp. 95–117.

[62] Chakrabarty (2005).

Dhruv Raina, and S. Irfan Habib, 'The Unfolding of an Engagement: The Dawn on Science, Technical Education and Industrialization', *Studies in History*, Vol. 9, No. 1, 1993, pp. 87–117.
M. K. Gandhi, *Young India*, October 7, 1926, p. 286.
M. K. Gandhi, *Hind Swaraj*, Ahmedabad: Navjivan Publishing House, 1938 (First Published 1909).
M. K. Gandhi, *Young India*, December 8, 1927.
Dhruv Raina, 'The Young P.C.Ray and the Inauguration of the Social History of Science in India', *Science, Technology & Society*, Vol. 2, No. 1, 1997, pp. 1-40;
M. K. Gandhi, *Young India,* October 7, 1926.
Deepak Kumar, *Science and the Raj*, New Delhi: Oxford University Press, 1995, pp.35-36.
Shiv Visvanathan, *A Carnival for Science: Essays on Science, Technology and Development*, New Delhi: Oxford University Press, 1997, p. 217.
Thomas B. Johansson, Amulya K. N. Reddy, Rob Williams and Rob Goldemberg, *Energy for a Sustainable World*, New Delhi: Wiley Eastern, 1988.
Benjamin Zachariah, *Developing India: A Social and Intellectual History c. 1930–1950,* New Delhi: Oxford University Press, 2005.
Aldous Huxley, 'Economists, Scientists and Humanists', in Thomas Holland (ed.), *Science in the Changing World*, London: G. Allen and Unwin, 1933, p. 211.
Baldev Singh (ed.), *Jawaharlal Nehru on Science and Society: A Collection of His Writings and Speeches*, New Delhi: Nehru Memorial Museum and Library, 1988, pp. 15-18.
Raghavan Iyer, *The Moral and Political Writings of Mahatma Gandhi*, New Delhi: Oxford University Press, 1986, pp. 310-315.
M. K. Gandhi, *Young India: 1919-1921*, New York: B. W. Huebsch, 1923.
Bruno Latour, 'On Technical Mediation – Philosophy, Sociology, Genealogy', *Common Knowledge*, Vol. 3, No. 2, 1994, pp. 29-64.
Wilhelm E. Fudpucker, 'Postmodern Production and STS Studies: A Revolution Ignored', in Stephen H. Cutcliffe and Carl Mitcham (eds.), *Visions of STS: Counterpoints in Science, Technology and Society Studies*, New York: State University of New York Press, 2001, pp. 139-154.
Wiebe Bijker, Thomas Hughes and Trevor Pinch (eds.), *The Social Construction of Technological Systems*, Massachusetts: MIT Press, 1987.
Langdon Winner, *Autonomous Technology: Technics-out-of-control as a Theme in Political Thought*, Cambridge: MIT Press, 1977.
Thomas J. Misa, 'How Machines Make History, and How Historians (and Others) Help them to do so', *Science, Technology and Human Values*, Vol. 13, Nos. 3-4, 1988, pp. 308-331.
Stephen H. Cutcliffe, *Ideas, Machines and Values: An Introduction to Science, Technology and Society Studies*, Lanham, Md: Rowman & Littlefield Publishers, 2000.
Sunil Khilnani, 'Nehru's Faith', *Economic and Political Weekly*, Vol. 37, No. 48, 2002, pp. 93-99.
Ramchandra Guha, 'Mahatma Gandhi and the Environmental Movement', in Mahesh Rangarajan (ed.), *Environmental Issues in India: A Reader*, New Delhi: Pearson Longman, 2007, p. 112.
Mahadev Desai, 'Preface to Mohandas K. Gandhi', *Indian Home Rule or Hind Swaraj*, Ahmedabad: Navjivan Publishing House, 1938, p. 7–9.
I. A. Karawan, Wayne McCormack and Stephen Reynold (eds.), *Values and Violence: Intangible Aspects of Terrorism*, New York: Springer, 2008, pp. 17–18.
Akeel Bilgrami, 'Gandhi did not oppose science: interview with Akeel Bilgrami' February 26, 2007, https://www.rediff.com/news/2007/feb/26inter.htm (accessed on November 30, 2021).
Mulford Q. Sibley, 'In Praise of Gandhi: Technology and the Ordering of Human Relations', lecture delivered at the meeting of the American Political Sciences Association, https://universalistfriends.org/sibley2005.html (accessed on November 30, 2021).
Aant. Elzinga, and Andrew Jamison, 'Cultural components in the scientific attitude to nature: Eastern and Western modes', *Technology and Culture*, Occasional Report Series, No. 2, 1981.
Dhruv Raina, *Images and Contexts: The Historiography of Science and Modernity in India*, New Delhi: Oxford University Press, 2003.
A. K. N. Reddy, 'Reflections of a Maverick', *Seminar*, Vol. 409, September 1993, pp. 1–9.

Dhruv Raina, 'Technological Determinism embodied in an Appro-tech Programme: Small Science in a High-Tech Environment' *Journal of Scientific and Industrial Research*, Vol. 52, 1993, pp. 471–482.

A. K. N. Reddy, 'Technologies Appropriate to Rural Development', *INSA Seminar on Contributions of Science to Rural Development*, New Delhi, 1978.

E. F. Schumacher, *Small Is Beautiful*, New York: Harper Perennial Books, 1975.

M. K. Gandhi, 'Economic versus Moral Progress', in *Speeches and Writings of M. K. Gandhi with an Introduction by Mr. C. F. Andrews and a Biographical Sketch*, Madras: G. A. Natesan and Co., 1922, 3rd edition, pp. 290–291.

A. K. N. Reddy, 'What is Appropriate Technology' Mimeograph, 1978, Bangalore.

Dhruv Raina and Chanakya Hoysala, 'Des microbes à vocation gandhienne dans un digesteur à biogaz. Les objectifs mouvants de la recherche sur les biogas', *Techniques & Cultures*, Vol. 67, 2017, pp. 154–175.

David Hardiman, *Gandhi in his Time and Ours: The Global Legacy of his Ideas*, Ranikhet: Permanent Black, 2003, p. 9.

M. K. Gandhi, *Harijan*, April 29, 1933, p. 2.

Dipesh Chakrabarty, "Legacies of Bandumg: Decolonisation and he Politics of Culture", *Economic and Political Weekly*, November 12, 2005, pp. 4812–4818.

Chapter 15
Contours of Post-colonial Comparative Philosophy: Modernity, Philosophy and Science

Nirmalya Narayan Chakraborty

15.1 The Very Idea of Comparative Philosophy

Philosophical questions start their journey with the human urge to fathom the mysteries that man encounters in his surrounding environment. These questions constitute some of the fundamental mortal queries, questions like the origin of the universe, the nature of I, the destiny of man etc. These questions are manifestations of the human urge to know, exhibiting his love of wisdom as the etymology of the word 'philosophy' implies. Gradually with the flow of time, these questions were reformulated, framed and answered in varied ways. Many more related questions were raised and attempts were made to systematize them in a rigorous manner. Philosophy as an independent discipline marked by its own set of problems and methodology evolved. Philosophy does not start in a vacuum. Even asking a question requires background knowledge. And this background knowledge, for historical contingencies, was manifest in different ways in different geographical spaces. The different intellectual horizons of the demanding mind provided man with wide-ranging ideas and insights resulting in raising assorted queries, sometimes formulating the same question from a different perspective. And with the intensification of thought and analysis, philosophical systems expanded and developed in very many different ways. Even within a specific geographical boundary, people might have lived through varied experiences producing different intellectual motivations to formulate certain questions and respond to these questions from what one thinks to be his privileged positioning. This much of pluralism seems inevitable. In fact, the commentorial tradition of classical Indian philosophical systems betray this benign pluralism in the untiring effort of each of the systems confronting the other systems and constantly engaging in raising new problems and reformulating its earlier stances so as to be able to

N. N. Chakraborty (✉)
Department of Philosophy, Rabindra Bharati University, Kolkata, India
e-mail: nirmalya.narayan@rbu.ac.in

S. Ahmed and S. Sarkar (eds.), *Decolonizing Science and Modernity in South Asia*,
https://doi.org/10.1007/978-981-97-1829-0_15

successfully respond to the possible objection, sometimes even an imaginary one. This is a beautiful example of creative comparative philosophy.

With the arrival of colonialism in India, the term 'comparative philosophy' gained a new momentum.[1] Comparative philosophy in colonial India acquired a different connotation, different from the creative comparative philosophy that has been there in practice in classical period. One of the far-reaching consequences of colonial regime is manifest in its implementing English education. Due to various socio-political reasons, India that the East India Company met was in a mess. The social cohesiveness and the state power were in dire states. Many people in India at that time saw the arrival of the East India Company as a blessing. Soon the weighing scale of the business man turned into the sceptre of a ruler and both the ruler and his subjects thought it to be prudent to introduce English education. With fast speed the millennium old indigenous learning system collapsed making room for institutions imparting European education bringing along with it its philosophy, culture, religion etc. The civilizing role of the English education system overwhelmed the vast majority of the people. This complete rupture in the age-old traditional knowledge system has got some interesting consequences for the Indian academic scenario, especially with regard to philosophy.

It is to be noted that for the first time in history, Indian philosophers got in touch with a highly developed alternate philosophy. Enamoured by the great figures of the then European philosophy, many Indian philosophers engaged themselves into study of European philosophers. Interestingly these philosophers focussed on those very European philosophers who they thought might echo some Indian philosophical ideas. The projects of European philosophy became the projects of Indian philosophers. The lion's share of the curriculum of the education institutions founded on the European model was devoted to European philosophy, with hardly anything on the classical Indian philosophical commentaries especially when these commentaries are the real repositories of Indian philosophical insights. Both the content and the method of teaching and learning philosophy in Indian institutions changed dramatically.

With the rise of nationalist fervour, some philosophers started taking the classical Indian philosophical ideas seriously. Many institutions were founded to impart teaching of indigenous philosophical systems, independent of European philosophical ideas. But with the spread of English education institutions, exposure to European philosophy became inevitable. Some philosophers with a nationalist spirit embarked on a comparative philosophy with the agenda to prove that many of the ideas propounded in recent European philosophy have been well articulated much earlier in ancient Indian philosophy. This claim, they thought, would blunt the civilizing claim of European philosophy. Some others, within the largely nationalist framework, tried to assimilate some of the European philosophical ideas in the Indian philosophical trajectory in a nuanced manner. Philosophy, in the works of these thinkers, is neither Indian nor Western. This is not a comparative philosophy with a hidden agenda. This

[1] Some of the ideas expressed in the following paragraphs have been articulated in my essay 'Comparative Philosophy: Why and How?' in Basu (2019).

is a creative comparative philosophy which is marked by the originality of the author. And K. C. Bhattacharya is the most eminent representative of this school.

With India and the West coming into contact with each other, intellectual interaction stepped up. But with the civilizing idea being quite well spread, there are philosophers who, while well exposed to European thought, nonetheless articulated and explained the philosophical insights available in classical India. Indian philosophical ideas got reinterpreted and reframed in contemporary jargon. In the works of these authors Indian philosophy is never looked at through the prism of Western philosophy. Radhakrishnan is perhaps the most important figure in this genre.

It is interesting to note that with the exchange of philosophy between India and West, Indian philosophy never received the attention of the philosophers in the West. It is true that a large number of indologists in the West have made significant contribution to the studies of Indian languages, religion, history etc. But the great treasure of Indian philosophy hidden in the large number of commentaries never received the attention that they deserve from the Western philosophers. Apart from the socio-political reasons, there is a widespread impression among the Western philosophers that Indian philosophy is mystical, spiritual, other-worldly etc. and consequently Indian philosophy is not philosophy proper. To remedy this wrong and biased idea about Indian philosophy, some philosophers have engaged themselves in the study of Indian philosophy by focussing on those specific questions and debates in classical Indian systems that have their cousins in some parts of Western philosophy. These works are aimed at proving that if Indian philosophy contains the debates and queries that have their counterparts in Western philosophy, then Indian philosophy is at par with Western philosophy in so far as its status as philosophy is concerned. B. K. Matilal is the most important name in this attempt.

Thus, one sees that in post-colonial India 'comparative philosophy' has been pursued in different ways and with different motives: (1) Comparative philosophy with a parochial bias, (2) Creative comparative philosophy and (3) Comparative philosophy with a clamour to claim that one brand of philosophy is no less a philosophy than the other.

Philosophical concepts and debates come with a baggage. And if philosophy is an attempt, on the part of the humans, to explain and decipher his experiences, then in order to understand these questions and conjectures, we have to place them in the conceptual landscape against which the deliberations took place. In other words, what we are dealing with is not a singular question or a concept, it is rather a fabric of ideas and speculations. We are talking about a philosophical conceptual scheme. So, understanding one philosophical system is interpreting an alien conceptual system. Comparative philosophy then devotes itself into interpreting alternative conceptual systems. Let me try to unpack what is involved in interpreting a conceptual system. The interpreter tries to decipher a particular conceptual system. Notice, the interpreter is already working with a conceptual system. So, interpreting an alien conceptual system consists in translating the alien conceptual system in one's own conceptual vocabulary. Thus, there are two conceptual systems at work. For translating one into another, we need a manual that would map the target conceptual system onto the home conceptual system. But then the translation manual itself is couched in a vocabulary that falls outside the realm of the two conceptual systems or at least should be rich enough to encapsulate both the conceptual schemes. Unless we assume that the interpreter is already in possession of this translational conceptual scheme, any talk of interpreting the foreign conceptual scheme cannot take off the ground. Then the same question could be asked about the understanding of the translational conceptual scheme and we would be moving backward ad infinitum. We have to assume two things in order to make sense of our attempt to interpret the alien conceptual scheme: (1) We have to assume that the interpreter has some knowledge, even if rudimentary in nature, of the world that the interpretee talks about. There is an overlapping shared world that is accessible to both the interpreter and the interpretee (a move to realism); (2) The interpreter must also assume that a large number of statements found in the alternate conceptual system are believed to be true by the participants of the foreign conceptual system. Interpretation, including wrong interpretation, can work only on the assumption that the interpretee is sincere in her beliefs and that the participant is not lying etc. (a move to charity). In other words, a common public space and a mutual trust are necessary prerequisites in any interpretative project. None of these, of course, implies that interpretation does not fail. Radical disagreement and massive error would render the interpretative work impossible to start with.

If translation manual is indispensable for interpretation, then we must entertain the possibility of alternative translation manuals and so we must make room for alternative interpretations of the alien conceptual scheme. Pluralism is the natural outcome of this multiple interpretations. It is hardly possible to come up with a criterion of choice of one particular interpretation merely on the basis of some objective matter of facts, simply because there is no such vantage point. This is precisely why interpreting conceptual schemes is such a challenging work.

What this shows is that comparative philosophy does not have a monolithic, linear structure. There is a lot of boot strapping involved, sometimes moving ahead, some other times moving backwards. One should be circumspect of the very idea of comparative philosophy. Comparative philosophy, in the field of Indian academic philosophy however, is inescapable. Philosophy qua philosophy should be courageous enough to stand in front of novel interrogations and doubts and it is the onus of a philosopher to answer and if necessary to reformulate and revise his own standpoint. If philosophy is not to be reduced to a history of thoughts, if philosophy is to be living and thriving, if philosophy is to invigorate man's understanding of his surroundings, then creative comparative philosophy remains our desired goal. One philosophical system needs to confront the other, but this confrontation must not be understood in terms of one system competing with the other. One philosophical system need not emulate another. Let us be sincere pluralists, and not selective pluralists.

15.2 The Origin of the Idea of Modernity

Any discourse on modernity and Indian connection is bound to start with a discussion of the idea of modernity as it emerges in Europe. Like many other concepts, modernity has its baggage to carry. The very idea of Europe is contestable, because one could argue that Europe does not necessarily represent a single set of monolithic ideas, rather Europe is an assemblage of heterogenic varied ideas often crisscrossing each other. The history of Europe is history of many communities within Europe. It is also often difficult to demarcate European thoughts from non-European thoughts simply because Europe has established colonies in many non-European countries and also the geographical boundary of Europe has changed many times. The birth of Europe as a cultural paradigm is contemporaneous with the idea of modernity. Naturally these two are related. Still, one could argue that there is a set of ideas that taken together could give Europe a shape of its own.

It is to be remembered that there was no single power that ruled Europe for long time. Many small state powers came and disappeared after a while. There was exchange of ideas among them, sometimes through marriage, sometimes through state craft, sometimes through war. But no single overarching state power could emerge in Europe. After the fall of the Roman Empire, states of the Eastern Europe maintained a distance from the Western European states. As there has been war, there have also been attempts to establish a balance of power. As a result of all these, we have got the separation of power of church and state. Thus arise a kind of indeterminacy, constant change in the centre of power.

And this is closely related to the long tradition of civic society. Because of the frequent change in the power centre, the local communities often raised the issue of protecting their own rights against the powerful elites. This led to questioning the centralisation of power. The common people from different strata like the workers, the trading communities, the intellectuals again and again questioned the sovereign power of the state. This facilitated the birth of many powerful political struggles that had great influence on many countries. This constant questioning of the state power made it impossible for any state power to show its brute strength for long. In all these, civic society played an important role. The difference between legality and legitimacy gained importance. This led to the firm founding of democratic mind set.

There has also been rupture between the present and the past that has been noticed often in European history. There have been many disagreements among the different sects of Christianity with regard to Christian religious texts and its rituals, paving the way for the birth of Greek orthodox church and many other similar sects. Out of these constant churning of these sectarian disputes secular idea grew, which in its turn gave a concrete shape to the idea of rationality. Sometimes while questioning the near past, attempts have been made to accept the distant past, a kind of going back to the original.

Europe has been witness to the emancipatory urge both at individual and social level. Europe is perhaps the birth place of individual dignity. Though individualism of some sort could be found in many civilizations, Europe is the only place where the political implication of individualism has been felt most intensely. Liberal political philosophy and capitalist ethics are the products of this.

Europe also had an international interest. They were interested in many non-European civilizations. Because of this contact with others, Europe imbibed many elements from other civilizations. And through this exchange, as Europe got the opportunity to know others, similarly Europe also got to know herself. Because of trade with many countries, because of spread of Christianity through missionaries and education institutions, because of spread of printing technology, exchange of ideas among a large number of people has been in place for long time. All these resulted in multiplicity of ideas, self-examination, and the idea that Europe is a miniature of the world.

Against this landscape, the European mind came to realize that nothing is permanent; future is not pre-determined. With the appearance of a fissure with the near past, the new must be welcome. With the call for the new, the predetermined location of experience must be overcome. This idea of progress talks about a hope for better future. This transcends the present. People started believing that with new experience, with new knowledge human society can change itself. This trust on human capability is the mark of modernity. Modernity prompts us to constantly explore various manifestations of social and political organizations. This is an ongoing flow. Modernity thus gets manifest in epistemology, politics and economics. This is also an attempt to know one's own social self. Modernity makes its presence felt in the pursuit of science and social laws and practices based on science. In this scheme rationality becomes the torch light that shows the future path to mankind.

15.3 Modernity: Some Philosophical Concerns

The tug of war between tradition on the one hand and its progression through history on the other is especially evident in a country like India which boasts of being one of the most ancient and rich civilisations mankind has ever produced and at the same time due to constant interaction with 'other' cultures is subject to influences foreign to it. Thanks to economic globalisation, native tradition is all the more threatened. Naturally a debate crops up about the value of indigenous tradition and its survival in the face of turbulent and fast changing time. I would begin my discussion with a clarification of the term 'tradition' *vis-a-vis* the term 'modernity'. Tradition, in some significant sense, is treated as the 'other', something that is situated far off from the modern man, something which the modern man finds obscure and at best esoteric, and hence is of little value and perhaps dispensable. This concept of the 'other' needs clarification. The idea of radical otherness will be critically assessed.

There is a simple and rather trivial way of understanding tradition as being opposed to modernity. Every age that is latest to arrive is modern and that which precedes the modern is the tradition. Whatever is temporally present is modern and that which comes before the present slips back into memory and contributes to the reservoir of the tradition. On this naïve view, with the constant flow of time the contents of tradition and modernity also constantly shift. Tradition viewed in this way actually ignores the deeper philosophical issues involved. If modernity is what is simply temporally present, then tradition being present in my memory can also be temporally present through my act of remembering. Consequently, the distinction between tradition and modernity collapses. Nonetheless, one important lesson to be learnt is that temporality is an important dimension of tradition. Tradition is not an a-temporal idea. So, any discussion on tradition would bring with it a temporal perspective. This is nothing unusual if we remember that one of the elements involved in modernity distancing itself from tradition is a big temporal gap. However, tradition and modernity have special significance in the Indian context, which cannot be fully apprehended by mere temporal distancing.

J. N. Mohanty rightly distinguishes orthodoxy from tradition.[2] Orthodoxy turns tradition into a fossilised, unchanging phenomenon. Orthodoxy cannot tolerate any questioning or self-criticism. And if philosophy is primarily a critical analysis, then it would be obvious that orthodoxy does not leave any room for philosophical activity. Tradition, as opposed to orthodoxy, is living in the sense that it constantly re-creates itself through self-questioning and responding to new situations and challenges. It is only when tradition loses its dynamicity and self-questioning capacity that it lapses into orthodoxy. It is only a dead tradition that needs orthodoxy to protect it from possible attacks or perversions. So, a living tradition, in this sense, leaves quite a bit of room for openness and flexibility. The opposition that we are talking about is not orthodoxy versus modernity, for they belong to two different categorical frameworks. Philosophically interesting and insightful opposition worth discussing would rather be tradition versus modernity.

If modernity means simple rejection of tradition, then the tension mentioned earlier would immediately cease to exist. And then any dialogue or mediation between the two would be out of place. This interpretation of modernity, to me, really avoids the difficult issues involved. Viewed this way, modernity would exhaust itself in accepting the present only because it is present. And this attitude is philosophically barren.

An interesting way of understanding modernity consists in addressing oneself to what is contemporaneous.[3] Modernity, looked at this way, does not simply stand for uncritically accepting what is in fashion. Modernity addresses itself from contemporary perspectives and challenges. It reorients itself with contemporary questions. Here modernity enters into a dialogue with what is contemporaneous. Since philosophy, in an important sense, is a kind of dialogue within oneself or inter-subjective, maybe even inter-cultural, discourse on modernity does contribute significantly to philosophical thought. But dialogue involves questioning and scepticism, seeking clarifications and facing challenges. And it is here that tradition can play the role of interlocutor in the dialogical activity of modernity. Modernity can very well enter into a dialogue with tradition, for tradition, as distinguished from orthodoxy, is ready

---

[2] Mohanty (1992).

[3] Ibid., p. 12.

to face possible challenges from modernity and is contemporaneous with it. The fact that tradition can pose challenges to modernity, the fact that tradition indulges in self-criticism and becomes reflective, suggests that tradition and modernity after all can be partners in the advancement of human society. This is especially important in the context of Indian philosophy where the traditional questions and issues, after facing healthy challenges from modern philosophical deliberation, are reinterpreted, recreated, getting different rendering. So, tradition and modernity need not be understood as involving an unbridgeable gulf.

A widely accepted way of understanding modernity is in terms of the idea of critical spirit. Uncritical acceptance, blind faith in a set of beliefs is something which modernity opposes with all its strength. This point demands a cautious understanding especially in view of various clichés in terms of which a comparison of Indian and Western philosophical speculations is often made. We must remember that any attempt to compare two 'paradigms' assumes a third paradigm on the basis of which the comparative estimate is being made. And naturally the third paradigm must first face the tribunal of modernity. Avoiding the risk of being a dogmatist one must ensure the justification of this third paradigm. Of course, in this justification one must not rely, in any way, on the paradigms to be compared, for that would be philosophically quite un-explanatory. How this could be done, however, is a million-dollar question in philosophy. Of course, one could apply justification to the third paradigm involving critical norms internal to that paradigm. If critical tools applied to the third paradigm are taken from inside the paradigm itself, then this of course would contribute to the inner strength of the paradigm. But then why not apply the same internal justification to the first and the second paradigms? One could indulge in critical appraisal of the first and second paradigms borrowing the critical apparatus from within those paradigms themselves. Then the very act of comparing would lose its rationale. In fact, inter-paradigm comparison could not even take off the ground.

If, on the other hand, we keep on pressing for comparative activity, then we have to think of some critical tools that are applicable across the paradigms. In other words, we have to think of a set of universal critical tools applied indifferently to any mode of thinking. The problem is to formulate such a universal set of critical norms. What happens is that often in the name of giving universal set of critical norms, one unwittingly follows one paradigm or another. Consequently, what is apparently universal is really internal to a paradigm. Thus arise the familiar characterisations of Indian philosophy as being 'other worldly' or lacking 'logic'. One could, in fact, go to the other extreme in arguing that any critical appraisal should be done from within the paradigm, and the basic framework of that paradigm cannot be questioned by that critical appraisal. So, any questioning or appraisal is always internal to the paradigm.

Even if the underlying message of this stance is worth considering, one must not be swept away by this zealous internalist attempt. It would be too narrow a view to think of paradigms as completely self-sufficient, shut off from the rest of the world. A cursory look at the historical development of Indian philosophical schools seems to suggest that even if each school has its own epistemological, logical and axiological baggage, still they closely interacted with each other, often resulting in refinements and revisions in their own respective theories. How else could one explain the emergence of varied philosophical systems from the same Vedic corpus? The established schools of Indian philosophy with their enormous number of philosophical arguments and counter-arguments went far beyond what original Vedic seers anticipated. It is of course undeniable that different systems of Indian philosophy share their concern with, if not theories of, a group of concepts like self, knowledge, action, liberation, etc. But one should also not lose sight of the historical fact that dissent in Indian philosophy has not been uncommon. Buddha and his followers opposed the tradition of self. Nāgārjuna criticises the logo-centrism of different metaphysical doctrines. Madhusūdana Saraswatī challenged the Nyāya pluralism. Raghunātha Śiromaṇi, being himself a Naiyāyika, challenged the validity of some inferences held by early Naiyāyikas and thereby set an example of internal criticism in the Indian philosophical context.

Given our understanding of modernity as engaged in a dialogue with what is contemporaneous and as being reflective, it would not perhaps be unjust to think of Indian philosophical tradition as modern in nature. Constant interaction with alternative paradigms and refining and revisioning own theories have been ineliminable features of Indian philosophical enterprise. Viewed in this manner tradition, as distinguished from orthodoxy, and modernity are wedded together. They need not be opposed to each other. In fact, that they have a meeting ground is well exemplified in the history of the development of Indian philosophical systems. I do not, of course, intend to eulogise Indian philosophical pursuit. Radical questioning, i.e., questioning the basic assumptions of a paradigm, including its historical, cultural and cognitive doctrines is intrinsic to philosophising. And we must prepare ourselves to confront Indian philosophical tradition as a critical philosopher does. But at the same time, one must maintain a cautious optimism with regard to modernising tradition. Radical questioning does not take place in vacuum. Alternative paradigms are welcome, but they must also be ready to face philosophical scrutiny. One must bear in mind, by the way, that the expression 'Indian Philosophy' does not stand for any homogeneous system of thought. The vast corpus of Indian philosophical literature, from the materialists to the Advaitins, from Sāṃkhya denial of God to those philosophers who advocate devotion to personal God as a way to liberation, from Mīmāṃsā hermeneutics to Nyāya realism implies that the subtle arguments and counter-arguments advanced in favour of each of these positions signify a set of heterogeneous doctrines, loosely connected though here and there.

A person keen to maintain universality of critical norms might argue that since such universality is nearly established in abstract sciences like formal logic, why couldn't we do the same in epistemological and metaphysical theories? My response to this would consist of two points. First, even if we use the term 'logic' in cases of Aristotle, Medieval Scholastics, Frege, Husserl, etc., what they were doing in their logical enterprises were poles apart from each other. No wonder then that there was so much misunderstanding amongst these philosophers or even among their followers accusing each other of not doing 'logic'. Phenomenological explanation of logical categories has been branded as psychologistic in nature, whereas many Fregean scholars have disputed the attempt to interpret Frege's logic as a theory of meaning at the bottom. Even Quine has decried modal logic as conceived in sin. So, first of all, it is not obvious that universality has been attained in formal logic. Secondly, if universality is so distant even in abstract sciences, then what to speak of universality in epistemological or axiological theories that are concerned not with any abstract universality but with the concrete content that are attempting to achieve a generality. So instead of putting everything in a Procrustean bed, the open endedness of philosophical theories is to be welcome, not something to get rid of. Moreover, psychologism is not necessarily a folly. A growing number of researches are being done in cognitive science where heavy dose of theories of psychological mechanism has been injected in explaining mental events like reasoning, thinking, imagining, expecting, etc. The anathema to psychologism is probably going down day by day. This is important to remember, for quite often Indian 'logic' has been branded as psychologistic.

We must however concede the universalist's claim that from Aristotle to Husserl, in so far as all of them are *doing* logic, they must share some common features or satisfy common requirements. And this again shows that not all critical norms are internal to the paradigm. To this if we add the idea of tradition as I have spoken of earlier, then no tradition, worth its name, can set a limit to the bound of critico-philosophical thinking. If a tradition is what it is, then in its reflective stance it must address various challenges and re-orient itself in the face of these questions.

15.4 Understanding Philosophy and Science: The Modern Indian Philosopher's Way

Philosophy is perhaps one of the most self-critical disciplines. Throughout its history, philosophers have had long debates on the nature, methodology and the proposed solutions to the questions that go by the name of philosophy. This long history of conjectures and refutations regarding the very nature of the activity that philosophers engage themselves in need not necessarily be looked at disparagingly. In fact, the very debate about the nature of philosophy has given rise to interesting areas of discourse that have enriched philosophical repertoire. Needless to mention, the idea of philosophy entertained by the professional philosophers is somewhat different from that held by the amateur ones, although they are not unrelated. Philosophy, being the mother of many disciplines that emerged later, has cast its shadow on many of the major areas of human enquiry. And this explains why the non-professionals have their interests in and expectations from philosophy, whatever philosophy stands for them. The confluence of ideas of varied kinds has generated interesting deliberations among the philosophers as to the nature of philosophy. The present section is an attempt to decipher the concept of philosophy as advocated by Rasvihary Das.[4] I shall try to expand and defend Rasvihary's ideas by taking cues from Kalidas Bhattacharyya, a junior contemporary of Rasvihary. At the end of the discussion one can see the

[4] 'My Conception of Philosophy', Das (1968).

rationale behind the prescription that one finds both in ancient Greece and India - know thyself.

Rasvihary starts his analysis by trying to distinguish science from philosophy. If, as usually claimed, science and philosophy provide us with the knowledge of reality, and if reality is one, then knowledge of reality is also required to be one. As a result, science and philosophy must coincide and the distinction between the two seems to collapse. This is indeed the view of some philosophers, but Rasvihari, though does not explicitly reject this view, is more keen on analyzing the attempts that try to distinguish science from philosophy.

According to some, science gives us the knowledge of reality in piecemeal, whereas philosophy harps on the reality as a whole. It is true that different branches of science deal with different aspects of reality. It is equally true that the knowledge of the whole requires the knowledge of the parts. And so, if philosophy is concerned with the reality as a whole, philosophical knowledge requires the knowledge of different branches of science. Rasvihari thinks that it is absurd to claim that knowledge of the different branches of science is a prerequisite to philosophical knowledge. It is simply practically impossible to master all the branches of science and then try to acquire philosophical knowledge.

Another attempt to distinguish science from philosophy consists in proposing that they differ in their respective methods. While science relies on observation and experiment, philosophy hinges on rational thinking. Scientific enquiry is based on experiment, philosophical pursuit rests on pure ratiocinative argumentation. Rasvihary is sceptical of this attempt. The reality that is grasped by observation and experience is not the reality that is apprehended in pure reason and vice versa. What is the object of pure thought can never be the object of empirical investigation. Pure thought encompasses abstraction of a certain level that is not amenable to observation. May be science gives us the knowledge of one aspect of reality and philosophy is concerned with another aspect of reality. But then how is one going to determine that science and philosophy are concerned with different aspects of the *same* reality? What is the source of the knowledge of this reality whose different aspects are said to be the objects of investigation of science and philosophy? Moreover, if philosophers claim to strive at the knowledge of reality whereas science remains content with focusing on appearance, then science counters by asserting that philosophers engage in fanciful speculation. At any rate, the debate goes on. Rasvihary concedes that it is not possible to deny that science gives us knowledge and if science exhausts itself in producing knowledge, then what we get from philosophy is not knowledge, but something different. One could, of course, explain the difference between science and philosophy in terms of producing qualitatively two different kinds of knowledge, one provisional and the other, absolutely certain.

Even if this is not a satisfactory explanation, Rashvihary asserts that "it is not necessary to confine the application of the term knowledge to scientific knowledge only".[5] It is undeniable that we get to have knowledge from different sources

[5] Ibid, p. 33.

like history, listening to reports from people etc. But the difference between scientific knowledge and the knowledge that we get from philosophy consists in science exhausting itself to the analysis of the object ignoring the role of the subject. But in production of knowledge the role of the subject is crucial. Rasvihariy contends that in knowing the object it is always possible to ask whether the object is there at all "and in this sense objective knowledge is always doubtful".[6] But in a knowledge situation where I am the knower, it cannot be meaningfully asked whether I am there. For there to be knowledge the subject is presupposed. Philosophy takes the subject seriously. While deciphering an epistemic experience philosophy looks at the knowledge *as known* by the subject. Object, independent of any relation to the knower, is a misnomer in philosophy. For something to be an object of knowledge, that something must be placed in relation to a subject. This "subjective bias" is what distinguishes science from philosophy, according to Rasvihary. This of course implies that the role of subject is of tremendous importance in so far as philosophical analysis of knowledge is concerned. Assuming that there could be varieties of understanding regarding the idea of subject, Rasvihary alludes to the woman saint of the Upaniṣad who refused to be swayed by the enticements of wealth in the thought that this mundane affluence would not bring her the immortality suggesting that she could have equally refused to be impressed by scientific knowledge which does not satisfy the demand of the spiritual need of the soul. Classical India entered into philosophical speculation with the aim of removing the spiritual *malaise* that man suffers from.

This subjective orientation of philosophy colours the way the object is understood in philosophy. Since science is blind to the role of subject, the object in science is pure in the sense that this object is understood as completely unaffected by the subject's understanding. Philosophical object, on the other hand, is heavily coloured by the subject. It is true that for a scientist, philosophical subject is not the object proper and hence falls short of reality, but since the problem of reality is something that makes us feel concerned, scientific object appears to be an abstraction that, even if interesting in itself, would fail to enlighten us on the nature of reality.

If philosophical understanding of the object is always via the subjective path, this subject is also not an abstraction, rather a concrete subject consisting of knowing, feeling and willing capacities. In the epistemic relation to the object, the emotive and affective dispositions also play their role. All the three ways of countering the object contribute to the making of the object, the object as known by the subject. The subject in action is what interests philosophers and in that modality object enters into the field of the subject and the subject also contributes to the uncovering of the object through its various dispositions. And in this way, philosophy could be viewed as contributing to self-knowledge. "The pivotal point of self-knowledge seems, in my view, to be the most important aspect of philosophical knowledge, so much so that a demand for philosophical knowledge may well be equated with a demand for

[6] Ibid.

self-knowledge and vice versa".[7] This is well in tune with the Greek and classical Indian prescription to know one's own self.

But if philosophy is to give us the knowledge of reality, and if reality is made of both self and not-self, then how could one equate philosophy with self-knowledge? Does not this leave out the not-self part of the reality? Rasvihary's reply is that self-knowledge cannot be divorced from the knowledge of reality consisting of both self and not-self. The attempt to know self by cutting off all relation to the world would land us in mysticism which would lie outside the scope of philosophy. Rasvihari argues that such a view of self as wholly distinct from and opposed to world "is quite unrealistic".[8] Just as object is what it is because of its relation to the subject, the subject is also to be understood in terms of its relation to the object. The reality, in so far as it is in relation to the subject, is all the way strewn with the remnants of the self. To search for the self is to look for it in the reality in all its facets. Self-knowledge, in this way, leads one to the knowledge of reality. Self-knowledge divorced from the world would be utter empty. Focusing on the knowledge of the self does not bar it from taking the world into consideration. Self as subject is in the state of forward looking to the object and so knowledge of the subject encompasses the knowledge of the world. Philosophical knowledge, being reflective, brings out the presuppositions of objective knowledge that includes the subject-object totality in concrete.

Wherefrom does philosophical activity start? What are the data that philosophical analysis could get to reflect on? According to Rasvihary, commonsense provides the data. Commonsense beliefs and ways of interpreting the world are what we inherit from the community that we are born and brought up in. Commonsense gives us a ready- made philosophy to start with. Our everyday life is directed by commonsense. This untutored commonsense is the beginning of philosophy. Commonsense beliefs are enough for the vast majority of people to deal with in their everyday lives. But the critical mind is not content with these commonsense beliefs. He seeks to question, clarify and if possible, comes up with an alternative understanding of the reality. The task of philosophy is two-fold: 1. Critical and 2. Constructive. Critical part consists in unearthing the inconsistencies and lack of clarities resulting in rupture in the commonsense edifice. The constructive part is made of building an alternative view of the reality that is claimed to be free from discrepancy and ambiguity. This is what all the major philosophers and philosophical systems have tried to do in the course of history.

It is true that there are sceptics and agnostics who argue against the success of any such attempt of philosophers to arrive at a conception of reality and by their criticism they also engage themselves in philosophical activity. It is not required of philosophers that they will come up with a complete knowledge of reality, but at least they can be expected to offer clarification of some obscure and opaque ideas about reality. Philosophy is not concerned with discovering new facts and in this sense philosophical knowledge is not cumulative. But philosophical knowledge grows by relentless clarification of the ideas that determine the very knowledge of facts. This

[7] Ibid, p. 35.

[8] Ibid.

explains why philosophical debates are never ending. Discovery of an objective fact could have an end somewhere. Philosophy, being concerned with object as it is related to the subject, is amenable to multiple interpretations of the subject-object dynamics. This explains why there is no final refutation of any philosophical theory. A particular philosophical stance, long after its refutation, may rise again from the ashes. Philosophising "is essentially critical reflection, reflection on vital aspects of experience with a view to gaining relative clarity and rational satisfaction",[9] concludes Rasvihary.

Thus, we can see, according to Rasvihary, the uniqueness of philosophy consists in taking the subject's role seriously in an account of knowledge of reality. The "subjective bias", as Rasvihary calls it, is the mark of philosophical knowledge. One must not take it as leading to vulgar subjectivism implying that philosophy is nothing but a record of individual history of encountering the world. Philosophy is not concerned with the subjective history of the individual's understanding and interpretation of the world. Doing justice to the role of the subject in the production of the knowledge of reality does not necessarily lapse into descriptive psychological study.

Let me clarify this as I understand it. When philosophy grasps the reality in terms of the subject-object totality, what philosophy yields could be called thought. So, philosophy aims at arriving at thoughts about reality. How are we going to understand thought? Thought is not mere representation (linguistic or otherwise) without having anything corresponding to it. If thought is the result of the subject-object dynamics, then thought contains the different modalities in which different objects are present to the subject. In other words, what we have, as the content of thought, can be called "objective ideality".[10] Modal assertions are the most important elements of philosophical knowledge and the objective idealities are the fundamental ingredients of modal assertion. As mentioned earlier, philosophy is not interested in discovering facts, it is rather keen on discovering the existence and categorial features of the reality as they are manifest to the subject. Factual details are not of any relevance to philosophy. It is the structural encapsulation of the facts that gets manifest in thought.

Doesn't thought grasp what is real? The answer is 'yes' but with a caveat. If we follow Rasvihary's model of philosophical knowledge, then the very idea of object bereft of any relation to the knowing subject does not make sense. So 'real' in philosophical discourse is never something that is isolated and independent of the cogniser. What thought grasps is real in the sense that they are sense-given, but they are not real in the sense of concrete existents lying independent of the subject. What is real is never distinguishable in itself. Real gold can hardly be found, it is always either a lump of gold or a necklace or some other ornament. In the model of philosophy that Rasvihary advocates, thought does not discover bare facts. This is why what thought grasps is called 'objective ideality'. This ideality constitutes reality as bare gold constitutes a necklace. This ideality is not fictitious. This ideality is what is discovered in thought. The ideality has its loci in the real but then it does

[9] Ibid, p. 39.

[10] 'The Business of Philosophy', Bhattacharyya (1965).

not exhaust itself in being real. Even our common sense understanding of the world is sophisticated enough to grasp this ideality. And this is precisely why philosophical reflection can very well regard common sense experience as its take off ground. The ideality is there in the real and thought discovers it implying that ideality was there in the real "as undistinguishably fused with reals".[11] So, ideality was there in the real, only not as ideality, rather in a generic shape or form. Thus, what thought grasps has its home in the real but itself is not real and so it is called objective ideal. In philosophical reflection these objective idealities are made to face rational scrutiny.

The objective part of objective ideality points to the real which is the loci of objective ideality. The ideality part of objective ideality implies that the objective ideality bears an inseparable relation to the knower; "they all necessarily hang on to the subject".[12] The subject discovers them with the application of his mental capacities. At the same time the objective idealities have an independent status in the sense that we can frame laws, derive significant conclusions from these idealities. Objective idealities have a "demand to reality" both in the sense of having the real as their loci and also that they are subject to laws. The reality of objective idealities does not reduce them to the real, nor the ideality of the objective idealities consists in rupturing its relation to the real. The ideal springs from the real and if philosophy is concerned with thought that grasps objective ideality, then philosophy can very well claim to give us knowledge of the real.

15.5 Conclusion

What the present essay shows is how the anglophone modern Indian philosophers located themselves in the cosmopolitan space created following the introduction of European Philosophy in the Indian academia. All of these philosophers were keen observers of the developments that took place in European philosophical tradition and addressed these issues in their own ways. And the nature of these interventions is important because here the modern Indian philosophers articulated and responded to these issues by freely borrowing from the conceptual repertoire of both classical Indian and modern European philosophical traditions. The articulation of the very nature of philosophy as distinct from science (discussed in the earlier paragraphs) shows that the ancient wisdom concerning philosophy, as the discipline revealing the knowledge of self (*ātman*), gets reinterpreted and receives a new cloak under the idea of objectivity. This idea of objectivity is one that is proposed by modern day philosophy of science. The very urge to unravel the relation of philosophy to science itself is an example of modernity reinterpreting the tradition. If tradition, as I have explained earlier, is to be distinguished from orthodoxy, then tradition has the onus of responding to the emerging novel questions and if necessary, reorienting itself in the light of these. The idea of creative comparative philosophy that I have talked about

[11] Ibid, p. 26.

[12] Ibid, p. 30.

earlier is a welcome development in post-colonial Indian philosophy. This model of comparative philosophy is different from either provincializing classical Indian philosophy or mindless mimicry of an alien tradition, and thus gives birth to new ways of looking at philosophy in twenty first century.

References

Durga Basu (ed.), *Sister Nivedita's Interpretation of Swami Vivekananda and Cross-Cultural Multidisciplinary Philosophy*, Kolkata, Ramakrishna Mission Institute of Culture, 2019, pp. 415–421.

Kalidas Bhattacharyya, *Philosophy, Logic and Language*, Calcutta, Allied Publishers, 1965, p. 21.

Rasvihary Das, *The Concept of Philosophy* (editor not mentioned), Varanasi, Banaras Hindu University, 1968, p. 31

J. N. Mohanty, *Reason and Tradition in Indian Thought*, Oxford, Clarendon Press, 1992, p. 11.

GPSR Compliance

The European Union's (EU) General Product Safety Regulation (GPSR) is a set of rules that requires consumer products to be safe and our obligations to ensure this.

If you have any concerns about our products, you can contact us on ProductSafety@springernature.com

In case Publisher is established outside the EU, the EU authorized representative is:

Springer Nature Customer Service Center GmbH
Europaplatz 3
69115 Heidelberg, Germany

Batch number: 10725823

Printed by Printforce, the Netherlands